Essentials of Bio-organic Chemistry

R. W. Hanson

Edward Arnold

First published in Great Britain 1984 by
Edward Arnold (Publishers) Ltd, 41 Bedford Square, London WC1B 3DQ

Edward Arnold, 300 North Charles Street, Baltimore, Maryland 21201, U.S.A.

Edward Arnold (Australia) Pty Ltd, 80 Waverley Road, Caulfield East, Victoria 3145, Australia

British Library Cataloguing in Publication Data

Hanson, R. W.
Essentials of Bio-organic Chemistry.
1. Biological chemistry
I. Title
574.19'2 QP514.2

ISBN 0–7131–3500–X

Printed in Great Britain by
Thomson Litho Ltd, East Kilbride, Scotland

Preface

The explosive development of molecular biology during the last twenty years illustrated by, for example, the elucidation of the structure and mode of action of haemoglobin makes it imperative for students of biological subjects to have a sound knowledge of organic chemistry. Unfortunately, this fact is not evident to many students at the beginning of their course and they must, therefore, be introduced to organic chemistry in a manner which makes its relevance to biology apparent. This book is an attempt to achieve this objective.

Students of biological subjects do not need to know the host of synthetic reactions which fill most text books of organic chemistry. However, in order to understand biological phenomena at the molecular level, particularly the metabolic pathways by which materials are interconverted in cells, they do need to know selected reactions of the functional groups which are present in molecules which occur in nature. Consequently a major part of the book (chapters four, five and six) is devoted to an account of such reactions using, where possible, a comparative approach (for example, halogeno, hydroxy and amino compounds are discussed collectively). I feel that this type of approach minimises the effort a student must make in order to assimilate the material and also places the role of certain biochemicals, for example *S*-adenosylmethionine, in clearer perspective than if they were discussed individually. Because the nature of the reactions of functional groups depend on the electronic structure of the groups the accounts of chemical reactivity are prefaced by a survey of chemical bonding (chapter 2) in which emphasis is placed, logically, on covalent bonding and by an account of some general aspects of chemical reactions (chapter 3). Stereochemistry, a topic which is of great importance in any account of biochemical reactions, but one which many students find difficult to grasp, is the subject of chapter seven of the book.

The final chapters of the book deal with enzymes and with co-enzymes. In my opinion enzymes and co-enzymes provide the most logical connection between the laboratory reaction of the classical organic chemist and its counterpart in a living cell. Most of the main types of organic reaction which have been discovered by chemists during the last 150 years must have been utilised by living cells since pre-Cambrian times. However, whereas the chemist has comparatively little control over the stereochemical course of a reaction and may have to use vigorous conditions and reaction times measured in minutes or hours in order to obtain adequate yields of products, enzyme catalysed reactions occur in high yield, usually with complete stereospecificity, during a few minutes or less at room temperature. What the chemist struggles towards, nature achieves with consummate ease and has done so for the past five hundred and seventy million years!

I have assumed that readers will have previously studied biology and chemistry to GCE 'A' level or its equivalent and have access to a textbook of biochemistry. I have not included accounts of the physical methods used to purify and to characterise carbon compounds, neither have I detailed any of the evidence on which the reaction mechanisms quoted are based. These omissions are necessary in order to keep the book to a reasonable length; additionally I am of the opinion that students would get better value by studying these topics from specialised monographs.

I am indebted to my colleagues Dr. Christine King and Dr. Rod Smith who read various sections of the manuscript and made suggestions for its improvement. Finally I must acknowledge the fact that the subject matter of a book of this type cannot be original. Here I feel I can do no better than to follow the example of Field Marshal Earl Wavell* by quoting Montaigne:

> 'I have gathered a posie of other men's flowers and nothing but the thread that binds them is my own.'

I hope that the choice of flowers and their manner of arrangement meets with approval and that none of them have been blemished by mis-interpretation of fact.

R. W. Hanson
Plymouth Polytechnic
Plymouth, UK

* Other Men's Flowers. A. P. Wavell. The Penguin Poets. Penguin Books Limited, 1960.

Contents

1

Introduction

The original definition of an **organic** compound was that it was one produced by a living organism. The early chemists were unable to synthesise such compounds in the laboratory and believed that they were formed under the influence of some 'vital force' which operated only in living cells. This was in marked contrast to **inorganic** compounds which could be prepared relatively easily from materials of mineral origin. However, the synthesis of urea (the nitrogenous excretion product of most land-dwelling vertebrates) from ammonium cyanate by Wöhler in 1827 and the laboratory synthesis in subsequent years of many other organic compounds have proved the 'vital force' theory to be incorrect.

$$KCN + PbO = Pb + KCNO$$

$$2KCNO + (NH_4)_2SO_4 = K_2SO_4 + 2NH_4CNO$$

$$NH_4CNO \xrightarrow{\text{heat}} (H_2N)_2CO$$

ammonium cyanate — urea

All organic compounds contain carbon and chemists have been able to synthesise not only carbon compounds which occur in living cells but also those which do not. The adjective organic has therefore come to have a slightly different meaning than it had originally. It now describes all carbon compounds other than those, such as carbonates, which are of mineral origin. The scope of organic chemistry, using the adjective in its modern sense, is obviously immense. Besides natural products, organic chemistry embraces synthetic dyestuffs, pharmaceuticals, agrochemicals (pesticides and herbicides), polymers of all types and many more specialised types of compound. Since many of these topics are outside the direct interest of the student of the life sciences this book uses the adjective organic in its original sense. It deals with those reactions of functional groups* which are relevant to biochemistry in that similar reactions occur in living cells and with the chemistry of the enzymes, the compounds which catalyse these reactions in living cells. Before developing any of these topics it will be useful to classify carbon compounds and to list and name the functional groups which they contain so that readers will be familiar with the structures of any compounds or groups which they encounter subsequently in the text.

* A functional group is a group of atoms which imparts characteristic properties to a compound, see Table 1.1.

1.1. Classification of Carbon Compounds

Carbon compounds can be divided into four main types:

(i) *Aliphatic* compounds which contain 'open' (non-cyclic) carbon chains, e.g. pentane

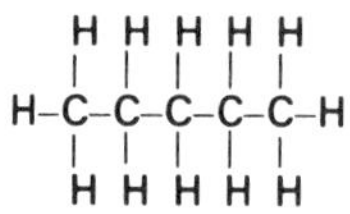

or $CH_3CH_2CH_2CH_2CH_3$

(ii) *Alicyclic* compounds which are the cyclic analogues of aliphatic compounds, e.g. cyclopentane

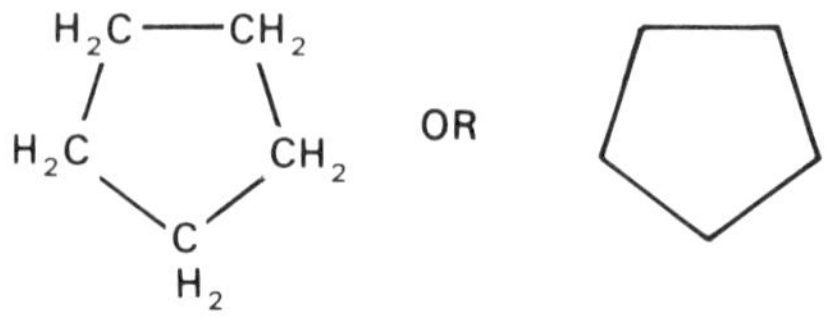

(iii) *Aromatic* compounds which contain an 'aromatic' or 'benzenoid' ring. The simplest compound of this type is benzene:

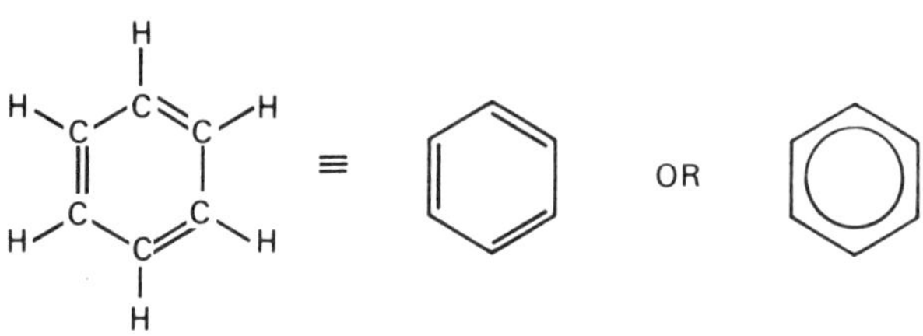

(iv) *Heterocyclic* compounds which are alicyclic or aromatic compounds in which one or more carbon atoms has been replaced by an oxygen, nitrogen or sulphur atom—i.e. by a 'hetero' atom.

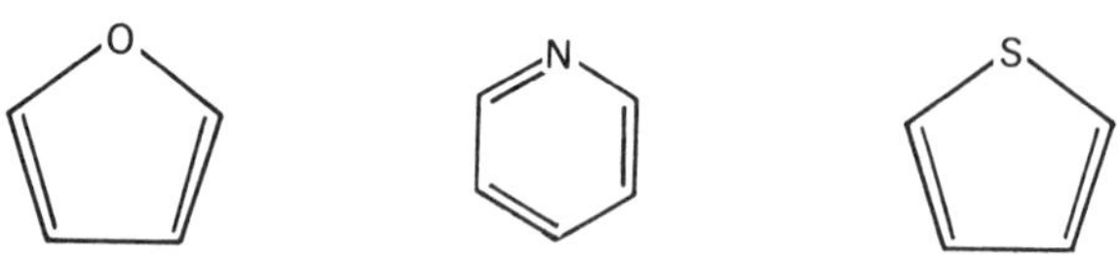

e.g. furan pyridine thiophene

1.2. The structures and nomenclature of functional groups

The four main types of organic compound may contain functional groups which give rise to characteristic properties. The names and structures of those functional groups which frequently occur in organic compounds are given in Table 1.1.

Table 1.1 Functional groups and their nomenclature

Structure of functional group	Name of functional group	Compound containing functional group
>C=C<	double bond	alkenes
–C≡C–	triple bond	alkynes
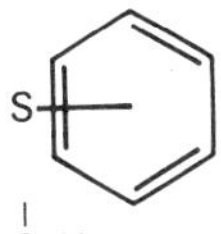	benzene ring (may be substituted-S, indicates a substituent in an undefined position)	arenes
–C–X (with two further bonds on C)		
X = halogen	halogeno	halogeno-alkanes halogeno-arenes
–OH	hydroxy	alcohols (aliphatic) phenols (aromatic)
–OR	alkoxy or aryloxy	ethers
–SH	sulphydryl	thiols or mercaptans
–SR	alkylthio or arylthio	thioethers
$–NH_2$	amino	primary amines
–NHR	alkylamino or arylamino	secondary amines
$–NR^1R^2$	dialkylamino, diarylamino or alkylarylamino	tertiary amines
>C=X with substituent Y		
X = 0, Y = unspecified	carbonyl	see below
H–	aldehydo or formyl	aldehydes
alkyl or aryl	oxo or keto	ketones
HO–	carboxyl	carboxylic acids
RCOO–	acyloxycarbonyl	acid anhydrides
halogen–	halocarbonyl	acid anhydrides or acyl halides
$(HO)_2P(=O)–O–$	phosphoryloxycarbonyl	acid anhydrides or acylphosphates
RO–	alkyloxycarbonyl or aryloxycarbonyl	esters
RS–	(alkylthio)carbonyl or (arylthio)carbonyl	thiol esters
X = NH Y = alkyl or aryl	imino	azomethines
X = NR Y = alkyl or aryl	substituted imino	Schiff bases
–C≡N	carbonitrile or cyano	nitriles or cyanides

2

Bonding in Organic Compounds

In addition to carbon, organic compounds invariably contain hydrogen and usually oxygen. Nitrogen is often present; sulphur and phosphorus occur less frequently, bromine occasionally and chlorine and other elements only rarely. The physical and chemical properties of an organic compound are determined by the number and arrangement of the atoms of these elements in its molecules and by the nature of the chemical bonds between them. An outline of some aspects of chemical bonding is therefore presented in this section as a prelude to the discussions of chemical reactivity contained in chapter 3.

The following important generalisations should be borne in mind when attempting to rationalise chemical bonding, viz: it is an experimental fact that energy must be supplied in order to break **any** chemical bond. For example, the dissociation of one mole of molecular hydrogen into atomic hydrogen requires an input of 431 kJ (Table 2.3). Although it is only homonuclear diatomic molecules which can be cleaved to give their constituent atoms (more complex molecules give correspondingly complex fragments) it is clear that bound atoms, or ions in an ionic lattice, are more stable than their atomic constituents. **Chemical bonding is an exo-energetic process.** Following from this it should be noted that the stabilisation which results when two atoms form a chemical bond is a reflection of the fact that, by forming the bond, each atom effectively attains the electronic configuration of the inert gas nearest to it in the periodic table. As their name implies, the inert gases, each of which possesses the maximum number of electrons in its 'valency' shell, are exceptionally stable; they are the only elements (other than mercury in the vapour state) which exist as atoms at normal temperature and pressure. Two types of bonding process allow atoms to effectively achieve the electronic configuration of an inert gas.

(i) Pairs of electrons may be 'shared' by two atoms provided that the electronegativities of the latter are not too dissimilar. Each electron of the pair interacts electrostatically with both nuclei and hence bonds them together. When each atom formally supplies one electron of the pair the bond is known as a covalent bond; when both electrons are formally supplied by a single atom the bond formed is known as a co-ordinate bond. The number of covalent bonds which an element can form is determined by the number of unpaired electrons which it possesses in its valency shell in the ground state, (or the number which it can give rise to by 'uncoupling' of electron pairs, see later), together with the total number of electrons with which its valency shell can be associated. Finally, note that electron pair bonds are strongly directional, the bonding forces being localised in the internuclear region.

(ii) Strongly electropositive elements may, formally, lose electrons to strongly electronegative ones giving rise to ordered three-dimensional arrays (lattices) of oppositely charged ions held together by strong, non-directional electrostatic forces. This type of bonding is referred to as ionic bonding.

Covalent and co-ordinate bonds will be discussed in more detail subsequently in this section.

2.1. The covalent bond

The covalent bond, in which two atoms are held together by the electrostatic interaction of two electrons with both nuclei is characteristic of organic compounds. Unfortunately electrons cannot be accurately described as point charges with a precise location without contravening Heisenberg's Uncertainty Principle. This principle, which is now widely accepted as one of the fundamental bases of physics, states that 'it is not possible to predict, at the same time, both the position and momentum of a particle of the size of an electron'. If Δx is the uncertainty in the position co-ordinates and Δp the uncertainty in the momentum then

$$\Delta x \Delta p = h/2\pi, \text{ where } h \text{ is Planck's constant}$$

However, this problem can be overcome by the use of wave mechanics. It can be demonstrated experimentally that electrons possess some of the properties of electromagnetic waves and this leads to the fact that an electron can be described in terms of a specific wave function, ψ, the value of which depends on the position of the electron in space and upon its energy. Each wave function is a solution of Schrödingers Wave equation. This generalised differential equation relates the amplitude of an electron wave at any point in space to its total and potential energy. Furthermore ψ is such that the value of its square (ψ^2) at a particular point in space is a measure of the probability of finding the electron at that point in space. Each wave function can be divided into two parts. First, the radial part of the wave function, the square of which determines the probability of finding the electron at any given distance from the nucleus irrespective of the orientation of the electron with respect to the nucleus and second, the angular part of the function the square of which determines the probability of finding the electron in a particular orientation with respect to the nucleus. Plots of the angular parts of wave functions are commonly referred to as **electron orbitals** and are frequently used to give pictorial representations of electrons associated with both free and bound atoms.

The angular part of the wave function of an s electron has spherical symmetry (that is, the probability of finding an s electron at any point in space is independent of the orientation of the electron with respect to the nucleus) and is represented in two dimensions by a circle (2.1a). In contrast the angular part of the wave function of a p electron is not constant but varies with cos θ, where θ is a measure of the orientation of the electron with respect to the nucleus and an appropriate axis. The best way to plot cos θ in this case is as the length of a line inclined at an angle θ with respect to an appropriate axis and (2.1b) illustrates the way in which p_x oribitals are depicted. The diagram indicates that a p_x electron is most likely to be found somewhere (the radial part of the wave function determines where) along the x co-ordinate; p_y and p_z orbitals differ from p_x orbitals only in that their maximum extensions co-incide with the y and z co-ordinates respectively.

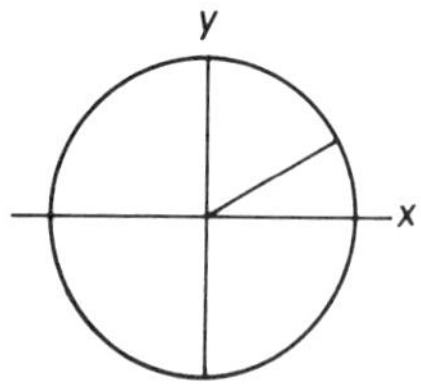

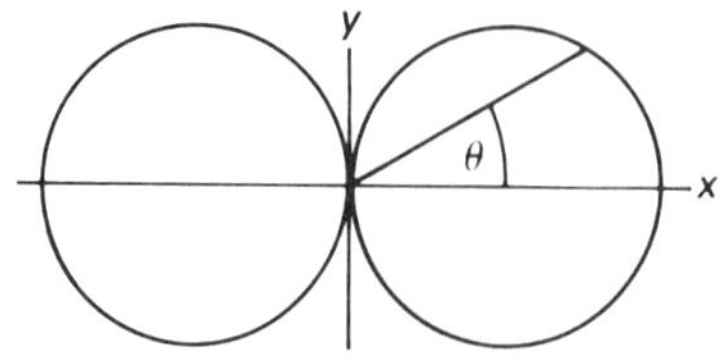

2.1 (a) Plot of the angular part of ψ_{1s} in the *xy* plane; the value of the function corresponds to the length of the radius vector.

(b) Plot of the angular part of the wave function of a p_x electron in the *xy* plane; the value of the function for any value of θ corresponds to the length of the line from the origin to the curve, inclined at an angle θ to the abscissa.

If the value of the square of an electron's wave function at any point in space is a measure of the probability of finding the electron at that point in space then it should be possible to determine the probability of finding each of the two electrons which constitute a covalent bond at every point within the inter-nuclear region. A summation of these probabilities will then give a measure of the 'charge density' in the inter-nuclear region and hence of the bond strength. In other words wave mechanics provides a method of describing covalent bonds which does not contravene Heisenberg's Uncertainty Principle. The mathematical summation of probabilities involved in the wave mechanical description of a covalent bond involves a term known as the 'overlap' integral which contains the wave function of both electrons and it is for this reason that covalent bonds are depicted, graphically, in terms of two overlapping electron orbitals. Note that the angular parts of wave functions are always used in graphical representations of covalent bonds rather than the radial parts of the wave functions. This is because organic molecules are three-dimensional objects and it is only the angular part of a wave function which can give information about the directional characteristics of a bond.

Single covalent bonds in organic compounds

The electronic configurations of the atoms commonly found in organic compounds are as shown in Table 2.1.

It follows that hydrogen is invariably monovalent forming σ bonds (in which the electron density is cylindrically symmetrical about the bond axis) which are **described** in terms of overlap of its 1s orbital with a suitable low energy orbital of another atom (2.2).

Carbon is invariably tetracovalent and in its saturated compounds the four bonds are disposed tetrahedrally. Such compounds cannot, therefore, be described in terms of the electronic configuration of carbon in the ground state which would predict divalency with bond angles of 90°. Wave mechanics describes saturated tetracovalent carbon as follows. The spins of the 2s electrons are 'uncoupled', one is 'promoted' and now described by the $2p_z$ orbital. It can be shown that if ψ_{2s}, ψ_{2p_x}, etc. are separately solutions of

Table 2.1 The electronic configurations of elements which occur in organic compounds

H	$1s^1$						
C	$1s^2$,	$2s^2$,	$2p_x^1$,	$2p_y^1$			
N	$1s^2$,	$2s^2$,	$2p_x^1$,	$2p_y^1$,	$2p_z^1$		
O	$1s^2$,	$2s^2$,	$2p_x^2$,	$2p_y^1$,	$2p_z^1$		
P	$1s^2$,	$2s^2$,	$2p^6$,	$3s^2$,	$3p_x^1$,	$3p_y^1$,	$3p_z^1$
S	$1s^2$,	$2s^2$,	$2p^6$,	$3s^2$,	$3p_x^2$,	$3p_y^1$,	$3p_z^1$

Schrödinger's wave equation (that is, acceptable atomic orbitals) then various linear combinations of them are also solutions.

$$\psi_1 = \tfrac{1}{2}(\psi_s + \psi_{p_x} + \psi_{p_y} + \psi_{p_z})$$
$$\psi_2 = \tfrac{1}{2}(\psi_s + \psi_{p_x} - \psi_{p_y} - \psi_{p_z})$$
$$\psi_3 = \tfrac{1}{2}(\psi_s - \psi_{p_x} + \psi_{p_y} - \psi_{p_z})$$
$$\psi_4 = \tfrac{1}{2}(\psi_s - \psi_{p_x} - \psi_{p_y} + \psi_{p_z})$$

In particular the combinations shown are found to describe four equivalent orbitals whose maximum angular probability vectors are directed towards the apices of a regular tetrahedron as in (2.3). These orbitals are known, logically, as sp^3 hybrid orbitals. Throughout this book, full lines in perspective diagrams of this type indicate orbital axes (or bond axes where appropriate) which lie in the plane of the paper, broken line axes directed behind the plane of the paper and wedge axes directed out of the plane of the paper. A compound such as methane (CH_4) can be described in terms of overlap of the four hybrid orbitals with the 1s orbital of four hydrogen atoms (2.2). Readers should bear in mind that uncoupling, promotion and hybridisation are mathematical manipulations which are necessary in order to describe a molecule as it already exists but otherwise have no physical reality.

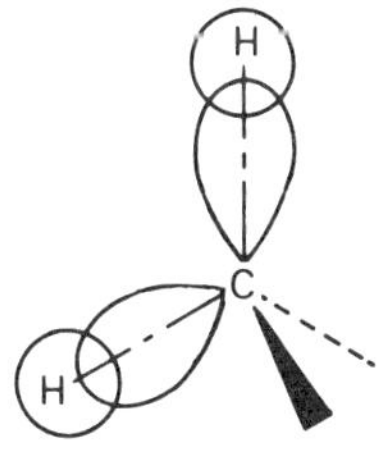

(2.2)
The methane molecule. Only two of the four σ bonds are shown for clarity.

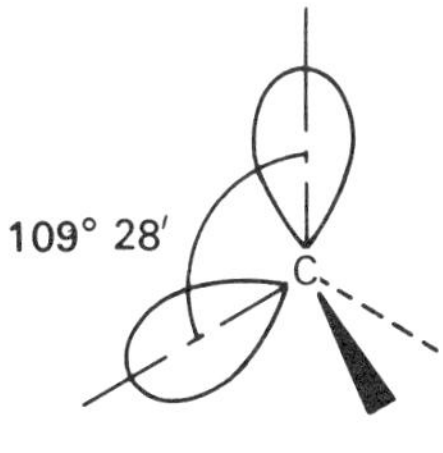

(2.3)
Angular probability diagrams for sp^3 hybrid orbitals. Only two shown for clarity.

Single covalent bonds to nitrogen, phosphorus, oxygen and sulphur can be described similarly in terms of sp^3 hybrid orbitals. In the case of trivalent compounds of nitrogen and phosphorus the three bonding electrons are described by three of the hybrid orbitals while the fourth describes a pair of non-bonding electrons (the 'lone pair'). Compounds of this type, for example ammonia and simple, symmetrically substituted, aliphatic amines are indeed pyramidal but the bond angles at the nitrogen atom are less than the tetrahedral angle. It appears that a lone pair of electrons, because it does not interact with a second nucleus, is held nearer to its 'parent' nucleus than are bonding electrons. The lone pair therefore repels the bonding pairs, decreasing the bond angles. Similarly in the case of dicovalent oxygen and sulphur the two bonding electrons are described by two sp^3 hybrid orbitals while the two remaining orbitals describe two lone pairs of electrons. The carbon–oxygen–carbon bond angle in compounds such as the ethers is approximately 110° while even smaller angles are observed in hydrogen sulphide and its organic derivatives.

The valence shells of phosphorus and sulphur, both of which are in Period III of the periodic table, can contain up to 18 electrons ($3s^2$, $3p^6$, $3d^{10}$). It is therefore possible for these elements to use all of their valence electrons for covalent bond formation by 'uncoupling' of one or two lone pairs. However, **saturated** derivatives of tetra- or hexavalent sulphur are of little or no interest in organic chemistry.

Before leaving the topic of single covalent bonds it is worth noting that the unique role of carbon in the formation of organic compounds is a reflection of its ability to form strong covalent bonds (Table 2.3) with itself and with other elements leading to an immense variety of linear, branched and cyclic structures.

Multiple covalent bonds in organic compounds

Double bonds

Carbon to carbon double bonds, as in compounds such as ethene ($H_2C{=}CH_2$), are described by another type of hybrid orbital. It is known, from physical measurements, that the six atoms in the ethene molecule are coplanar with H–C–H and H–C–C bond angles of 120° and that the methylene groups ($H_2C{=}$) cannot be rotated with respect to each other about the bond axis. These facts can be accommodated by a description which utilises three equivalent co-planar sp^2 hybrid orbitals (2.4a), each formed by a suitable linear, mathematical combination of ψ_{2s}, ψ_{2p_x} and ψ_{2p_z}, and the unhybridised $2p_y$ orbital of each carbon atom. The maximum angular probability vectors of the hybrid orbitals make angles of 120° to each other and the plane in which they all lie is perpendicular to the maximum probability vector of the $2p_y$ orbital. A carbon to carbon σ bond is described by overlap of one sp^2 hybrid orbital from each carbon atom (2.4b), the second carbon to carbon bond is then described in terms of sideways overlap of the unhybridised $2p_y$ orbital of each carbon atom (2.4c). This type of interaction between p electrons gives rise to an accumulation of charge above and below the inter-nuclear axis rather than symmetrically about it as in a σ bond and the bond which results is known as a π bond. It should be clear from (2.4c) that because the probability vectors in the region of overlap are smaller than the maximum values a π bond will be weaker than a σ bond (see Table 2.3); it is, however, strong enough to prevent rotation of the carbons with respect to each other about the σ bond. Finally two bonds between hydrogen atoms and each carbon atom are described in terms of overlap between hydrogen 1s orbitals and sp^2 hybrid orbitals of carbon (2.4d).

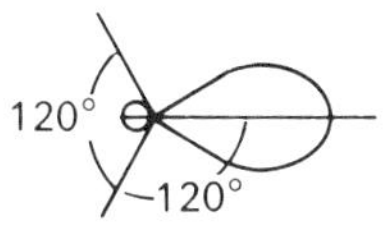

(a)

(a) angular probability function for an sp^2 hybrid orbital (only one shown, for clarity)

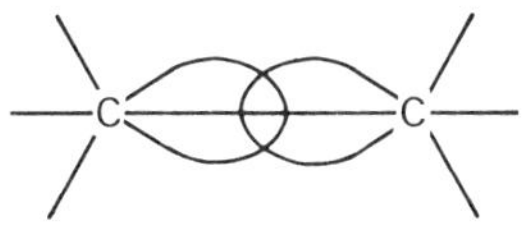

(b)

(b) carbon to carbon σ bond described by overlap of two sp^2 hybrid orbitals in the *xz* plane

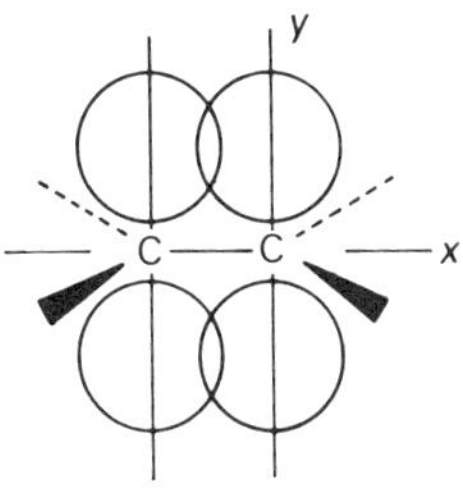

(c)

(c) carbon to carbon π bond formed by sideways overlap of two p_y orbitals

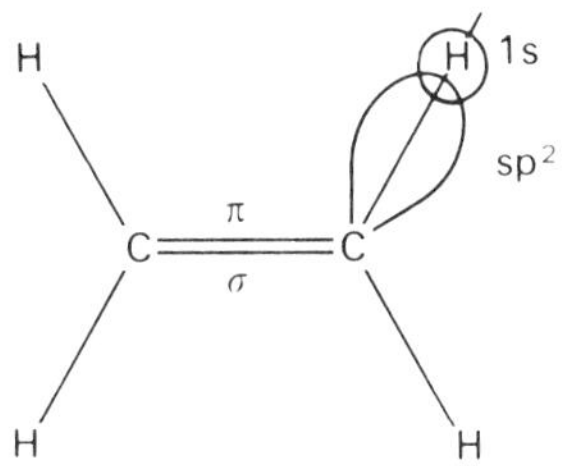

(d)

(d) carbon to hydrogen bonds in ethene. All atoms lie in the *xz* plane

(**2.4**) Orbital diagrams of bonding in ethene.

Double bonds to nitrogen, as in the imino group, (>C=N–), are described in a similar way to the carbon to carbon double bond, except that one of the sp^2 hybrid orbitals of nitrogen now describes a lone pair of electrons rather than a single bonding electron (2.5). Oxygen can also form double bonds, for example as in the carbonyl group (>C=O). Here, it is not necessary to invoke hybrid orbitals of oxygen to describe the bonding. One of the two bonds is described as a σ bond in terms of overlap of the carbon sp^2 hybrid orbital with the $2p_x$ orbital of oxygen (2.6a) and the other as a π bond in terms of sideways overlap of the carbon $2p_y$ and oxygen $2p_y$ orbitals (2.6b).

Triple bonds

Carbon can also form triple bonds as, for example, in the linear ethyne (HC≡CH) molecule. The bonding in this molecule is described in terms of a third

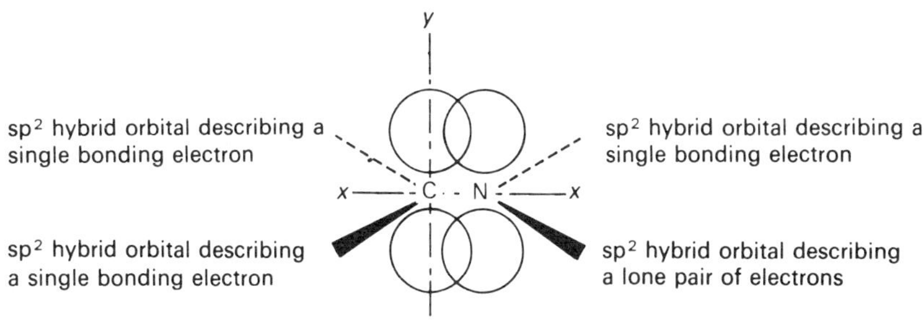

(**2.5**) Orbital diagrams for a >C=N– bond.

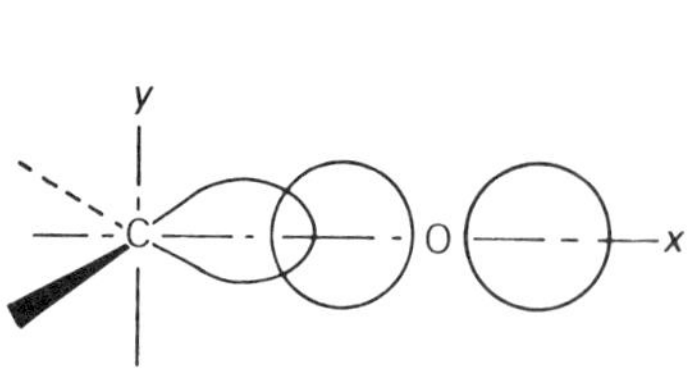

(a)

(a) σ bond described by overlap of carbon sp^2 and oxygen $2p_x$ orbitals.

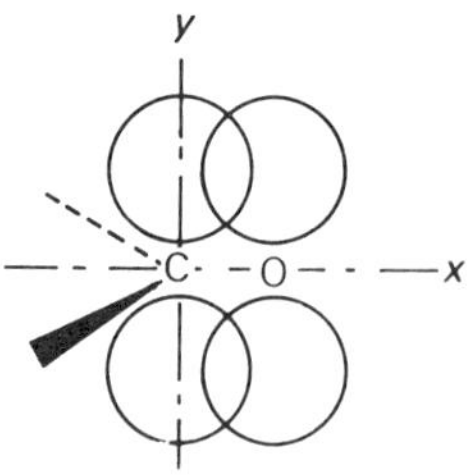

(b)

(b) π bond described by sideways overlap of carbon $2p_y$ and oxygen $2p_y$ orbitals

(**2.6**) Orbital diagrams for a >C=O bond.

type of hybrid orbital. Two appropriate linear, mathematical combinations of ψ_{2s} and ψ_{2p_x} are found to describe two equivalent, colinear, sp hybrid orbitals whose maximum angular probability vectors coincide with the *x* co-ordinate (2.7a). The two remaining valence electrons of each carbon atom are described by the usual $2p_y$ and $2p_x$ orbitals. A σ bond is described by overlap between one of the hybrid orbitals of each carbon atom (2.7b) and **two** π bonds are described in terms of sideways overlap of the $2p_y$ and $2p_z$ orbitals on each carbon atom (2.7c,d). The one hybrid orbital remaining on each carbon atom is used to describe a bond to a hydrogen atom. In keeping with experiment the description is of a linear system.

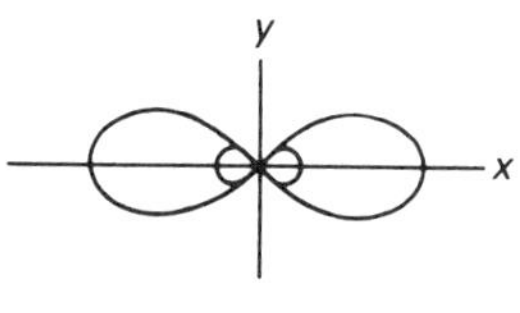

(a)

(a) co-linear sp hybrid orbitals

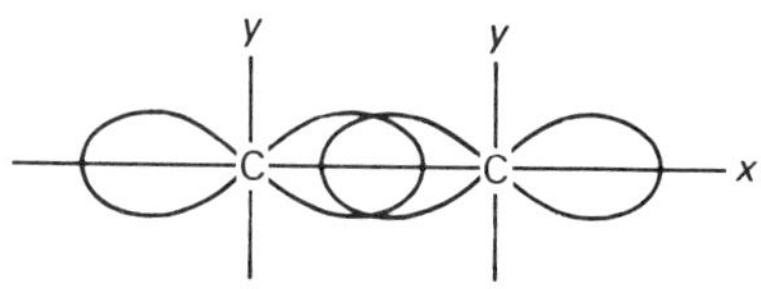

(b)

(b) carbon to carbon bond formed by overlap of sp hybrid orbitals

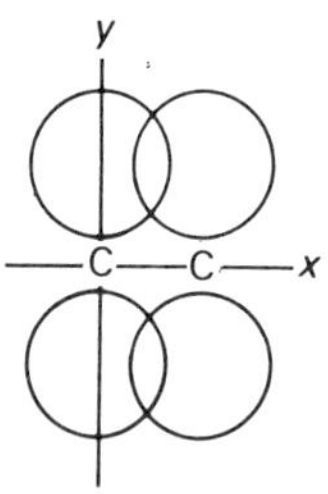

(c)

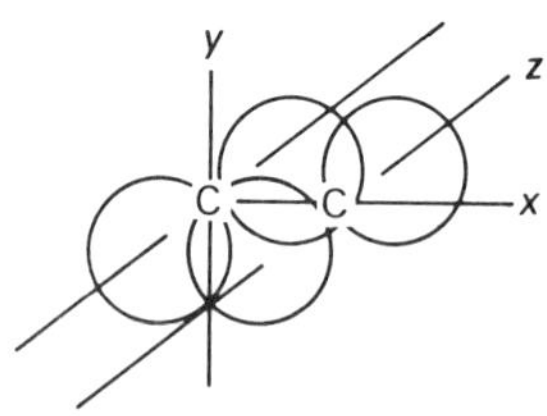

(d)

(c), (d) carbon to carbon bonds formed by overlap of $2p_y$ and $2p_z$ orbitals

(**2.7**) Orbital diagrams for a –C≡C– bond.

A triple bond to nitrogen, as in the carbonitrile (cyano) group, $-C\equiv N$, is described in the same way as a triple bond to carbon. The lone pair of electrons of the nitrogen atom is described by the sp hybrid orbital which is not involved in bonding to the carbon atom.

Multiple bonds involving phosphorus and sulphur

Silicon, phosphorus and sulphur, in contrast to their Period II cogeners carbon, nitrogen and oxygen, form relatively few compounds containing multiple bonds. Tricovalent derivatives of phosphorus containing double or triple bonds are extremely rare and of little or no interest to biochemists. Unsaturated compounds in which phosphorus is formally pentavalent are abundant and, of these, derivatives of orthophosphoric acid (2.8a) are very important in biochemistry. Orthophosphoric acid derivatives are tetrahedral in shape; it is debatable whether the phosphorus to oxygen bond is a true double bond (2.8a) or whether it is better described as a co-ordinate bond (2.8b, see Section 2.2).

$$\underset{(a)}{(HO)_3P{=}O} \qquad \underset{(b)}{(HO)_3P \rightarrow O} \qquad \textbf{(2.8)}$$

Dicovalent sulphur can form conventional double bonds, for example in the thiocarbonyl group ($>C{=}S$) but compounds containing such groups are of little importance in biochemistry. Additionally sulphur participates in multiple bonds in compounds in which it is formally tetra- or hexacovalent. For example sulphur in organic sulphoxides (2.9a) is formally tetracovalent while in sulphones (2.9b), in derivatives of sulphonic acids (2.9c) and in sulphuric acid (2.9d) it is formally hexacovalent. Again, it is debatable whether the double bonds shown in these compounds are true double bonds or whether they are better represented as co-ordinate bonds.

(a) (b) (c) (d) **(2.9)**

Polarity in covalent bonds

The foregoing account of covalent bonds suggests that the electrons in such a bond interact equally with both bound nuclei. This is true only if the bound atoms are identical. In contrast, consider a carbon to oxygen bond. The electronegativity of oxygen is considerably greater than that of carbon. Consequently the electrons of the bond interact more strongly with the oxygen nucleus than with the carbon nucleus giving rise to a small excess of negative charge associated with the former atom and a corresponding deficiency of charge associated with the latter atom.

As the bond is associated with small electrical poles it is said to be a **polar** covalent bond. Polarity in covalent bonds is a very important determinant of both the physical and chemical properties of an organic compound and will be discussed further in Sections 2.7 and 3.1.

2.2. Co-ordinate bonding in organic compounds

A co-ordinate bond is a special case of covalency where one atom (the donor atom) supplies **both** electrons of the bond. The acceptor atom must, of course, be such that it can form a stable association with the electron pair. There is a temptation to say that the acceptor atom must have an empty orbital in which the electrons can be accommodated—readers are reminded of the mathematical nature of orbitals. The simplest examples of co-ordinate bond formation which are relevant to organic chemistry involve the formation of 'onium ions'. Ammonia and the organic amines which are derived from it are basic because they can utilise the lone pair of electrons of the nitrogen atom to form a co-ordinate bond with a proton.

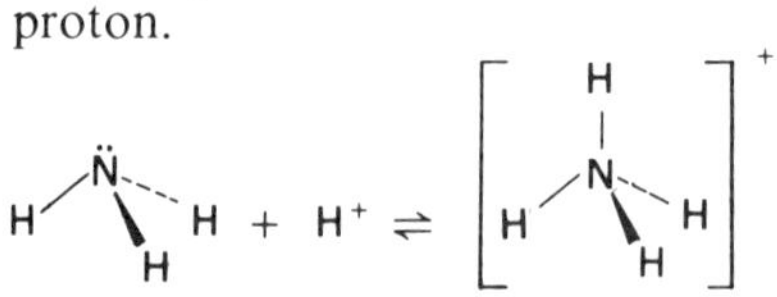

The four bonds in the ammonium ion which is formed are identical; in other words the co-ordinate bond, once formed, is no different from a normal covalent bond. Oxygen, phosphorus and sulphur can increase their covalency in a similar manner by forming oxonium, phosphonium and sulphonium salts. (See Chapter 5.)

It should be noted that a co-ordinate bond is invariably associated with some degree of polarity. The acceptor atom effectively gains control of one electron at the expense of the donor atom. The former is therefore often shown as bearing a negative charge and the latter a positive charge.

$$A{:} + B \rightleftharpoons \overset{+}{A}{:}\overset{-}{B}$$

In practice however, the degree of polarity will depend on the relative electronegativities of the bound atoms. If A is more electronegative than B then there will be less than one unit of negative charge associated with the latter atom. Consequently, it is usual to denote a co-ordinate bond, associated with an unspecified degree of polarity, by an arrow, the tail originating on the donor atom and the head on the acceptor atom, see (2.10).

Co-ordinate bonding is most highly developed in derivatives of the transition elements. The electronic configurations of the first short transition series is shown in Table 2.2 from which it can be seen that the elements are characterised by possession of partially filled d orbitals. The energy of a 3d electron is not very different from that of a 4s or 4p electron, consequently comparatively little energy is required to re-distribute the valency electrons of the transition elements among these energy levels and the elements exhibit variable valency. These valence states are stabilised by the formation of coordination complexes in which 'empty' 3d, 4s or 4p orbitals of the transition metal accept electron pairs from donor ions or molecules, known as ligands. For example, water molecules stabilise cupric ions in hydrated copper sulphate.

$$CuSO_4.5H_2O \equiv [Cu(H_2O)_4]^{2+}\,SO_4^{2-}.H_2O$$

Table 2.2 Electronic configurations of the first series of transition metals

	3d					4s
Sc	↑					↑↓
Ti	↑	↑				↑↓
V	↑	↑	↑			↑↓
Cr	↑	↑	↑	↑	↑	↑
Mn	↑	↑	↑	↑	↑	↑↓
Fe	↑↓	↑	↑	↑	↑	↑↓
Co	↑↓	↑↓	↑	↑	↑	↑↓
Ni	↑↓	↑↓	↑↓	↑	↑	↑↓
Cu	↑↓	↑↓	↑↓	↑↓	↑↓	↑
Zn	↑↓	↑↓	↑↓	↑↓	↑↓	↑↓

The four electron pairs from water are formally described by a 3d, 4s and two 4p orbitals. Experimentally, however, the four ligands are found in a square plane; the bonds to them from copper are therefore described in terms of four equivalent spd^2 hybrid orbitals* (Although hybrid orbitals can be used to describe the shapes of transition metal complexes they do not give a satisfactory explanation of the spectra of the compounds. The use of hybrid orbitals has now been superseded by the Ligand Field Theory to describe the bonding in co-ordination complexes.) Nitrogen ligands can partially, or completely, replace water in the tetra-aquo-copper II complex. Thus, when excess ammonia is added to copper sulphate a deep blue solution containing the tetra-aminocopper II ion is obtained.

$$[Cu(H_2O)_4]^{2+} + 4NH_3 \rightleftharpoons [Cu(NH_3)_4]^{2+} + 4H_2O$$

Many organic ligands contain more than one donor group. They are known as di-, tri-, tetra- or multidentate ligands depending on the number of electron pairs available for bonding. For example 2-amino acids (5.52) exist as anions in alkaline solution and two of these anions can replace the ammonia in the tetra-aminocopper II ion (2.10).

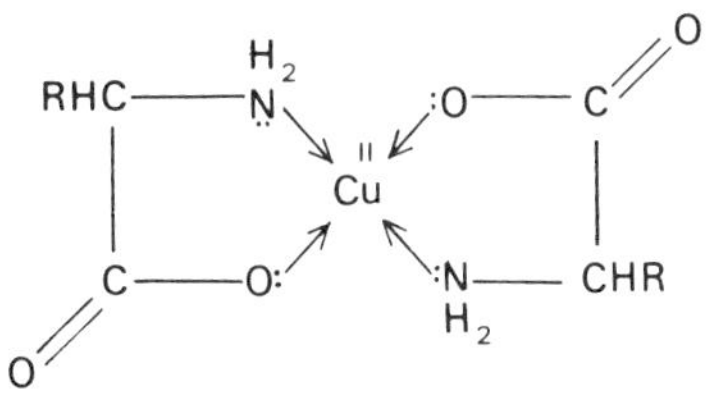

(2.10)

Ligands such as these which give rise to ring formation are often described as **chelating** agents: the metal ion is said to be **chelated** or **sequestered**. One of the commonest and most useful synthetic multidentate ligands is ethylenediamine tetra-acetic acid (EDTA), which is frequently used in analytical chemistry for assaying a number of metals. It is often added to enzyme assays to sequester

* Not all copper complexes are planar. Tetrahedral complexes are also known; the bonding in them is described in terms of sp^3 hybrid orbitals.

metals which would otherwise inhibit the enzyme.

EDTA 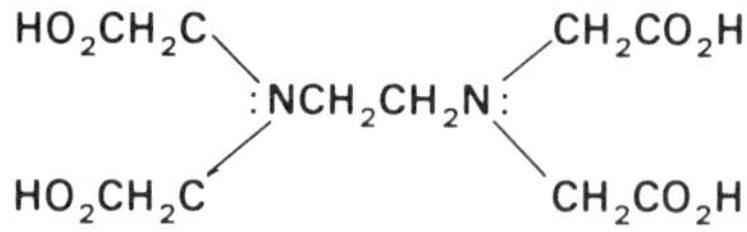

Co-ordination complexes are of prime importance in biochemistry. One of the most thoroughly studied examples is undoubtedly haemoglobin, the red, iron containing protein which transports oxygen in the blood of most vertebrates and some invertebrates. The oxygen binding site in the oxygenated form of haemoglobin contains octahedrally hexa co-ordinated ferrous iron with oxygen forming one of the six ligands. Many enzymes contain co-ordinately bound metal ions which are essential for their activity; an example is carboxypeptidase A which is discussed in Section 8.3.

2.3. Ionic bonding in Organic Compounds

Ionic bonds, which result from the complete transfer of an electron from one atom to another, are characteristic of crystalline salts, such as sodium chloride, and not, in general, of organic compounds. They are, of course, present in crystals of the salts of organic acids ranging from low molecular mass acids such as ethanoic acid to polymers such as the nucleic acids and in crystals of quaternary ammonium compounds such as choline $(HOCH_2CH_2N^+(CH_3)_3\ :\bar{O}H)$.

Electrostatic forces of the same type as those involved in ionic crystals contribute to the aggregation and folding of protein chains. Here a protonated basic group in one chain interacts with an anionic centre, usually a carboxylate group, in another chain or in another part of the same chain.

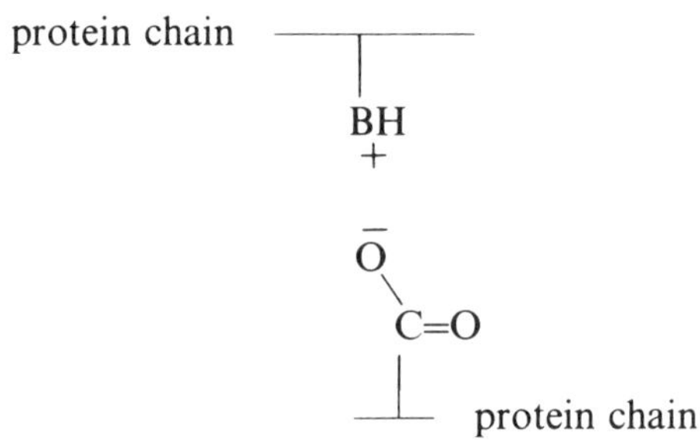

This type of bond is usually referred to as a 'salt bridge'.

2.4. Resonance Hybrids

The concept of resonance hybrids stems from attempts to describe the hydrogen molecule using wave mechanics. Early attempts to calculate the bond energy of the molecule in terms of the purely covalent model (2.11a) did not provide a satisfactory value. It was then found that considerably better agreement between the calculated and observed bond energies was obtained if it was assumed that the electron pair is **not completely localised** between the two nuclei but can interact, at least momentarily, with one or other nucleus exclusively. As the calculation of the bond energy involved a mathematical analogy to resonating pendulums it was suggested that the hydrogen molecule could be described as a **resonance hybrid** in terms of the three **hypothetical** structures (2.11 a, b, c).

H : H	$H:^-$ H^+	H^+ $:H^-$	**(2.11)**
(a)	(b)	(c)	

The use of the term 'resonance hybrid' is unfortunate in that it has connotations with interaction, or oscillation, between physically distinct structures. There is, however, only one form of hydrogen molecule or of any other molecule which is described as a resonance hybrid. The situation is that no one classical valence diagram can be used to describe such molecules adequately. Better descriptions can be obtained by taking a number of diagrams together, each describing certain physical or chemical properties of the molecule, and each having a weighting in the overall description in proportion to the property which it describes. For example, in the case of hydrogen the hypothetical ionic structures have only a small weighting in the overall description of the molecule which is reasonably well, but not exactly, described by the covalent formula.

Although resonance theory may seem somewhat artificial (and confusing!) at first sight it has proved to be of great value in chemistry. Many ions and molecules, both inorganic and organic, are known which cannot be described adequately by a single valence diagram. For example, the two carbon to oxygen bonds in the carboxylate ion of the salts of organic acids are identical. Diagram (2.12a) is therefore an inadequate description of the ion since it suggests that one bond is different from the other. However, (2.12a and b) equally weighted and taken together do give an adequate description of the ion.

(a) (b) **(2.12)**

An alternative way of describing the carboxylate group would be to draw atomic orbitals to depict the π bond and the lone pair of electrons responsible for the negative charge.

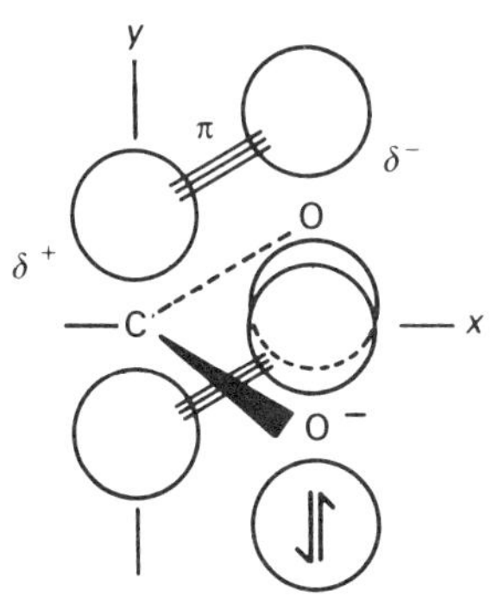

The carboxylate ion.
(lateral dimensions exaggerated for sake of clarity)

The carboxyl group, in the absence of other factors, would be polarised as shown. Consequently charge described by the $2p_y$ orbital of the 'anionic oxygen' interacts with the electron deficient carbon atom and continues to be **delocalised** until the charge density in both bonds is equal. Benzene is yet another example of a molecule which is best described as a resonance hybrid and in which bonding electrons are delocalised over several nuclei (Section 4.3). In fact, as these examples suggest, the necessity to describe a molecule, group or ion as a resonance hybrid is always a reflection of the fact that electronic charge is delocalised to some extent within the structure. This phenomenon will be returned to in Section (3.2).

Diagram (2.12) illustrates the usual way of depicting resonance hybrids. The appropriate structural formulae are interconnected with double-headed arrows (not to be confused with $\rightleftharpoons$ which indicates a reversible process) and enclosed in brackets to emphasise that they describe a single species. The nature of the diagrams which are used to describe a structure as a resonance hybrid depend on the physical and chemical properties of the molecule. However, some generalisations can be made:

(i) The atoms must always have the same relative orientation in space with respect to each other (otherwise the diagrams would not describe the same species).

(ii) the direction of 'formal movement' of electrons within the diagrams is generally (but not invariably, see page 26) in accord with the relative electronegativities of the bound atoms. Such formal movements are only possible if the acceptor atom possesses a suitable, vacant,* low-energy, orbital in which the electrons can be accommodated.

(iii) each diagram must be associated with the same number of unpaired electrons and each must possess the same net charge (otherwise they do not describe the same species).

2.5. The hydrogen bond

Hydrogen bonds are weak (4–25 kJ mol^{-1}) electrostatic bonds operating over a distance of about 0.12 to 0.15 nm. The electric charges which generate the forces of attraction are always associated with a system of the type shown in (2.13) where the electronegativity of X (X = 0, N or F) is much greater than that of H or Z (Z = H or C).

$$-\overset{\delta-}{X}-\overset{\delta+}{H}\cdots\cdots\cdots\overset{\delta-}{X}-\overset{\delta+}{Z}- \qquad \textbf{(2.13)}$$

The involvement of hydrogen in this type of bonding reflects the fact that removal of charge from a hydrogen atom partially exposes a small, bare nucleus with an associated intense electrostatic field. Although hydrogen bonds are only weak the formation of a large number of them can markedly stabilise a system. Thus many of the properties of water can be explained as being due to the existence of an extensive system of hydrogen bonds, in the solid, liquid and vapour states. In ice each water molecule is hydrogen bonded to 4 others arranged tetrahedrally.

* See p. 12.

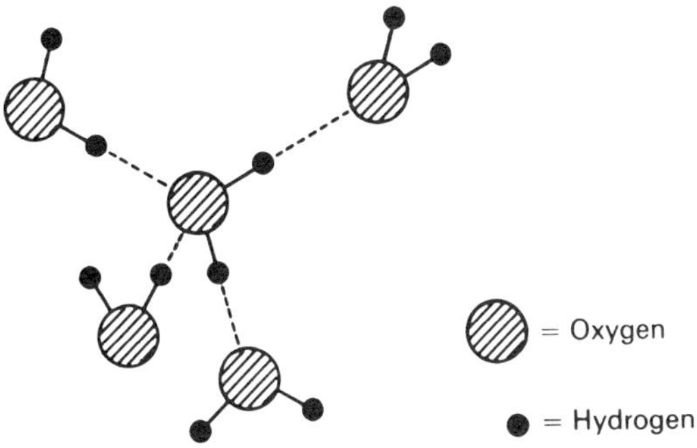

Hydrogen bonded water molecules

In liquid water at 0° C much of the ice structure persists—each water molecule is hydrogen bonded to, on average, 3.4 other molecules and hydrogen bonding persists in the vapour phase up to about 600°C. Both the high boiling point and high latent heat of vapourisation of water can be attributed to hydrogen bonding. This type of bonding is also of great importance in determining and maintaining the structure of both proteins and nucleic acids.

2.6. Bond strengths and lengths

The strength of a chemical bond is measured by the **bond dissociation energy** (*D*), which is defined as the energy absorbed when a bond is caused to dissociate into neutral fragments in the vapour phase. As can be seen from the first four entries in Table 2.3 the dissociation energy of a particular bond varies with the chemical environment of that bond. In many situations average bond energies (*E*) as listed in the remainder of Table 2.3 are more useful. As would be expected a triple bond between two given atoms is stronger and shorter than a double bond, which in turn is shorter and stronger than a single bond between the same atoms. Note, however, that a double bond is not twice as strong as a single bond, indicating that a π bond is weaker than a σ bond, as would be expected on the basis of the wave mechanical description of the bond (page 8).

Table 2.3 Some representative bond energies and lengths

Bond	Energy ($kJ\,mol^{-1}$)	Length (nm)
CH_3–H	427	
CH_3CH_2–H	406	
$(CH_3)_2CH$–H	393	
$(CH_3)_3C$–H	374	
$-\overset{\vert}{\underset{\vert}{\mathrm{C}}}-\mathrm{H}$ (av)	414	0.109
>N–H	389	0.103
–O–H	464	0.097
–S–H	347	
$-\overset{\vert}{\underset{\vert}{\mathrm{C}}}-\overset{\vert}{\underset{\vert}{\mathrm{C}}}-$	347	0.154

Table 2.3 (Contd.)

Bond	Energy (kJ mol^{-1})	Length (nm)
$\rangle C{=}C\langle$	607	0.140
–C≡C–	803	0.121
–C–O–	335	0.143
$\rangle C{=}O$	694	0.120
$-\overset{\vert}{\underset{\vert}{C}}-S-$	272	0.182
$-\overset{\vert}{\underset{\vert}{C}}-N-$	284	0.147
–C≡N	866	
$-\overset{\vert}{\underset{\vert}{C}}-F$	447	0.142
$-\overset{\vert}{\underset{\vert}{C}}-Cl$	326	0.177
$-\overset{\vert}{\underset{\vert}{C}}-Br$	284	0.191
$-\overset{\vert}{\underset{\vert}{C}}-I$	213	0.213
H–H	431	0.074
–H - - - X–	4–25	0.12–0.15

2.7. Some physical properties of organic compounds in relation to structure

Solubility

Non-polar covalent compounds are soluble in solvents of a similar nature but are insoluble in water. However, the introduction of a polar or an ionic group into a molecule often causes the latter to be soluble in water because electrostatic interactions between the group and the solvent makes the process of solution energetically favourable. For example, whereas propane (2.14a) is insoluble and hexanoic acid (2.14c) is only sparingly soluble in water, propanone (2.14b) and sodium hexanoate (2.14d) are freely soluble.

$$\underset{(a)}{CH_3CH_2CH_3} \qquad \underset{(b)}{CH_3\overset{\overset{O}{\|}}{C}CH_3} \qquad \underset{(c)}{CH_3(CH_2)_4CO_2H} \qquad \underset{(d)}{CH_3(CH_2)_4CO_2^-Na^+} \qquad \mathbf{(2.14)}$$

It should be borne in mind that the solubility of a carbon compound reflects the balance between hydrophilic and hydrophobic groups within the molecule. This point is illustrated by the homologous aliphatic alcohols. (A homologous series of compounds is one in which the structure of any compound differs from that of adjacent ones by only a methylene group, $-CH_2-$.) Methanol (CH_3OH) and ethanol (CH_3CH_2OH) are miscible with water in all proportions, solubility then decreases as hydrophobic methylene groups are added and hexanol ($CH_3(CH_2)_4CH_2OH$) is almost completely insoluble. Most true organic compounds are soluble in water (many exist in aqueous solution in the cytoplasm of cells) despite the fact that they may have very high molecular masses. This is because their molecules contain several (often a large number) of polar or ionic groups which can interact with water.

Volatility, boiling point and melting point

Low molecular mass, non-polar carbon compounds are either gaseous at room temperature or exist as volatile liquids with low boiling points. As the molecular mass increases volatility decreases, boiling points increase, and many non-polar carbon compounds of moderate relative molecular mass (e.g. naphthalene, $M_r = 80$) exist at room temperature as solids with low melting points. High volatility, low boiling points and melting points are a reflection of the fact that the forces of attraction between non-polar molecules in the liquid or solid state are weak. They are, in fact, weak ($\approx$ 5 kJ mol^{-1}) electrostatic forces between dipoles in adjacent molecules (dipole–dipole interactions) and are known as Van Der Waal's forces.

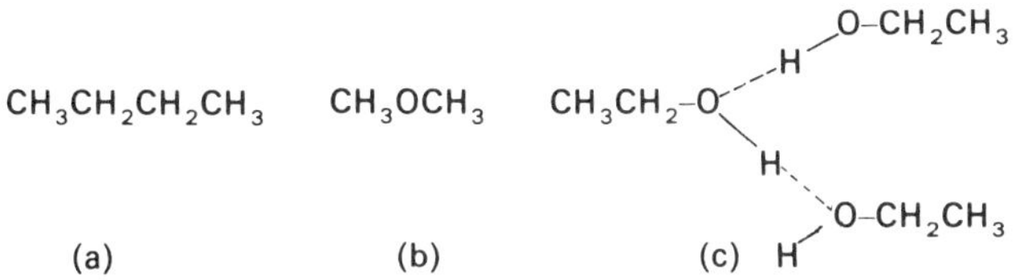

(2.15)

This situation changes markedly if electrical polarity or, more particularly, ionic centres are present in a molecule. For example butane (2.15a) has b.p. -5°C whereas propanone (2.14b) boils at 56.5°C although they have the same molecular mass. The carbonyl group is polar and appreciable dipole–dipole interactions have to be overcome in order to vaporise the liquid. The change in boiling point is even more marked if the polarity in a molecule is such as to give rise to intermolecular hydrogen bonding as, for example, in the simple alcohols. Methoxymethane (2.15b) boils at -24°C whereas ethanol (2.15c) with which it is isomeric boils at 78.5°C. More energy has to be supplied to vaporise the alcohol because the hydrogen bonds between its molecules must be broken. Similarly the melting point of hexadecanoic acid (palmitic acid, $CH_3(CH_2)_{14}CO_2H$) is 63°C whereas its sodium salt does not melt, but instead decomposes at about 150°C, the exact temperature depending on the rate of heating. Only weak Van Der Waal's forces have to be broken to disrupt crystals of the acid whereas strong ionic bonds ($\approx$ 100 kJ mol^{-1}) must be broken in the case of the salt.

Again, true organic compounds are possibly atypical of carbon compounds in general. Their role in life processes usually demands that they should be non-volatile at physiological temperatures. The same polar and ionic groups which confer water solubility on many organic compounds also confer negligible volatility and most organic compounds decompose at moderately high temperatures when heated, rather than melt.

3

Chemical Reactions

Chapters 4, 5 and 6 describe reactions of functional groups which are relevant to biochemistry. It is appropriate to preface these descriptions with the discussion, below, of some general aspects of chemical reactions.

3.1. General aspects of chemical reactions

Chemical reactions involve collisions between molecules, atoms or ions of the reactants. The collisions cause some of the existing bonds in the reactants to break and new bonds to form thus creating products. Since chemical reactions are complex processes, they are usually studied and discussed in terms of their thermodynamics, kinetics and mechanism rather than as a whole. The discipline of **chemical thermodynamics** enables the energy changes accompanying a reaction to be analysed in terms of a number of the properties of the reacting system. The results of such an analysis allows one to predict whether the reaction will occur under a given set of conditions and, if so, how far it will proceed before equilibrium is reached. **Chemical kinetics** is concerned with the rates of reactions and the factors which determine them. (An introductory account of both disciplines as applied to biological problems is available in 'A Biologist's Physical Chemistry', J. Gareth Morris, Edward Arnold Ltd., London, 1968.) The **reaction mechanism** describes the order in which bonds are made and broken and the nature of the intermediates which are formed during the reaction. It is this aspect of organic reactions which is, perhaps, of most importance to students of the biological sciences, since knowledge in this area should allow metabolic reactions to be rationalised.

3.2. Reaction mechanisms

The starting point in any attempt to determine the mechanism of a reaction is to postulate a pathway based on the electronic properties of the reactants and which is compatible with the kinetics of the reaction. Experiments are then designed to produce evidence in support of the postulated mechanism. It is not the purpose of this section to detail the experimental methods which are used to investigate reaction mechanisms but rather to outline how the electronic structures of the reactants determine, at least in part, the mechanism of a reaction. As a first step towards this end it is useful to classify both reactions and reagents.

Classification of reactions

Four basic types of reaction are known, involving:

(i) substitution, or displacement, of an atom or group in a molecule by another atom or group. Both saturated aliphatic and aromatic compounds undergo substitution (3.1, 3.2).

$$\underset{\text{bromomethane}}{CH_3Br} + H\bar{O} \overset{\text{heat}}{\rightleftharpoons} \underset{\text{methanol}}{CH_3OH} + \bar{Br} \qquad \textbf{(3.1)}$$

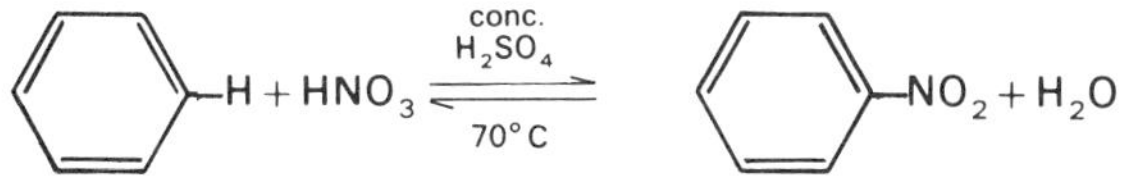

benzene nitrobenzene **(3.2)**

(ii) Elimination of the elements of a small molecule, usually of the type HX, with concomitant production of a multiple bond. The atoms, or groups, which form HX are most frequently eliminated from adjacent carbon atoms; the process is then said to involve 1,2- or β-elimination (3.3, 3.4).

$$\underset{\text{ethanol}}{CH_3CH_2OH} \underset{H_2SO_4}{\overset{\text{hot conc.}}{\rightleftharpoons}} \underset{\text{ethene}}{CH_2{=}CH_2} + H_2O \qquad \textbf{(3.3)}$$

$$\underset{\text{ethyltrimethylammonium hydroxide}}{H\bar{O}\ CH_3CH_2\overset{+}{N}(CH_3)_3} \overset{\text{heat}}{\rightleftharpoons} H_2O + CH_2{=}CH_2 + \underset{\text{trimethylamine}}{N(CH_3)_3} \qquad \textbf{(3.4)}$$

(iii) Addition to multiple bonds. Addition reactions are characteristic of unsaturated compounds and are the converse of elimination reactions. For example (3.3) can be reversed by dissolving the alkene in concentrated sulphuric acid and then diluting the reaction mixture with water.

(iv) Molecular rearrangement. An example of this type of reaction is the Beckmann rearrangement in which an oxime is converted to an amide under the influence of hot acid (3.5).

$$\underset{\text{diphenylketoxime}}{Ph_2C{=}NOH} \underset{H_2SO_4}{\overset{\text{hot conc}}{\rightleftharpoons}} \underset{N\text{-phenylbenzamide}}{PhCONHPh} \qquad \textbf{(3.5)}$$

The steps in a complex reaction may involve two or more of the basic types of reaction in sequence. For instance acylation (3.6, see Section 6.2) involves addition followed by elimination, giving overall substitution.

$$(R)(X)C{=}O + HY \rightleftharpoons (R)(X)C(OH)(Y) \rightleftharpoons (R)(Y)C{=}O + HX \qquad \textbf{(3.6)}$$

All four of the basic types of reaction are encountered in living cells; relevant examples are discussed in subsequent sections of the text.

˙assification of reagents

Ↄne of the reactants in any reaction, usually the simplest, is regarded as the eagent. For example in (3.1) the reagent is hydroxide ion and in (3.2) nitric acid. ⁚he distinction, however, is somewhat arbitrary. If the reactions of a given type of compound were being discussed then the materials with which they reacted would be reagents, irrespective of their relative size. Three types of reagent are known; their involvement in bond formation is shown in (3.7) to (3.11).

A = reagent

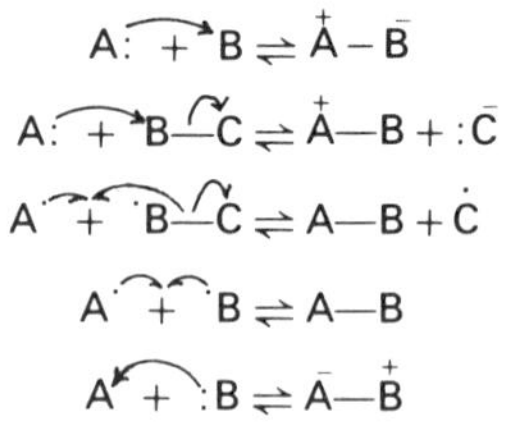

(3.7)
(3.8)
(3.9)
(3.10)
(3.11)

Before discussing the reagents, some general points relating to (3.7) to (3.11) should be noted.

(i) an arrow with a double barb is used to indicate the movement of an electron pair; the tail of the arrow originates on the source of the charge and the head terminates on the destination of the electrons. An arrow with a single barb is used in the same manner to indicate the movement of a single electron.

(ii) Bond cleavage in which both electrons of the bond are transferred to one of the products (3.8) is said to involve **heterolysis** of the bond, whereas if one electron is transferred to each product (3.9) the process is said to involve **homolysis** of the bond.

(iii) electrical charges must be conserved throughout a reaction. Thus if an atom or molecule such as A: (3.7) effectively loses control of an electron to B it must assume a formal positive charge and the acceptor atom a negative charge.

(iv) B in (3.7) and A in (3.11) must be co-ordinately unsaturated. That is, they must possess an 'empty' orbital in which the electron pair can be accommodated. On the other hand B in (3.8) and (3.9) is assumed to be co-ordinatively saturated and hence, as a new bond to B is formed the B–C bond must break.

Nucleophiles

Reagents such as A: in (3.7) and (3.8), which supply **both** electrons for formation of a new bond, are known as **nucleophiles** (nucleus loving, or seeking) because they attack electron deficient species. Some common nucleophiles are listed in Table 3.1.

It might be expected that the nucleophilic strength of a reagent would be related to the ease with which the nucleophilic atom can release its electron pair—i.e. to its electronegativity and, in general terms, this is so. Thus negatively charged reagents are stronger than their neutral analogues and nitrogen nucleophiles are usually stronger than oxygen nucleophiles which in turn are stronger than halogen nucleophiles. Furthermore because basic activity is a specific example of nucleophilic activity a general parallelism between the two would be expected. This is observed when elements in the same period of the periodic table are compared. Thus both basic and nucleophilic strength decrease in the sequence amines > alcohols > fluorides. However, the order of basicity and nucleophilicity is often reversed when elements within the same group of the periodic table are compared. Thiols are weaker bases than alcohols but are stronger nucleo-

Table 3.1. Some common nucleophiles

$H\ddot{O}H$	water
$R\ddot{O}H$	alcohols
$R^1\ddot{O}R^2$	ethers
$H\bar{O}$:	hydroxide ion
$R\bar{O}$:	alkoxide ions
$H\ddot{S}H$	hydrogen sulphide
$R\ddot{S}H$	thiols
$R^1\ddot{S}R^2$	thioethers
$HS\bar{:}$	bisulphide ion
$RS\bar{:}$	thiolate ions
$H_2\ddot{N}H$	ammonia
$R\ddot{N}H_2$, $R^1R^2\ddot{N}H$, $R^1R^2R^3\ddot{N}$	primary, secondary and tertiary amines
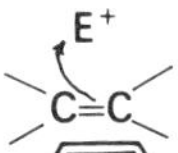	double bonds (E^+ represents an electrophilic reactant, see page 24)
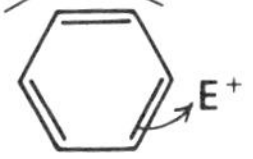	aromatic rings (E^+ represents an electrophilic reactant)
	carbonions (reactive intermediates formed during reactions)
$:X^-$	halide ions

philes. Polarisability and the efficiency with which the nucleophile can be 'de-solvated' appear to be important factors in determining nucleophilic strength.

Free radicals

Reagents which supply only **one** electron for covalent bond formation (3.9 and 3.10) are called **free-radicals.** They are neutral species which contain one, or more, **unpaired** electron. Free radicals are usually (but not invariably, see below) characterised by extremely high reactivity; it is not surprising, therefore, that normally they are produced only as transient intermediates during reactions. Their formation usually involves homolysis of a relatively weak, electrically symmetrical bond. For example, many free radical polymerisations can be initiated by adding an organic peroxide to the reaction mixture which is then warmed.

$$\underset{\text{dibenzoyl peroxide}}{PhCOO{-}OCOPh} \xrightleftharpoons{\text{heat}} \underset{\text{benzoyloxl free radical}}{2PhCOO^{\cdot}}$$

There is one free radical, however, which occurs naturally in great abundance and which is of paramount importance in biochemistry—oxygen. Oxygen exists as a **di-radical**, (fortunately, one of low kinetic reactivity under normal conditions) a fact which can be explained when the electronic structure of the molecule is described in terms of molecular orbitals. Many oxidations involving molecular oxygen show all the characteristics of radical reactions—they are *extremely* rapid, once initiated, and are self-sustaining whilst any appreciable quantities of reactants are present.

Electrophiles

Reagents which accept an electron pair in covalent bond formation (3.11) are known as **electrophiles**. Some common examples of this type of reagent are shown in Table 3.2. In general all positively charged electrophiles are more powerfully electrophilic than neutral compounds; carbonium ions (species containing positively charged carbon) are particularly effective reagents. The commonest electrophile in carbon chemistry is probably the proton. Metal ions also often act as electrophiles, particulalry in enzyme catalysed reactions. Alkyl halides, protonated alcohols and ethers, tertiary oxonium and sulphonium ions and quaternary ammonium ions are electrophiles (Section 5.1) as are compounds containing carbonyl groups (Section 6.1).

Table 3.2. Some common electrophilic groups

H^+	proton
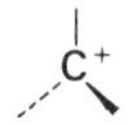	carbonium ion
NO_2^+	nitronium ion
metal^{n+}	metal cation
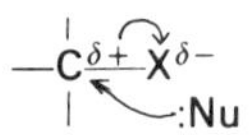	X = halogen, $-\overset{+}{O}H_2$, $-\overset{+}{O}HR$, $-\overset{+}{O}R^1R^2$, $-\overset{+}{S}R^1R^2$, $-\overset{+}{N}R^1R^2R^3$:Nu is a nucleophilic reactant
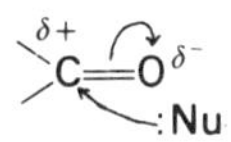	carbonyl groups :Nu is a nucleophilic reactant
	halogens :Nu is a nucleophilic reactant

Charge distribution in organic molecules

A nucleophilic reagent must, by virtue of its nature, attack an electrophilic centre in a reactant and an electrophilic reagent a nucleophilic centre; free radicals attack electrically symmetrical bonds or other free radicals. In order to determine where an organic molecule might be attacked by any of the three types of reagent defined above, it is necessary to be able to determine where the charged centres in the molecule are located. The distribution of charge in a molecule is described in terms of a number of **polarisation effects**, the most important of which are the **inductive** and **mesomeric** polarisations.

Inductive Polarisation

Inductive polarisations exist only in **single** bonds and are due to the electrical dissimilarity of two bound atoms. This type of polarisation has already been outlined, but not named, on page 11. The name of the effect derives from the fact that the polarity in, for example, a carbon to chlorine bond can **induce** similar, but smaller polarisations in adjacent bonds which are formally non-polar.

$$-\overset{\delta'''+}{\mathrm{C}}-\overset{\delta''+}{\mathrm{C}}-\overset{\delta'+}{\mathrm{C}}-\overset{\delta-}{\mathrm{Cl}} \qquad \delta'+ > \delta''+ > \delta'''+$$

Due to the intervention of the atomic nuclei the effect falls off rapidly with increasing distance from the site of the initial polarisation and can be considered to be negligible after it has traversed three bonds (in practice it is often necessary to consider only the initial polarisation). The inductive effect of an atom or group can be classified on the basis of whether it releases charge towards, (+I effect), or withdraws charge (−I effect), from the rest of the molecule when compared with a hydrogen atom. The sign of the polarisation effect usually reflects the electronegativities of the atoms concerned, relative to hydrogen, making it unnecessary to discuss individual cases here. Alkyl groups, however, constitute an exception to this statement. For reasons which are outside the scope of this book alkyl groups exhibit +I effects, decreasing in the order $(CH_3)_3C-$ $> (CH_3)_2CH- > CH_3CH_2- > CH_3-$

Mesomeric polarisation and π electron delocalisation

Mesomeric polarisations, or effects, are analogous to inductive polarisations except that they occur in **π bonds**. For example, oxygen is more electronegative than carbon and hence the charge in the π bond of a carbonyl group will be displaced towards the oxygen atom. The usual way of describing the polarisation is shown in (3.12).

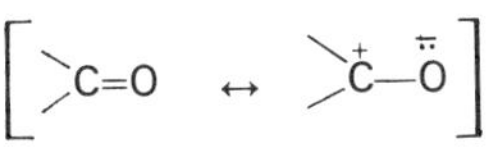

(3.12)

Here the left hand diagram represents an unpolarised carbonyl group, while the right hand one represents a completely polarised group. Both are hypothetical and the real structure of the carbonyl group lies somewhere between the two extremes. This situation is reflected by the name given to this type of polarisation—**mesomeric** means 'between the parts'. (Note that an inductive polarisation will also exist in the σ bond, but is not shown).

In contrast to inductive polarisations mesomeric polarisations can be transmitted very efficiently through a system providing that it contains conjugated bonds. Consider, for example, 2-butenal (crotonaldehyde, $CH_3CH = CHCHO$, see (3.13)).

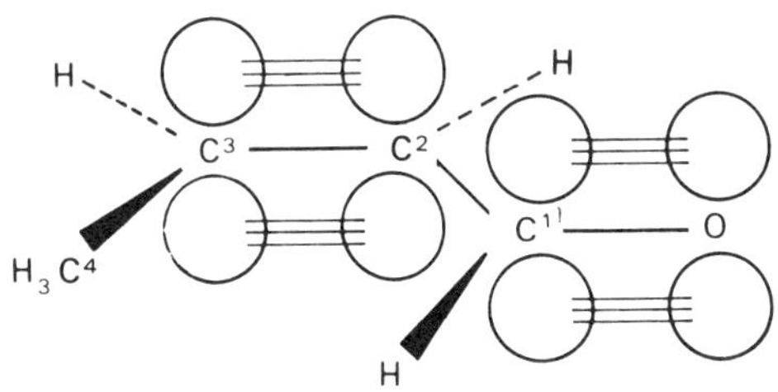

(3.13) Orbital diagrams of the π bonds in 2-butenal

In the absence of other effects the carbonyl group in this compound would be polarised as shown in (3.12) leaving the 'carbon end' of the bond electron deficient. However, the charge described by the p orbital of C^2 can interact just as efficiently with that of the p orbital of C-1 as with that of C-3. Consequently charge is withdrawn from the C-2 to C-3 π bond towards the carbonyl group, a situation which can be described in terms of the three structural formulae shown (3.14), taken together.

$$[CH_3CH{=}CHCH{=}O \leftrightarrow CH_3CH{=}CH\overset{+}{C}H{-}\bar{O} \leftrightarrow CH_3\overset{+}{C}H{-}CH{=}CH{-}\bar{O}] \qquad \textbf{(3.14)}$$

The actual structure of the molecule will lie somewhere between the extremes depicted by these diagrams; crotonaldehyde is said to be a **mesomeric molecule**. The rules for constructing the diagrams used to describe mesomeric polarisations are the same as those used to construct the diagrams to describe a molecule as a resonance hybrid (p. 16).

The mesomeric effect of an atom or group, like an inductive effect, can be classified on the basis of whether it withdraws (– M effect) or releases charge (+ M effect) to the rest of the molecule when compared with hydrogen (see Table 3.3). It may seem anomalous that the X – groups in the left side of Table 3.3 should exhibit + M effects when X is always more electronegative than carbon. However, the evidence for the classification given is overwhelming.

Table 3.3 Mesomeric effects of groups

+ M	H	– M
$[\ddot{X}{-}CH{=}CH{-} \leftrightarrow \overset{+}{X}{=}CH{-}CH{=}]$		$[O{=}X{-}CH{=}CH{-} \leftrightarrow \ddot{O}{-}\overset{+}{X}{-}CH{=}CH{-} \leftrightarrow \bar{O}{-}X{=}CH{-}\overset{+}{C}H{-}]$
X = häl		
$R\ddot{O}-$		X = O←N– (//)
$R\ddot{S}-$		
$R^1R^2\ddot{N}-$		
R = alkyl or H		=C<

Consider, for example, chlorobenzene (3.15). The molecule is relatively inert towards nucleophiles but is attacked by electrophiles at the 2, 4 and 6 positions. The carbon to chlorine bond is shorter than the carbon to chlorine bond in, say chloroethane, and the dipole moment of the molecule is much smaller than would be expected on the basis of its dimensions and the electronegativities of its constituent atoms. All these facts can be explained if it is assumed that a lone pair of electrons of the chlorine atom interacts, to some extent, with the π orbitals of the benzene ring.* The polarisation in the molecule can be described by (a) or (b), (3.15). The delocalisation of charge occurs, despite the fact that chlorine is strongly electronegative, because it leads to increased stability of the molecule. An analogy can be made to the potential energy of a charged, hollow, spherical condenser. The larger the surface area of the condenser, the lower the potential imparted by a given charge. The same appears to be true of molecules. The greater the extent of π electron delocalisation, the greater the stabilisation which results. Note that for π electron delocalisation to be most effective the conjugated bonds must be coplanar and bond angles must be normal.

* The π electrons in the benzene ring are completely delocalised – see section (4.3).

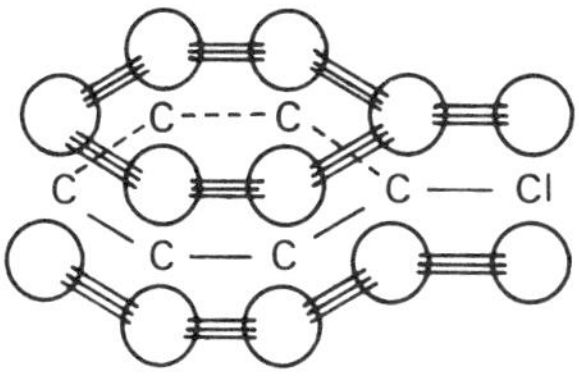

(a)

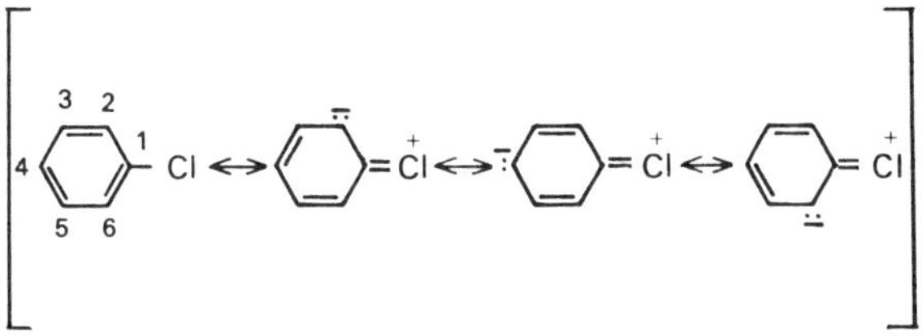

(b)

(3.15)

(a) Orbital diagram of π bonds in chlorobenzene, ≡ indicates sideways overlap of p orbitals.
(b) description of chlorobenzene as a mesomeric molecule.

π Electron delocalisation and resonance

It should be clear that the descriptions of 2-butenal (3.14) and chlorobenzene (3.15) as mesomeric molecules are identical with their descriptions as resonance hybrids. This is usually the case where molecules containing conjugated bonds are concerned. Note, however, that benzene (Section 4.3) can be described as a resonance hybrid but **not** as a mesomeric molecule because no polarity exists in the molecule. The difference between resonance theory and mesomerism is that the former stems from attempts to calculate bond energies in terms of the distribution of the total charge in a molecule and is therefore applicable to both saturated and unsaturated molecules while the latter stems from attempts to explain chemical reactivity in terms of any polarisations which exist in the π bonds of a molecule; mesomerism can only be applied to polarised, unsaturated molecules.

3.3. Rates of reaction and Catalysis

Equation (3.16) describes a generalised chemical reaction involving a transition state A···B* which might occur in solution.

$$A + B \rightleftharpoons A \cdots B^* \rightleftharpoons C + D \qquad (3.16)$$

Thermodynamics predicts that the reaction will occur i.e. there will be net formation of C and D, provided that the change decreases the free energy, G, of the system. In quantitative terms this means that the change in free energy per unit extent of reaction, ΔG, accompanying the reaction, as given by equation (3.17), must be negative.

$$\Delta G = \Delta G^{\ominus} + RT \ln \frac{[C][D]}{[A][B]} \qquad (3.17)$$

In equation (3.17) $\Delta G^{\ominus}$ is the standard free energy change for the reaction—the free energy change per unit extent of reaction when all reactants are present at unit concentration—and [A], [B], [C] and [D] are the initial concentrations of the reactants. The reaction will proceed until the free energy change per unit extent of reaction is zero, when the system will be in equilibrium. Thermodynamics, however, gives no indication at all as to how fast the reaction will go; rates of reactions, as noted previously, are the province of chemical kinetics.

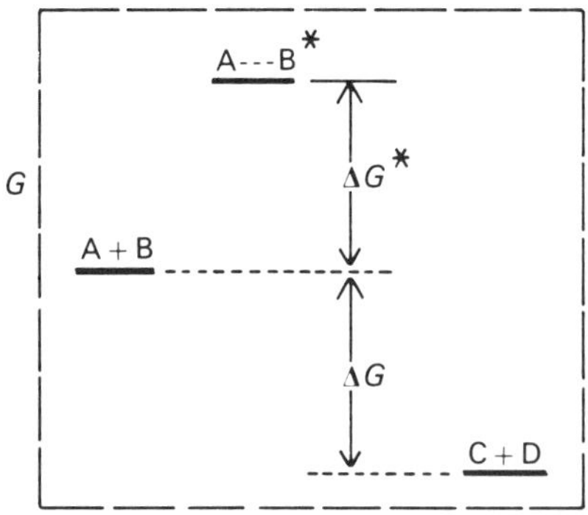

(3.18)

Plot of the relative free energies of starting materials, activated complex and products of reaction (3.16)

Figure (3.18) compares the free energies of stoichiometric amounts of the starting materials, transition complex and products of reaction (3.16). Here ΔG is the **total** change in free energy which accompanies the conversion of one mole each of A and B completely to C + D. ΔG^* is the free energy of activation per mole of transition complex. A simple physical interpretation of ΔG^* is that it is a measure of the energy which must be associated with a collision between A and B in order for the transition complex (and hence products) to be formed. Chemical kinetics predicts that the rate constant of a reaction, k_r, is determined by the value of ΔG^* for the reaction.

$$k_r = \frac{k_b T}{h} e^{-\Delta G^*}$$

The smaller ΔG^* the more rapid the reaction. This conclusion allows valuable rationalisations of kinetic reactivity to be made in terms of chemical structure. In particular, if the transition complex is charged it is possible to deduce, qualitatively, the relative rates of reaction of two reactants in the same process. Since delocalisation of charge increases stability (p. 26) the reactant which possesses structural features which can delocalise charge in the transition complex will undergo the faster reaction, because ΔG^* for that reaction will be decreased. For example, as noted in Section 6.2, decarboxylation of carboxylic acids proceeds via an intermediate carbanion (3.19).

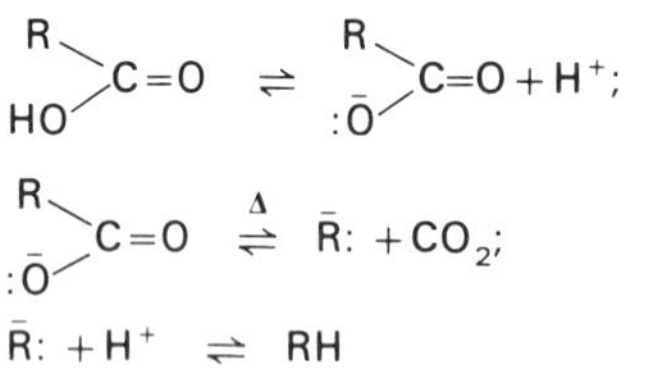

(3.19)

Acetic acid can be de-carboxylated by fusing anhydrous sodium acetate with soda-lime.

$CH_3CO_2^- + OH^- \rightarrow CH_4 + CO_3^{2-}$

Nitroacetic acid, however, reacts much more rapidly—it can be decarboxylated merely by heating an aqueous solution of the sodium salt to 80°C.

$$O_2NCH_2CO_2^- + OH^- \rightarrow \underset{\text{nitromethane}}{CH_3NO_2} + CO_3^{2-}$$

The difference in rates reflects the fact that the nitro group in nitroacetic acid can de-localise charge in the intermediate carbanion and hence, also, in the transition complex leading to it.

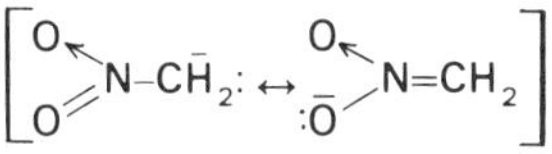

Catalysis

A catalyst is a substance which accelerates a chemical reaction without itself appearing as a product of the reaction. Catalysts can be shown to accelerate both the forward and reverse components of a reaction. That is, they accelerate the attainment of equilibrium although they do not alter its position. In general terms catalysts of all types produce their effect by lowering the free energy of activation of the reaction which they catalyse. Many laboratory reactions are catalysed by species such as the proton and virtually all cellular reactions are catalysed by enzymes. Subsequent reactions of this book contain many references to enzyme catalysed reactions and to be able to appreciate these reactions some properties of enzymes are introduced here, although a detailed discussion of them is postponed to Chapter 8.

The reactants in a reaction catalysed by an enzyme are known as the **substrates** of that enzyme. The substrate(s) is (are) bound non-covalently to a region of the surface of the enzyme known as the active site or centre, where the catalytic reaction occurs. All enzymes are proteins; some, however, require the presence of a small non-protein molecule before they can exert their catalytic effect. These small molecules were formerly classified as co-enzymes or as prosthetic groups dependent on whether they could be removed from their apo-enzyme* by dialysis. **Diffusible co-enzymes** were claimed to be bound to their apo-enzyme non-covalently while **non-diffusible prosthetic groups** were claimed to be bound covalently. It has been discovered, however, that a number of materials which were formerly regarded as covalently bound prosthetic groups are, in fact, bound non-covalently to their apo-enzyme but bound so tightly as to be non-diffusible. In view of this it is best to designate all of these non-protein materials which are essential for the activity of various enzymes as co-enzymes or as co-factors. Some of the more important co-enzymes are discussed in Chapter 9. It is sufficient to note here that co-enzymes act in one of two ways. Some, such as nicotinamide adenine dinucleotide (NAD^+), act as obligatory substrates in the reaction catalysed by their apo-enzyme. For example in (3.20) and in other similar reactions catalysed by NAD^+ dependent dehydrogenases, the co-enzyme is the

* The enzyme to which a co-enzyme or prosthetic group binds is known as the apo-enzyme, while a co-enzyme-apo-enzyme or prosthetic group-apo-enzyme complex is referred to as a holo-enzyme.

oxidising agent, as is flavin adenine dinucleotide (FAD) in reaction (3.21).

$$\underset{\text{ethanol}}{CH_3CH_2OH} + \overset{+}{NAD} \overset{ADH}{\rightleftharpoons} \underset{\text{ethanal}}{CH_3CH{=}O} + NADH + H^+ \qquad \textbf{(3.20)}$$

ADH = alcohol dehydrogenase

$$\underset{\text{succinate}}{{}^-O_2CCH_2CH_2CO_2^-} + \text{Enz–FAD} \rightleftharpoons \underset{\text{fumarate}}{{}^-O_2CCH{=}CHCO_2^-} + \text{Enz–FADH}_2 \qquad \textbf{(3.21)}$$

Enz–FAD = succinate dehydrogenase

The two oxidising co-enzymes differ in that one, NAD^+, is diffusible while the other, FAD, is not. Other co-enzymes, such as pyridoxal phosphate and thiamine pyrophosphate, act by enhancing the reactivity of the substrate of their apo-enzyme. Although they, like the 'reagent type' co-enzymes mentioned above, also participate in the catalytic reaction they are regenerated and hence do not appear in its stoichiometry.

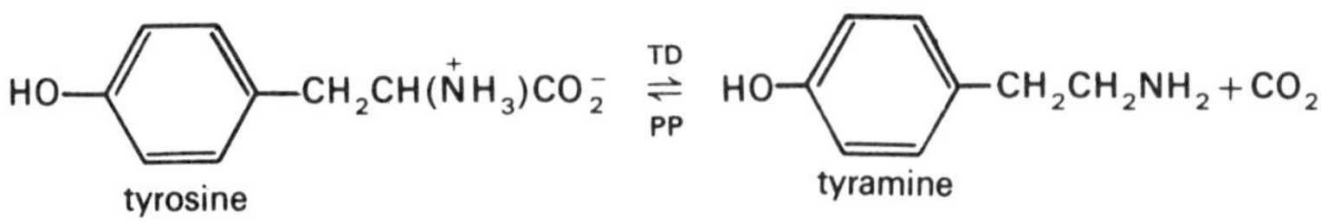

TD = tyrosine decarboxylase; PP = pyridoxal phosphate

4

Alkanes, the double bond and the aromatic nucleus

It is convenient to consider the reactions of alkanes, of the double bond and of the aromatic nucleus under a common heading. The unifying theme here is that compounds which do not contain a functional group (alkanes), those which contain only a double bond(s) (alkenes) and those which contain an aromatic nucleus, but no other functional group (arenes), are all **hydrocarbons.**

4.1. Alkanes

Alkanes are aliphatic hydrocarbons of general formula C_nH_{2n+2}; some representative members of the homologous series are shown in Table 4.1

Table 4.1. Some representative alkanes

Formula	b.p./° C	Name
CH_4	−161	methane
C_2H_6, CH_3CH_3	−88	ethane
C_3H_8, $CH_3CH_2CH_3$	−42	propane
C_4H_{10}, $CH_3(CH_2)_2CH_3$	0	butane
C_5H_{12}, $CH_3(CH_2)_3CH_3$	36	pentane
C_6H_{14}, $CH_3(CH_2)_4CH_3$	69	hexane
$C_{20}H_{42}$, $CH_3(CH_2)_{18}CH_3$	m.p.37	eicosane

Nomenclature of alkanes

The nomenclature of the alkanes is of great importance because it forms the basis of the system used for naming all other carbon compounds. The class suffix is **-ane**. Semi-trivial names are retained, by virtue of long usage, for the first four members of the series; thereafter a Greek prefix denotes the number of carbon atoms present (Table 4.1).

Structural isomerism

Accepting that hydrogen is univalent and carbon tetravalent there is only one possible structure for each of the alkanes having one, two or three carbon atoms. However, this is not so for higher homologues in the series. As shown in Table 4.2 there are two alkanes (4.1) and (4.2) with molecular formula C_4H_{10}, three with molecular formula C_5H_{12} (4.3 to 4.5) and five with molecular formula C_6H_{14} (4.6 to 4.10). Compounds such as (4.1) and (4.2) which have identical molecular formulae, but different structures, are **structural isomers** of each other. They have different physical properties, (e.g. b.p. of 0°C and −12°C respectively), slightly different chemical properties, and their molecules have quite different shapes (4.11).

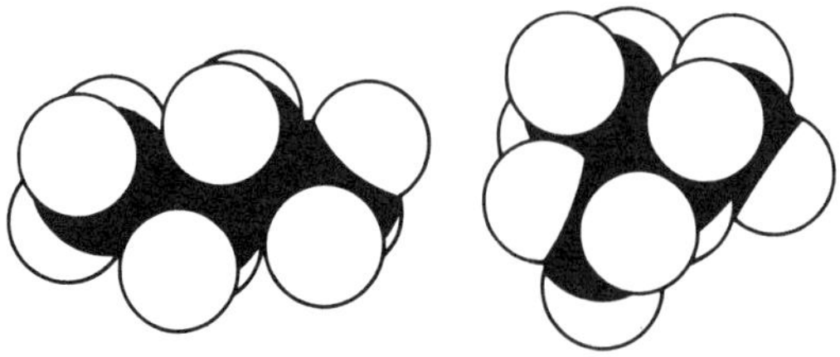

(a) (b)

(**4.11**) Space filling models of (a) butane, (b) isobutane

Table 4.2. Some examples of isomeric alkanes

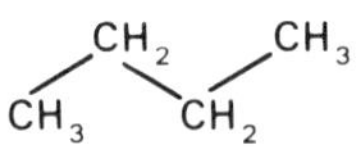 butane (**4.1**)	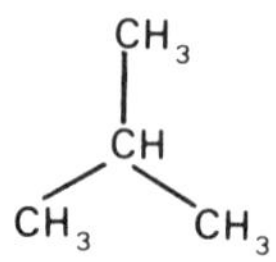 isobutane (**4.2**)	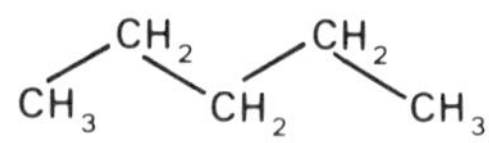 pentane (**4.3**)
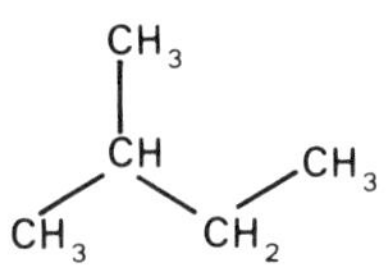 isopentane (**4.4**)	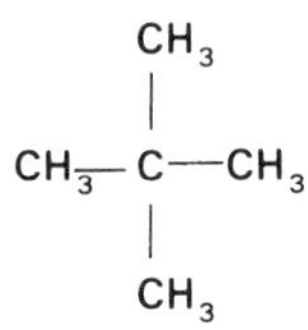 neopentane (**4.5**)	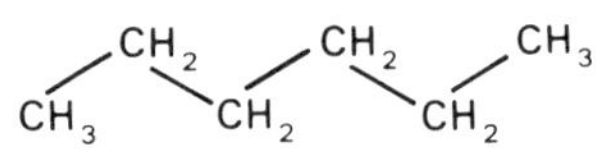 hexane (**4.6**)
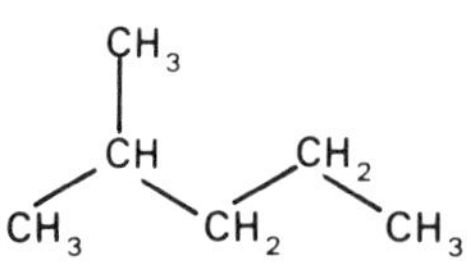 isohexane (**4.7**)	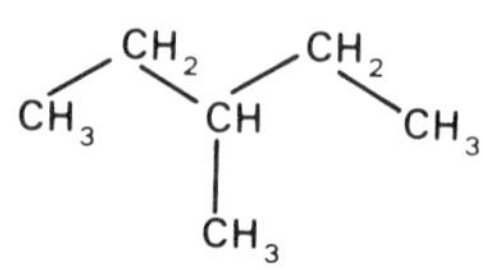 3-methylpentane (**4.8**)	
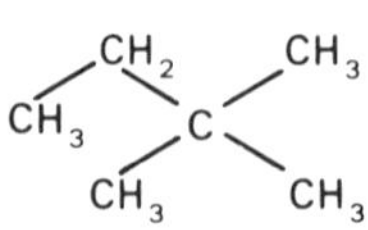 2,2-dimethylbutane (**4.9**)	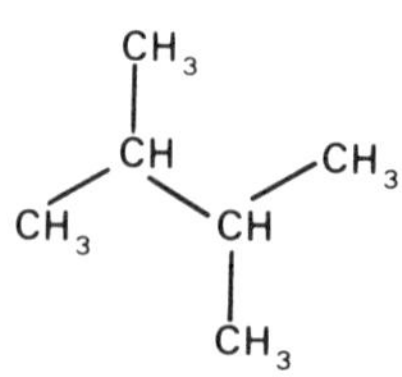 2,3-dimethylbutane (**4.10**)	

Nomenclature of branched chain hydrocarbons (Table 4.2)

The prefix iso- denotes the presence of a geminal dimethyl group, $(CH_3)_2CH$–. Thus, (4.2) is isobutane, (4.4) isopentane and (4.7) isohexane. The prefix neo is retained to describe (4.5). However, such prefixes are of only limited value for naming even more complex alkanes. In such cases it is preferable to use the following sequence of operations in order to derive the systematic name of the hydrocarbon:

(i) The parent hydrocarbon, which corresponds to the longest unbranched chain of carbon atoms present, is identified and named.

(ii) Any side chains present are named as radical substituents of the parent hydrocarbon. The names of monovalent radicals are derived by replacing the terminal **-ane** in the name of the corresponding hydrocarbon by **-yl** and divalent radicals by replacing **-ane** by **-ylidene** as indicated in Table 4.3

Table 4.3 Some radicals derived from alkanes

CH_3–	methyl	$CH_3CH_2CH_2CH_2$–	butyl
–CH_2–	methylene*		
CH_3CH_2–	ethyl	$(CH_3)_2CH$–CH_2–	isobutyl
CH_3CH_2<	ethylidene	$CH_3(CH_3CH_2)CH$–	*sec*-(or 2)-butyl
$CH_3CH_2CH_2$–	propyl		
$CH_3\dot{C}HCH_3$ or $(CH_3)_2CH$–	isopropyl	$(CH_3)_3C$–	*tert*-butyl

* a non-systematic name retained by virtue of long usage

(iii) The side chains are cited in alphabetical order. Appropriate multiplying prefixes (di, tri, tetra, etc) are used when two or more identical substituents are present but these prefixes are **not** considered when determining the alphabetical order of substituents, (for example triethyl precedes dimethyl).

(iv) The positions of the side chains are defined by locant numbers which are allocated by numbering the carbon atoms of the parent hydrocarbons so that the lowest series of locant numbers is obtained. The lowest of two series of locant numbers is not that with the lowest sum, it is that in which the lowest number occurs at the first point of difference. Note that **each** substituent, even if multiplied, must have a separate locant number.

Using these rules (4.12) is named as follows:

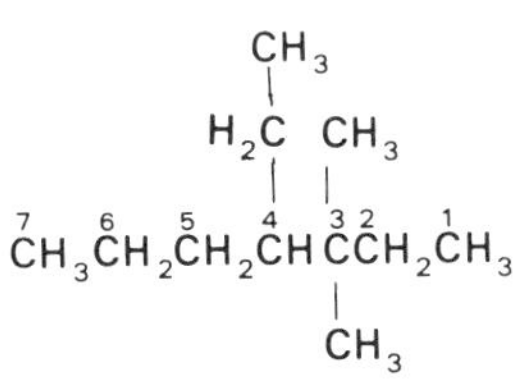

(4.12)

parent hydrocarbon: heptane
substituents: ethyl, dimethyl → ethyldimethylheptane
locant numbers: 3,3,4 → 4-ethyl-3,3-dimethylheptane

Similarly, (4.13) is named as 2,7,8-trimethyldecane,

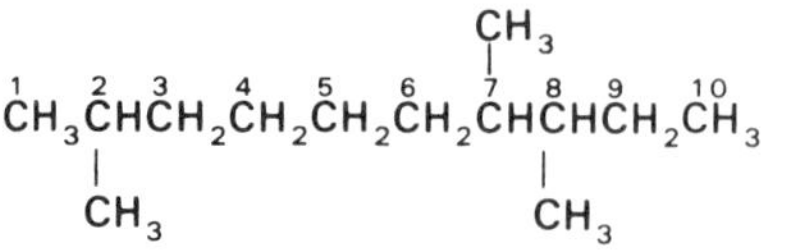

(4.13)

This example gives a good illustration of rule (iv). Numbering the hydrocarbon from the other end would give the series of locant numbers 3,4,9 which, despite having a lower sum, is the higher series since it possesses the higher number (3) at the first point of difference.

The nomenclature of the hydrocarbons in Table 4.2 should now be comprehensible.

Naturally occuring alkanes

Alkanes occur in large quantities in crude petroleum oils and in natural gas. These hydrocarbon deposits are the result of anaerobic decomposition of organic matter followed by compression under immense forces. However alkanes are not only associated with non-living matter but are also widely distributed in living organisms, although the amounts found are small. An exception to this last statement is pristane (2,6,10,14-tetramethylpentadecane, 4.14), which constitutes about 14% of the non-saponifiable fraction of the oil from the liver of the basking shark.

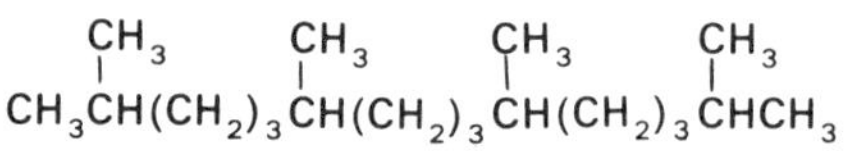

(4.14)

Beeswax contains about 10% of a mixture of alkanes in which hentriacontane, $C_{31}H_{64}$, predominates. Alkanes containing either an even or an odd number of carbon atoms in the approximate range $C_{24}-C_{36}$ occur in the leaf cuticle wax or fruit wax of many plants and also in the wax on fungal spores. Their role appears to be associated with prevention of loss of water from the organism. A number of bacteria use carbon dioxide, rather than oxygen, as their final oxidant and hence reduce it to methane. These 'methane producing bacteria' should not be confused with the methane oxidising bacteria which are discussed later in this section.

Properties of alkanes

As carbon and hydrogen have approximately equal electronegativities, alkanes exhibit properties which are typical of non-polar compounds. Low molecular mass alkanes are gases at room temperatures; pentane and higher homologues are volatile liquids or solids with low m.p. Alkanes are insoluble in water and in polar solvents such as ethanol but are soluble in non-polar solvents such as ether and benzene.

General reactivity

As was mentioned in the last paragraph, carbon to hydrogen bonds are essentially non-polar. It is therefore not surprising that alkanes react with free radicals, but not with polar reagents. A typical example of a reaction of an alkane involving free radicals is that represented by (4.15) in which methane is chlorinated under the influence of sunlight.

$$CH_4 + Cl_2 \overset{h\nu}{\rightleftharpoons} CH_3Cl + HCl \qquad \textbf{(4.15)}$$

The steps in the reaction are detailed in (4.16) below. In the **initiation** step chlorine molecules absorb radiant energy, the amplitude of the bond vibration increases and homolysis occurs producing chlorine atoms (free radicals). Each chlorine atom then immediately abstracts a hydrogen atom from methane, generating hydrogen chloride and a methyl free radical. The latter intermediate at once abstracts a chlorine atom from a chlorine molecule generating the second product of the reaction, methyl chloride, together with another chlorine atom which can continue, or **propagate**, the reaction providing that significant quantities of methane and chlorine are still present. Eventually, the concentrations of methane and chlorine become so small that the rates of the propagation steps become less than the rates of steps in which radicals combine and the reaction is **terminated**.

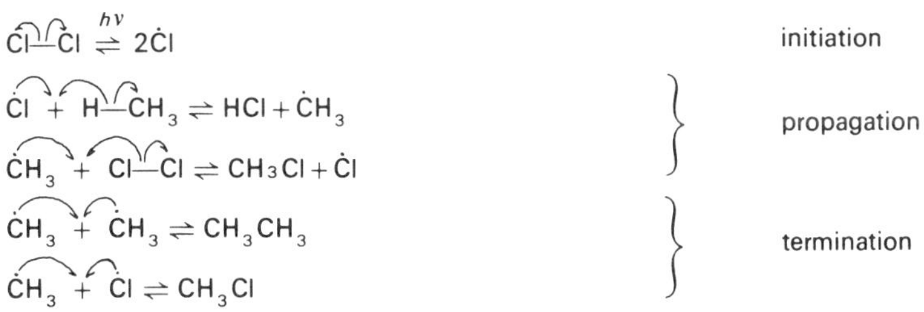

(4.16)

Reaction (4.15) is not of great relevance to biochemistry except in that it illustrates the three steps, initiation, propagation and termination, which are common to all reactions involving free radicals.

Oxidation of alkanes

(a) With molecular oxygen. The extensive use of natural gas and compressed propane or butane as fuels for heating or lighting reflects the fact that the complete oxidation of an alkane is a highly exergonic process (4.17) and (4.18).

$$CH_4(g) + 2O_2(g) \longrightarrow CO_2(g) + 2H_2O(l);$$
$$\Delta G^{\ominus}(298\ K) = -817\ kJ\ mol^{-1};$$
$$\Delta H^{\ominus}(298\ K) = -890\ kJ\ mol^{-1} \qquad \textbf{(4.17)}$$

$$C_4H_{20}(g) + 6.5O_2(g) \longrightarrow 4CO_2(g) + 5H_2O(l);$$
butane
$$\Delta G^{\ominus}(298\ K) = -2745\ kJ\ mol^{-1};$$
$$\Delta H^{\ominus}(298\ K) = -2879\ kJ\ mol^{-1} \qquad \textbf{(4.18)}$$

Once a reaction of this type has been initiated by ignition it proceeds **extremely** rapidly and is self-sustaining until either reactant has been consumed. These are characteristics which are typical of most radical reactions. Reactions (4.17) and (4.18) can be considered to be representative of the oxidation of organic compounds in general. Oxidations are invariably associated with a large, negative value of ΔG to which, in the case of compounds more complex than hydrocarbons, the $T\Delta S$ term makes a significant contribution ($\Delta G = \Delta H - T\Delta S$). This last is because oxidation usually produces a more disordered system both in energetic and often, also, in material terms. It should be noted at this point that life in all forms depends on oxidative processes for its continuance. Animals ingest fuel molecules (carbohydrates, fats and proteins) while plants synthesise them from simpler molecules using solar energy. Both then **oxidise** the fuel molecules. However, in plants and animals complete oxidation does not occur in one step and the chemical energy liberated during oxidation does not appear solely as heat. Instead, oxidations *in vivo* take place by multi-step processes and much of the chemical energy which would otherwise be liberated as heat is conserved using **coupled reactions** (Section 9.1) which produce materials necessary for the continued existence of the cell (proteins, carbohydrates, nucleic acids, lipids, etc.), or which lead to the performance of mechanical, electrical or osmotic work.

Hydrocarbons do not, in general, serve as fuel molecules for living cells. However a number of bacteria, for example **Methanomonas methanooxidans**, can oxidise methane directly. Additionally, a number of strains of bacteria and yeast have been found which are able to oxidise higher alkanes ($C_{10}-C_{18}$) and to incorporate alkane carbon into their cellular constituents. Considerable effort is currently being expended to develop such processes for the commercial production of 'single cell protein'. The oxidative sequence which is thought to occur is as follows:

$$RCH_3 \rightarrow RCH_2OH \rightarrow RCHO \rightarrow RCO_2H$$

The carboxylic acid which is produced can undergo further oxidation by the 'β-oxidation' pathway and the reactions of the tricarboxylic acid cycle, finally giving rise to carbon dioxide and water. (It is common practice, particularly in biochemical literature, to designate carbon atoms in a chain to which a functional group is attached as α, β, γ, δ etc. Thus the term 'β-oxidation' denotes a process in which a carbon atom β to a functional group, in this case a thiolester group, is oxidised.) The first reaction in the sequence above requires the presence of an oxidisable co-substrate (H_2X) and is catalysed by an enzyme known as a mixed function oxidase.

$$RCH_3 + O_2 + H_2X \rightarrow RCH_2OH + H_2O + X$$

Note that some micro-organisms contain enzymes which can catalyse the oxidation of both ends of an alkane chain concurrently.

(b) Oxidation by dehydrogenation. The 'cracking' of petroleum fractions to produce alkenes, particularly ethene, is a process of great industrial importance. The type of reaction which occurs is shown in (4.19). Following from this it is worth noting that several oxidative pathways *in vivo*, for example the β-oxidation of fatty acids, commence with a dehydrogenation step although this probably does not involve free radicals. The alkene which is produced during β-oxidation, or in similar oxidative pathways, then undergoes further oxidation as shown in (4.20).

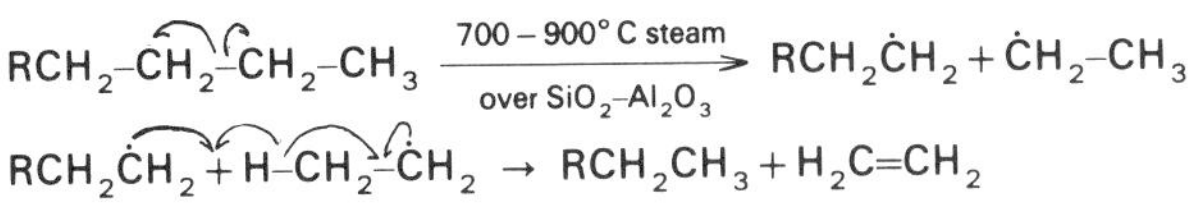

(**4.19**)

$$-CH_2CH_2- \xrightarrow{-2H} -CH{=}CH- \xrightarrow{+H_2O} -CH(OH)CH_2- \xrightarrow{-2H} -\overset{\overset{\displaystyle O}{\|}}{C}CH_2- \quad (4.20)$$

The position of the carbonyl group which is produced is such that it allows fragmentation of the molecule into smaller units which are then oxidised completely to yield carbon dioxide and water.

4.2. The double bond -alkenes

Hydrocarbons which contain one or more double bonds are known as **alkenes**; when only one double bond is present they have the general formula C_nH_{2n}.

Examples and nomenclature

The presence of a single double bond in a hydrocarbon is indicated by replacing the terminal **-ane** in the name of the parent hydrocarbon by **-ene**; when two double bonds are present the substitution is **-adiene**. The parent hydrocarbon corresponds to the longest carbon chain containing the maximum number of double bonds and is numbered so as to give the **double bonds** the lowest series of locant numbers, irrespective of what alkyl substituents are present (cf. alkanes). Some simple alkenes and their names are shown in Table 4.4.

Table 4.4 Some representative alkenes

$H_2C{=}CH_2$	ethene, (ethylene*)
$CH_3CH{=}CH_2$	propene
$CH_3CH_2CH{=}CH_2$	1-butene
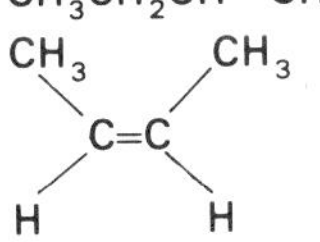	*cis*-2-butene‡
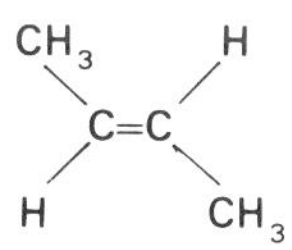	*trans*-2-butene‡
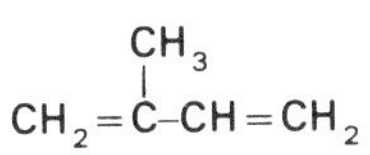	2-methylbuta-1,3-diene, (isoprene*)
Some common monovalent unsaturated radicals derived from alkenes:	
$H_2C{=}CH-$	ethenyl, (vinyl*)
$CH_3CH{=}CH-$	prop-1-enyl
$CH_2{=}CH-CH_2-$	prop-2-enyl, (allyl*)
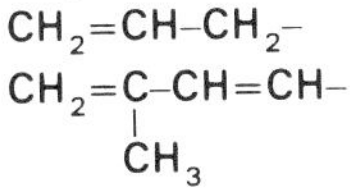	isoprenyl*

* a trivial name approved by the IUPAC commission on nomenclature
‡ see page 131 for an explanation of the meaning of the prefixes *cis* and *trans*.

Some compounds of biological interest which contain double bonds

It has been shown that ripening apples and tomatoes evolve ethene and propene in trace amounts. Terpenes, which are compounds that are derived by the head to tail polymerisation of isoprene units, are widely distributed in nature, being particularly abundant in the essential oils of plants. Thus in Table 4.5 examples (4.21) to (4.27) are all terpenes or are derived from terpenes. α-Pinene* (4.21) is the main constituent of terpentine, the oil produced by conifers. Limonene (4.22), a diene, can be extracted from lemon grass oil. β-Carotene* (4.23) is the principle pigment in carrots while the related Vitamin A (4.24) is obtained from fish liver oil. Natural rubber (4.25) is all *cis* polyisoprene while guttapercha (4.26) is the all *trans* compound. Many other compounds containing one or more double bonds in addition to other functional groups are found in living cells. Examples are cholesterol (4.27), an important sterol which is synthesised by animals, fumaric acid (4.28) which is an intermediate in the

Table 4.5. Some naturally occurring compounds which contain double bonds

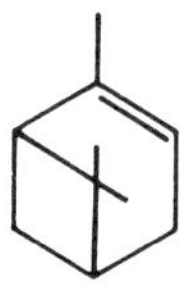

α-pinene
(4.21)

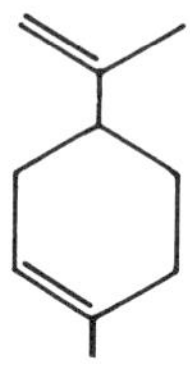

limonene
(4.22)

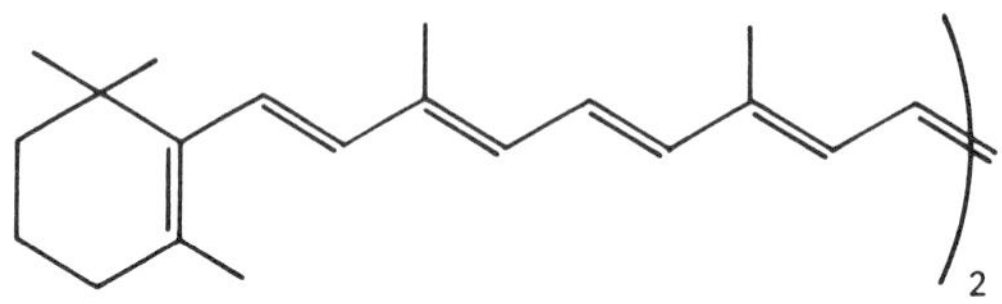

β-carotene
(4.23)

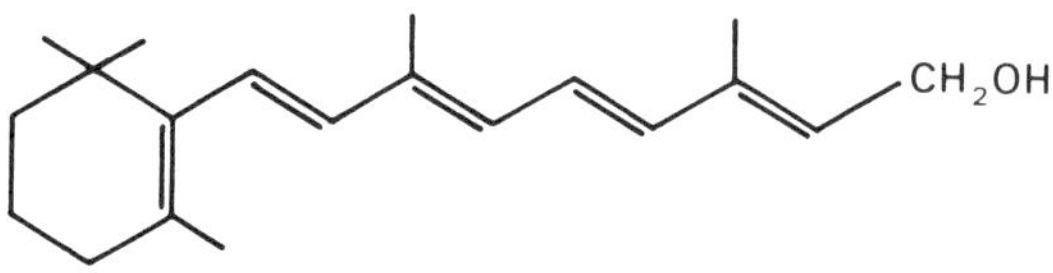

Vitamin A
(4.24)

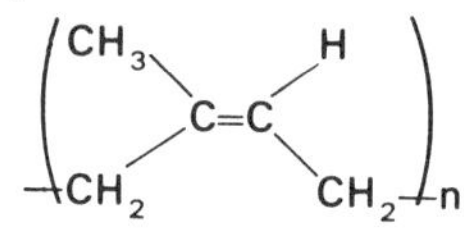

Natural rubber
(4.25)

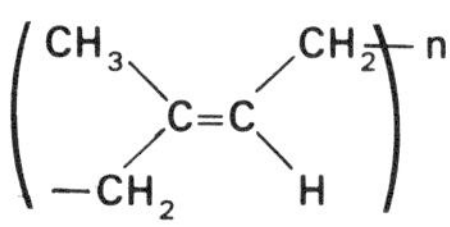

gutta-percha
(4.26)

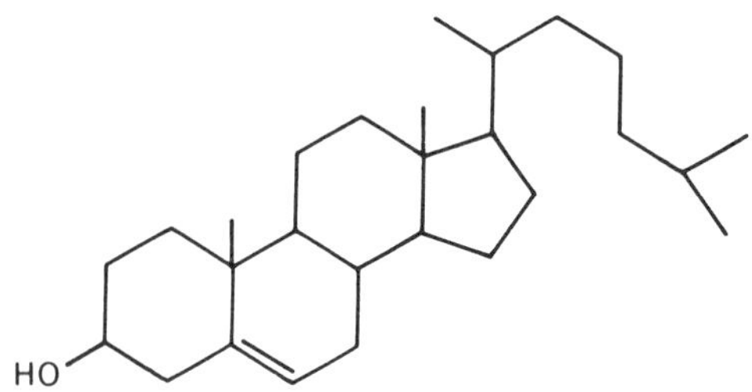

cholesterol
(4.27)

$HO_2CCH{=}CHCO_2H$

fumaric acid
(4.28)

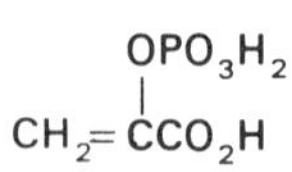

Phospho-enol pyruvic acid
(4.29)

* The greek prefix defines the position of a double bond.

tricarboxylic acid cycle of reactions and the important glycolytic product, phosphoenol pyruvic acid (4.29).

Properties of the double bond

The physical properties of alkenes are very similar to those of alkanes. They are relatively non-polar and exhibit solubility properties typical of such compounds. Low molecular mass alkenes exist as gases at room temperature while hexene and higher homologues are volatile liquids.

Typical reactivity

The double bond can be considered to be the functional group of alkenes, since it imparts characteristic properties to these compounds. Reactions of multiple bonds of whatever type (>C=C<, >C=O, >C=N–, –C≡C–, etc) typically involve **addition**. In very general terms this reflects the fact that π bonds are weaker than σ bonds. Consequently, addition, which involves forming two σ bonds at the expense of breaking one π and one σ bond is, in general, energetically favourable.

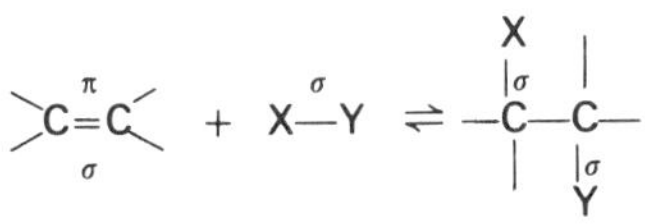

(4.30)

In the case of simple alkenes the double bond is attacked, **initially**, by an electrophile (X^+ in 4.31).

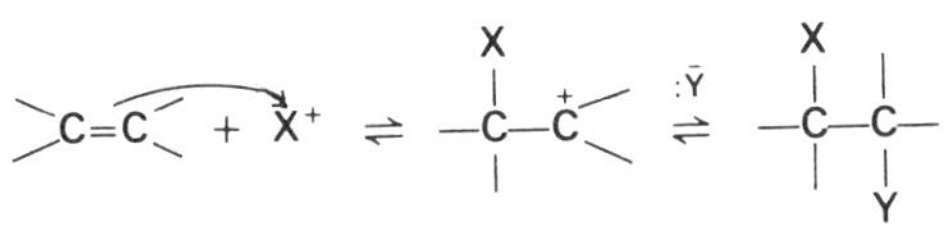

(4.31)

The carbonium ion which results from this step then reacts with a nucleophile (:$\bar{Y}$ in (4.31)) to give the adduct proper. The involvement of an electrophile in the first step of addition is explained by the fact that, because it is associated with four electrons, the internuclear region in a double bond constitutes a region of relatively high charge density. Moreover, the charge in the π bond is outside the internuclear axis and hence is easily accessible to the approaching reagent.

Addition of halogens

The mechanism by which halogens, for example bromine, add to double bonds is shown in (4.32).

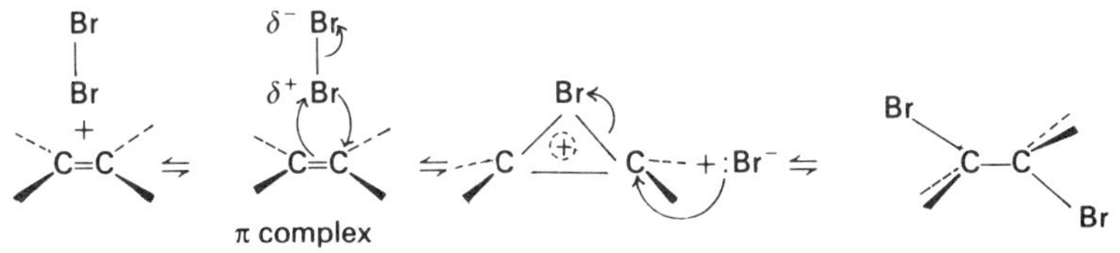

(4.32)

Formally, bromine is not an electrophile. However, as the reactants approach each other the region of high charge density associated with the double bond repels charge in the σ bond of the bromine molecule and a π complex is formed in which the bromine is loosely bound to the alkene by the electron pair of the π bond. Eventually the developing positive end of the bromine dipole captures the electron pair of the π bond and a cyclic bromonium ion is formed which then reacts with bromide to give the adduct. A bromonium ion is implicated in the reaction rather than the simpler, alternative, bromo-carbonium ion (4.33) on the basis of the stereochemistry of the products which are formed.

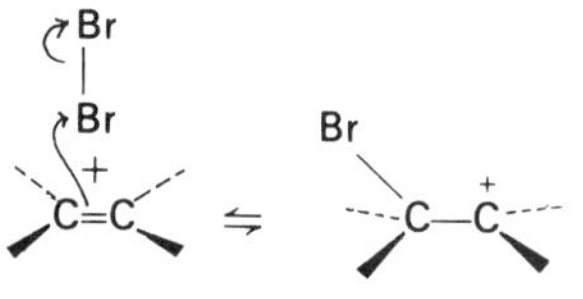

(**4.33**)

The bromine atoms in the adducts invariably occupy positions which indicate that they add to opposite faces of the double bond (*anti* addition). This can be explained if the last step in the sequence is essentially an S_N2 reaction (Section 5.1) in which bromide attacks the electrophilic carbon atom of a bromonium ion on the face opposite to that to which the leaving group (–Br–) is bonded. In contrast, the stereochemistry of the products cannot be explained in terms of a bromocarbonium ion. The three bonds formed by a positively charged carbon are co-planar. A carbonium ion would therefore be attacked, at least to some extent, on both faces giving rise to some product in which both bromine atoms have added to the same face of the double bond (*syn* addition). Since *syn* addition products are not observed a bromocarbonium ion cannot be an intermediate in the reaction. Bromination of an alkene is a reaction which is of little biochemical interest. It is included here because it is typical of the reactions of alkenes, because it is used as a diagnostic test for the presence of a double bond in a molecule (consumption of bromine, which is red brown, can be followed visually) and because it is sometimes used to purify alkenes, as shown in (4.34).

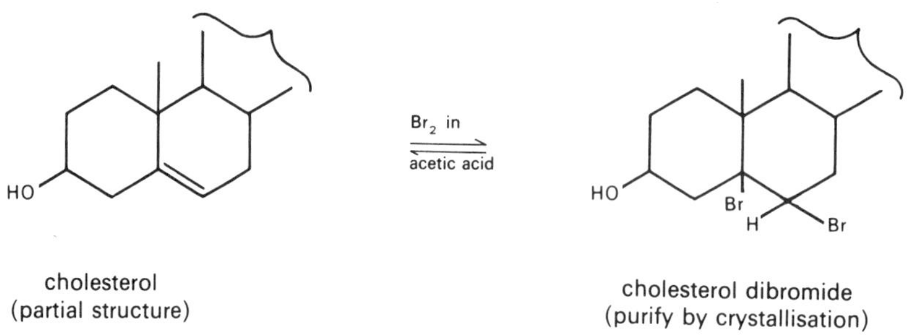

$$\text{cholesterol dibromide} \xrightleftharpoons{\text{Zn}} \text{cholesterol} + ZnBr_2$$

(**4.34**)

The addition of iodine to double bonds has been used for many years to determine the degree of unsaturation of fats.

Addition of water and related compounds

Hydration of a double bond is probably one of the most important reactions which it undergoes, both in the laboratory and in living cells. In the laboratory, the hydration of alkenes is used to produce alcohols. For instance, ethanol results from dissolving ethene in cold concentrated sulphuric acid and then diluting the solution carefully with water (4.35).

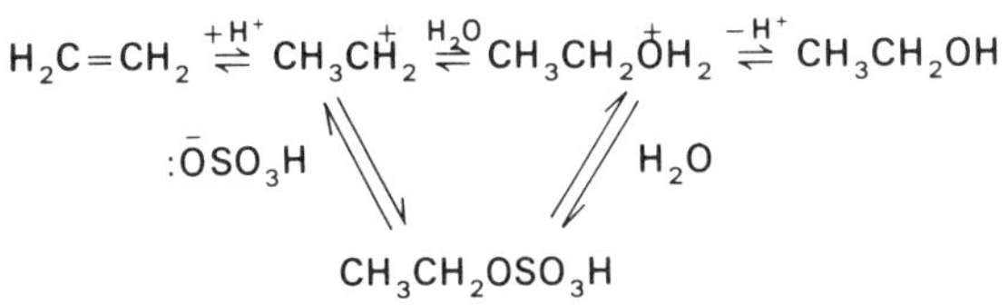

(4.35)

In this reaction the electrophile responsible for the initial attack on the alkene is the proton. The carbonium ion which is produced can be attacked by any nucleophile which is present in the reaction mixture. Consequently the protonated alcohol results both from direct attack by water and from displacement of bisulphate by water, (an S_N2 reaction, Section 5.1). Notice that because the last step regenerates a proton, the overall reaction is subject to acid catalysis.

Double bonds which are hydrated during biochemical reactions are almost invariably present in a position 2,3- (or α, β) relative to the carbonyl group of a carboxylic acid, ester or thiolester.

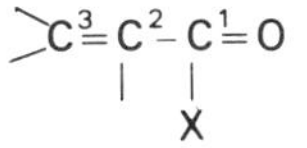

X = –OH, –OR, –SR

It is logical, therefore, to examine the reactions of this type of compound. In the laboratory powerful nucleophiles such as hydrogen sulphide, ammonia and their derivatives add across the double bond in the absence of a catalyst.

$$CH_3CH{=}CHCO_2H + NH_3 \rightleftharpoons CH_3CH(NH_2)CH_2CO_2H$$

but-2-enoic acid 3-aminobutanoate

Such reactions are explained on the basis of the existence of a mesomeric polarisation in the molecule (4.36) and of the mechanism shown in (4.37).*

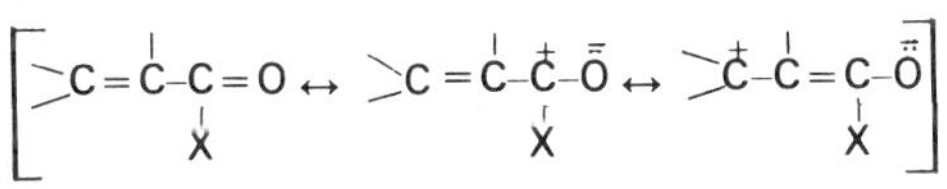

(4.36)

It can be seen that the last step in this reaction mechanism involves the migration of a proton between two positions in the structure with a concomitant rearrangement of the bonding electrons.

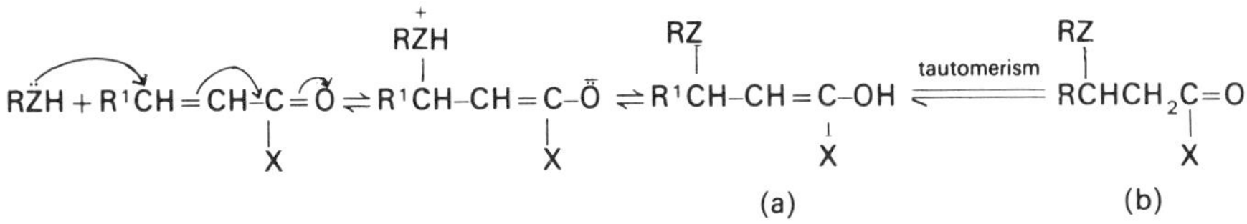

R = H or alkyl; Z = NH, S; X = –OH; $–OR^2$, $–SR^2$ **(4.37)**

* Note that the nucleophilic reagent could also attack the unsaturated carbonyl compound at C_1; the products of this mode of attack, however, are rarely stable.

This phenomenon is known as **tautomerism** and is explained as follows. An **enol**, that is a compound containing a hydroxyl group attached to a double bond, as in (a) in (4.37), is always very weakly acidic, the driving force for dissociation being the production of a relatively stable anion.

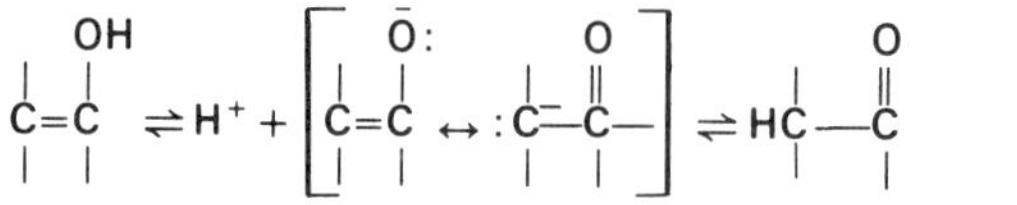

enol tautomer   keto tautomer

It can be seen that the anion contains two centres of high charge density, allowing the proton to return to one of two alternative sites. An **enol**, therefore, always exists in equilibrium with its corresponding **keto** form. The keto compound is generally considerably less acidic than the enol and therefore predominates in the equilibrium mixture, often almost to the exclusion of the latter for all practical purposes. In passing, it can be noted that tautomerism is encountered quite frequently in biochemistry. For example, the heterocyclic bases which are present in the nucleic acids are tautomeric compounds.

Water is too weak a nucleophile to be able to add directly to a 2,3-unsaturated carbonyl compound. Hydration is possible however, in the presence of an acid catalyst.

$$\underset{\text{propenoic acid}}{H_2C{=}CHCO_2H} + H_2O \underset{100^\circ C}{\overset{\text{conc. } H_2SO_4}{\rightleftharpoons}} \underset{\text{3-hydroxypropanoic acid}}{HOCH_2CH_2CO_2H}$$

The catalyst acts by protonating the carbonyl group* (4.38) and in so doing greatly increases the polarisation in the π bonds of the molecule (there will be a much greater tendency for charge to 'flow' to neutralise a positively-charged oxygen atom than for it to move towards uncharged oxygen). Attack by water at C-3, followed by loss of a proton, generates an enol which tautomerises to give the product proper.

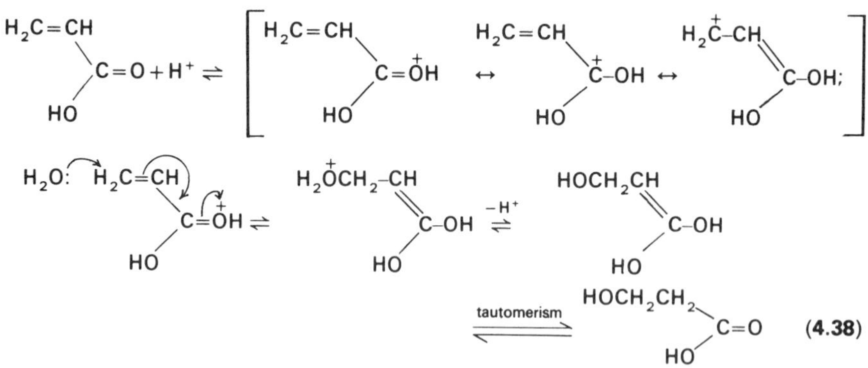

(4.38)

* The carbonyl oxygen is protonated in preference to the hydroxyl oxygen because the π bond is easily polarised.

Some examples of biochemical reactions which involve nucleophilic addition to 2,3-unsaturated carbonyl compounds or to 2,3-unsaturated acids or derivatives are as follows. Aspartase, an enzyme which is present in plants and in some micro-organisms but not in animals, catalyses the formation of aspartate by addition of ammonia to fumarate.

$$^{-}O_2CCH{=}CHCO_2^- + NH_4^+ \xrightleftharpoons{\text{aspartase}} {}^{-}O_2CCH(\overset{+}{N}H_3)CH_2CO_2^-$$

A considerable number of reactions involving the reversible hydration of a double bond adjacent to a carbonyl group are known to occur in living cells. For example one step in the *β*-oxidation of fatty acids involves the hydration of 2,3-unsaturated derivatives of coenzyme A and the catabolism of the branched chain amino acids, leucine, isoleucine and valine involves similar reactions. Enolase catalyses an important reversible hydration in glycolysis while aconitase and fumarase catalyse hydrations in the tricarboxylic acid cycle of reactions.

Oxidation and reduction of the double bond

Oxidation: In the laboratory, double bonds can be oxidised by a number of reagents, for example an alkaline solution of permanganate, which initially hydroxylates each carbon atom of the bond. The products of oxidation are normally carboxylic acids but the intermediate 1,2-diol can be isolated if the oxidation is performed at low temperatures using mild conditions (4.39).

$$\underset{\text{propene}}{CH_3CH{=}CH_2} \xrightarrow[KMnO_4/HO^-]{\text{cold dil}} \underset{\text{1,2-propanediol}}{CH_3\underset{|}{\overset{}{C}}H\text{-}\underset{|}{C}H_2} \xrightarrow[\text{more conc. reagent}]{\text{heat with a}} [CH_3CH(OH)_2 + (OH)_2CH_2]$$

(the two hydroxyls of 1,2-propanediol shown as OH OH beneath the carbons)

$$\xrightarrow{-2H_2O} \underset{\text{ethanal}}{CH_3CHO} + \underset{\text{methanal}}{HCHO} \longrightarrow \underset{\text{ethanoic acid}}{CH_3CO_2H} + CO_2 + H_2O \qquad \textbf{(4.39)}$$

More powerful reagents produce carboxylic acids directly.

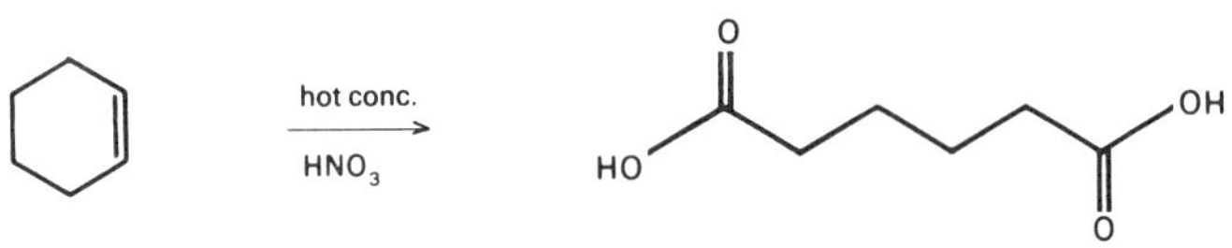

cyclohexene adipic acid

As was mentioned previously (4.20) on p. 37, compounds containing double bonds are intermediates in many oxidative metabolic pathways. Their further oxidation involves hydration and the resulting alcohol is oxidised to a carbonyl compound.

$$-CH{=}CH- \xrightleftharpoons{H_2O} -CH(OH)CH_2- \rightleftharpoons -\overset{\overset{O}{\|}}{C}CH_2-$$

Alkenes can also be oxidised by dehydrogenation. The products, compounds containing a carbon to carbon triple bond, are known as alkynes. Although many alkynes have been found to occur naturally, particularly in plants of the *Compositae* family, they are not involved in any of the major metabolic activities of plants or animals and consequently will not be discussed in this book.

Reduction: Isolated double bonds can be reduced by allowing them to react with molecular hydrogen in the presence of a finely divided metal catalyst such as 5 % palladium suspended on charcoal.

$$\underset{\text{fumaric acid}}{HO_2CCH{=}CHCO_2H} \xrightarrow[5\%\ Pd/C]{H_2 @ 1\ At.,\ R.T.} \underset{\text{succinic acid}}{HO_2CCH_2CH_2CO_2H}$$

The reaction occurs on the surface of the catalyst and has characteristics of a radical reaction in that homolytic bond breakage occurs. The reduction of double bonds in living cells, on the other hand, may proceed by an ionic mechanism. For example, 2,3-unsaturated acyl derivatives of acyl carrier protein (ACP) are reduced by reduced nicotinamide adenine dinucleotide phosphate (NADPH) as the last step in the biosynthesis of fatty acids (4.40).

$$\underset{\text{but-2-enoyl ACP}}{CH_3CH{=}CH\overset{O}{\overset{\|}{C}}SACP} + NADPH + H^+ \xrightarrow[\text{reductase}]{\text{enoyl}} \underset{\text{butanoyl ACP}}{CH_3CH_2CH_2\overset{O}{\overset{\|}{C}}SACP} + \overset{+}{NADP} \qquad \textbf{(4.40)}$$

ACP = acyl carrier protein

As will be explained in Chapter 9, NADPH acts as a donor of hydride ion; it is therefore possible that reaction (4.40) involves nucleophilic attack at C-3 of the thiolester.

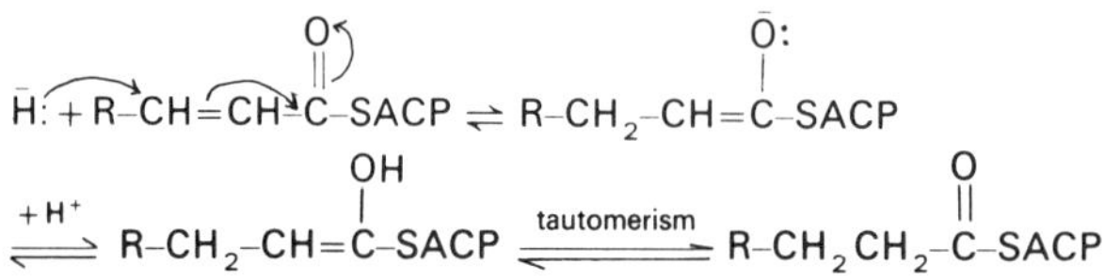

4.3. The aromatic nucleus -arenes and derivatives

The modern, systematic, class name for a hydrocarbon derived from benzene is **arene**. The name stems from the older description of compounds derived from benzene as 'aromatic compounds', the latter term relating to the fact that many such compounds are volatile liquids possessing characteristic, often pleasant, smells. The self-explanatory term 'benzenoid compound' is also commonly used in the chemical literature to describe arenes and their derivatives.

Examples and nomenclature

The structural formulae of benzene, some of its homologues, and of some simple monovalent aryl radicals are shown in Table (4.6). The prefixes *ortho*, *meta* and *para* are still in common usage for naming, respectively, 1,2-, 1,3- and 1,4-disubstituted benzenoid compounds. However, since no comparable system is available for more highly substituted compounds it is preferable to use locant numbers in all cases. The difference between tolyl and benzyl radicals should be noted.

Table 4.6 Some typical arenes and some derived radicals

benzene

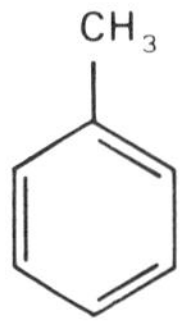

toluene
(methyl benzene)

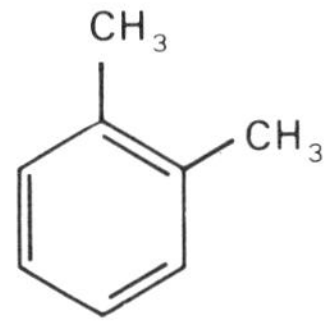

1,2-dimethylbenzene
(*o*-xylene)

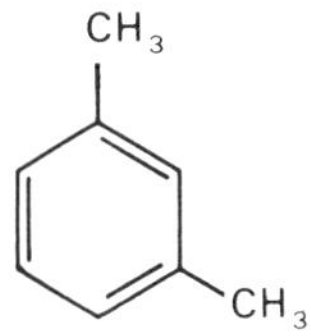

1,3-dimethylbenzene
(*m*-xylene)·

1,4-dimethylbenzene
(*p*-xylene)

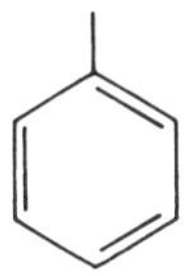

phenyl

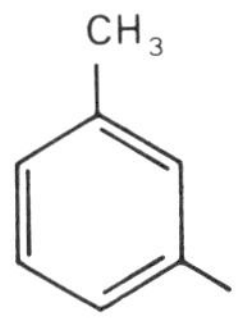

3-tolyl
(*m*-tolyl)

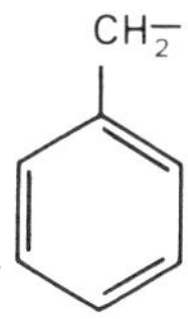

benzyl

Naturally occurring arenes and derivatives

Benzenoid hydrocarbons are not found to any significant extent in living organisms. However, a great number of compounds are found which contain one or more functional groups (particularly hydroxyl or carboxyl groups) substituted for hydrogen in a benzenoid ring; a selection of such compounds is given in Table 4.7.

Phenylalanine (4.41), tyrosine (4.42) and thyroxine (4.43) are 2- (or α-) amino acids, the first two being normal constituents of proteins. Thyroxine is a hormone which is essential for proper mental development of mammals and control of their basic metabolic rate. The hormone is synthesised in, and secreted by, the thyroid gland. Adrenalin (4.44) is also a hormone. It is produced and secreted by the adrenal medulla in response to a variety of stimuli such as excitement, fear or anger and prepares the animal for 'fight or flight'. Both phenylalanine and tyrosine are precursors of adrenalin. Anisole (4.45), eugenol (4.46), cinnam-aldehyde (4.47) and methyl salicylate (4.48) are examples of the wide variety of aromatic compounds which are found in the 'essential oils' of many plants. Oestradiol (4.49) is a female sex hormone (an oestrogen) produced by the ovaries and is responsible for development of female sexual characteristics and for the maintenance of the reproductive cycle in women. Indolylacetic acid (heteroauxin, 4.50) is a plant growth hormone.

Structure and reactivity of benzene

Although benzenoid compounds are in a minority compared with aliphatic compounds in plant and animal metabolism, it is nevertheless necessary for a biochemist to have some understanding of the reactions of aromatic structures.

Table 4.7 Some naturally occuring benzenoid compounds

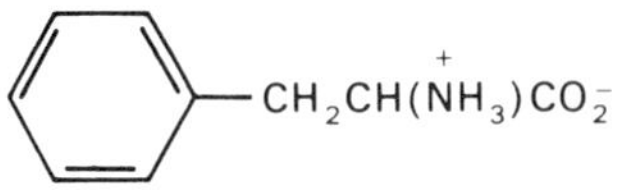

phenylalanine (**4.41**)

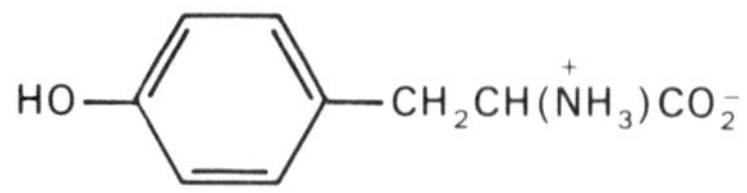

tyrosine (**4.42**)

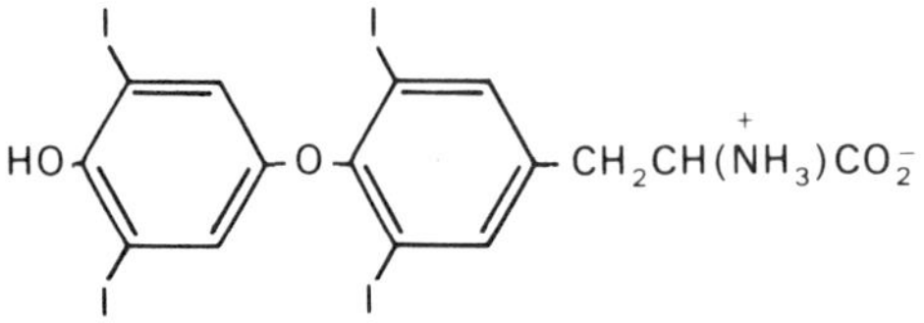

thyroxine (**4.43**)

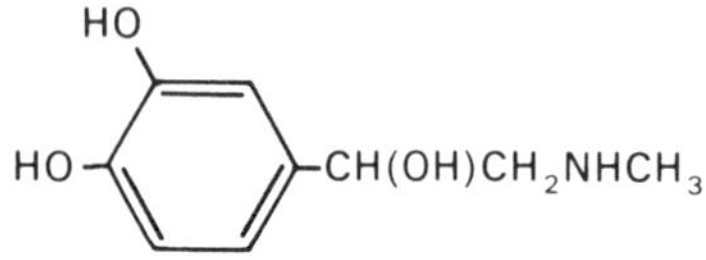

adrenaline (**4.44**)

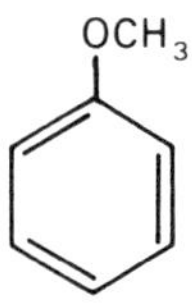

anisole (**4.45**)

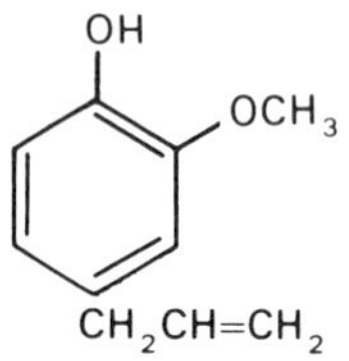

eugenol (**4.46**)

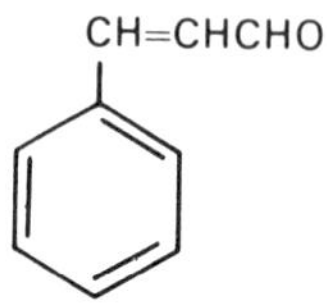

cinnamaldehyde (**4.47**)

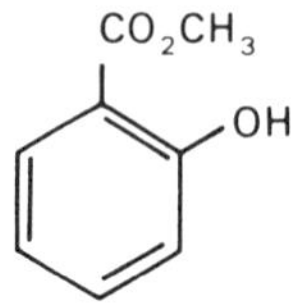

methyl salicylate (**4.48**)

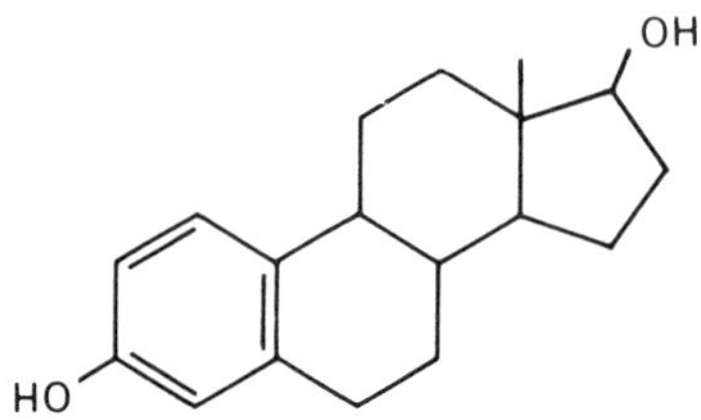

oestradiol (**4.49**)

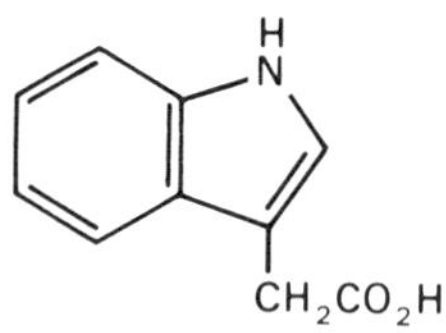

indolylacetic acid (**4.50**)

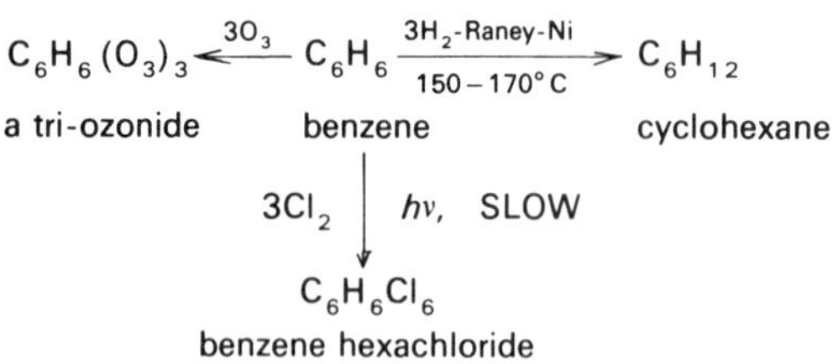

(**4.51**)

To this end, the remainder of this section discusses aromatic reactivity, mainly with reference to benzene. Obviously functional groups attached to an aromatic ring will modify its reactivity. Such modifications are generally outside the immediate interest of a biochemist and are not dealt with here.

The structure of benzene has already been mentioned, briefly, in relation to the concept of resonance hybrids (Section 2.4). A more complete discussion of the structure follows.

The molecular formula of benzene is C_6H_6. The hydrocarbon **adds** three molecules of chlorine, hydrogen or ozone under appropriate conditions (4.51) and hence appears to contain three double bonds. The molecule must, therefore, be cyclic (Kekulé, 1865) and is often represented by (4.52), the Kekulé formula.

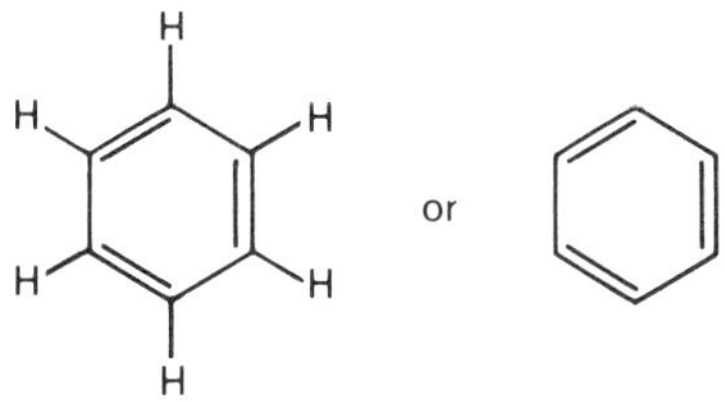

(**4.52**)

The Kekulé formula, however, is incompatible with several properties of benzene, viz:

(i) The double bonds are abnormally unreactive. The addition reactions quoted above require considerably more vigorous conditions than do similar reactions involving alkenes. Furthermore, hydrogen halides do not add to benzene as they do to alkenes and benzene is not readily oxidised by alkaline solutions of potassium permanganate as are alkenes.

(ii) Reactions of benzene typically involve **substitution** rather than addition; substitution reactions are not typical of alkenes.

(iii) Physical measurements show that **all** of the carbon–carbon bonds in benzene have the same length, 0.139 nm, intermediate between the lengths of isolated single and double bonds (Table 2.3). The Kekulé formula suggests that the bond lengths should alternate between 0.134 and 0.154 nm.

(iv) Benzene is considerably more stable than would be expected on the basis of the Kekulé formula. Studies of many simple alkenes show that the heat of hydrogenation of an isolated double bond is approximately constant at about 126 kJ mol^{-1}. In particular, the heat of reaction associated with the conversion of cyclohexene to cyclohexane by hydrogenation is 120 kJ mol^{-1}. One might expect, therefore, that the heat of reaction when benzene is converted to cyclohexane by hydrogenation would be 3 × 120 = 360 kJ mol^{-1}. The experimental value, however, is only 208 kJ mol^{-1}, indicating that benzene is 152 kJ mol^{-1} more stable than would be expected on the basis of Kekule's formula (4.53).

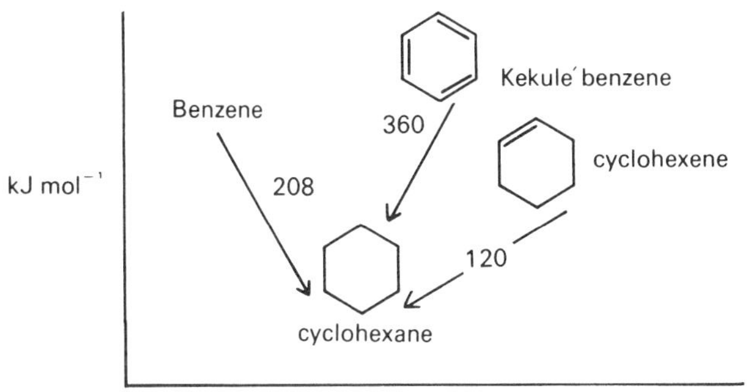

(**4.53**)

Bonding in the benzene molecule. Section 2.1 indicated that the four electrons of a doubly bound carbon atom can be described in terms of three sp^2 hybrid orbitals and one unhybridised p orbital. If this description is applied to the electrons of each carbon atom in benzene the following picture of the hydrocarbon results. Six coplanar carbon–carbon bonds, forming a regular hexagon, are described by overlap of sp^2 hybrid orbitals on adjacent carbon atoms and six coplanar carbon–hydrogen bonds are described by the overlap of the remaining sp^2 hybrid orbital of each carbon atom with the s orbital of a hydrogen atom (4.54).

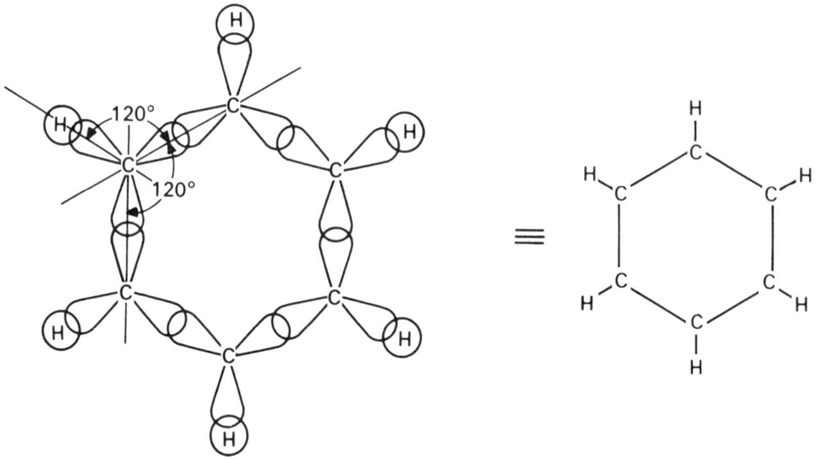

(**4.54**)

Inspection of the system now indicates that three π bonds can be described in terms of sideways overlap of the p orbitals. However, the geometry of the system is such that each p orbital can overlap with either of its neighbours with equal facility; π bonding as shown in (4.55a) is just as likely as that shown in (4.55b).

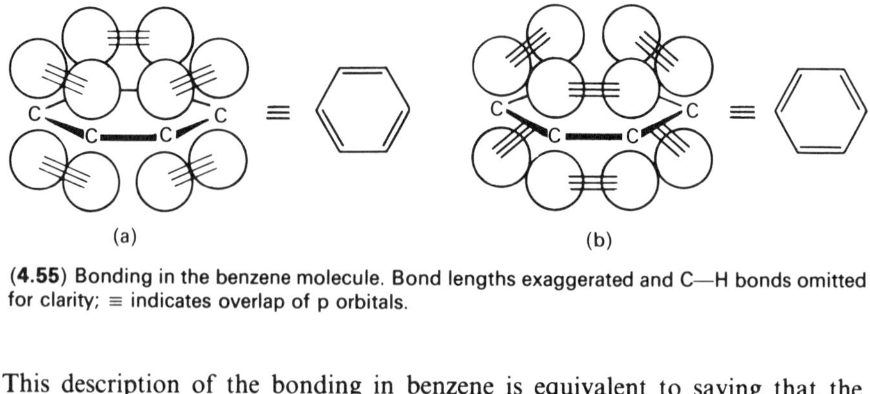

(**4.55**) Bonding in the benzene molecule. Bond lengths exaggerated and C—H bonds omitted for clarity; ≡ indicates overlap of p orbitals.

This description of the bonding in benzene is equivalent to saying that the hydrocarbon can be described as a resonance hybrid in terms of the two Kekulé structures shown in (4.55). It also indicates that the π electrons are completely delocalised, a fact often denoted by using symbol (4.56) to represent the molecule.

(**4.56**)

Complete de-localisation of the π electrons in benzene is the feature of its structure which leads to its stability. The description explains why the carbon to carbon bond lengths are all equal and intermediate between those of isolated single and double bonds and also why benzene does not display typical alkene-like reactivity—namely because it does not contain true double bonds.

Typical reactivity of the aromatic nucleus

Firstly, it should be noted that benzene usually reacts with **electrophilic** reagents for the same reason that the double bond reacts with electrophiles, namely because the charge concentrated above and below the plane containing the nuclei would repel a nucleophile but invites attack by an electron deficient species.

The stability which results from π electron de-localisation explains why reactions of benzene usually involve substitution rather than addition. Electrophilic attack of benzene by, for example, bromine would generate a monobromo-carbonium ion. It is energetically more favourable for this intermediate to lose a proton, giving rise to the substitution product in which the delocalised, low energy, benzenoid system is retained rather than to react with bromide to give the addition product in which the system over which π electrons are delocalised is considerably restricted (4.57).

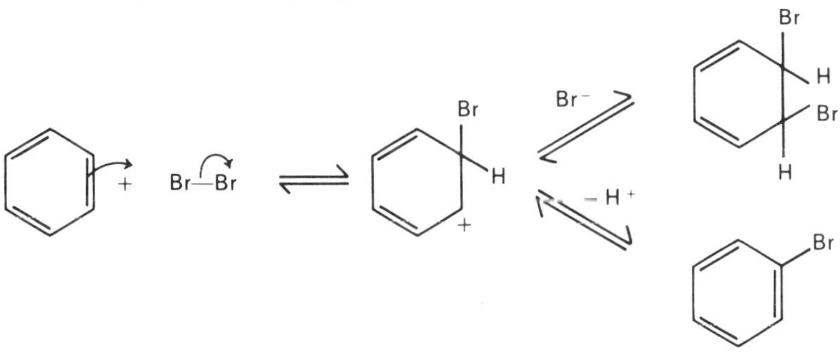

(**4.57**)

Five of the most important reactions of benzene which involve electrophilic substitution are summarised in Table 4.8. Inspection of the table shows that all of the reactions conform to the generalised equation; all the reactions are catalysed and in all cases the catalyst functions to produce an effective electrophile (sulphuric acid acts as both catalyst and reagent for sulphonation). Initially the electrophile binds loosely to the π electrons of benzene giving a π complex. This complex is then converted to a σ complex, the terminology indicating that the electrophile, having captured a pair of π electrons, is now bound to carbon by a σ bond. Finally the σ complex is 're-aromatised' by loss of a proton. These and analogous reactions involving derivatives of benzene have proved to be of

Table 4.8. Some Reactions of Benzene which involve Electrophilic Substitution

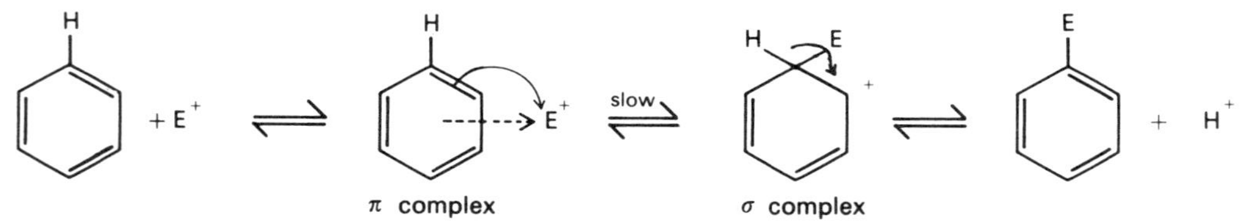

Reaction	Overall Reaction	Action of catalyst	Comments
Nitration	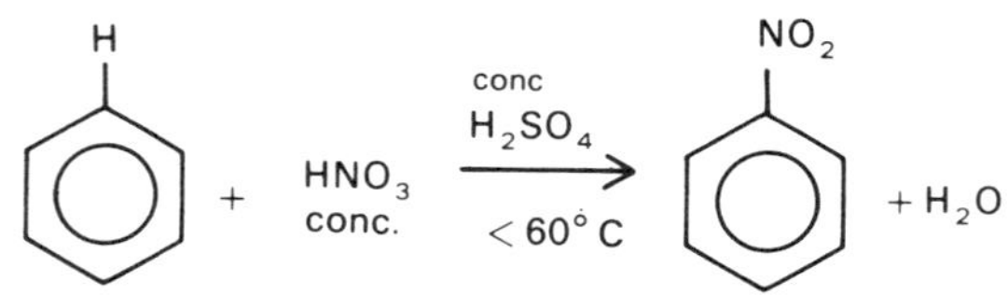 nitrobenzene	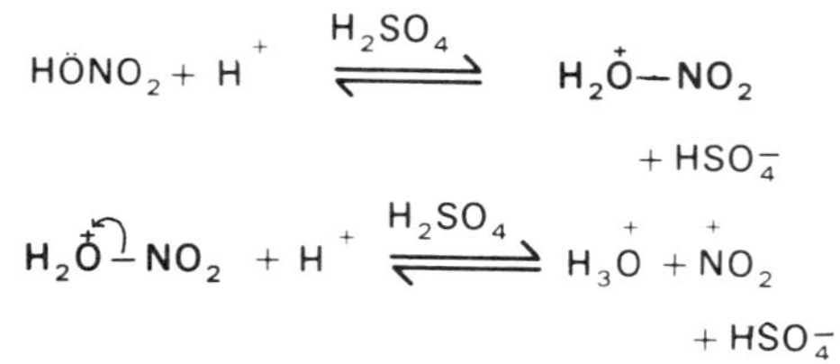	Temp kept below 60° C otherwise *m*-dinitrobenzene is produced.
Friedel-Craft alkylation	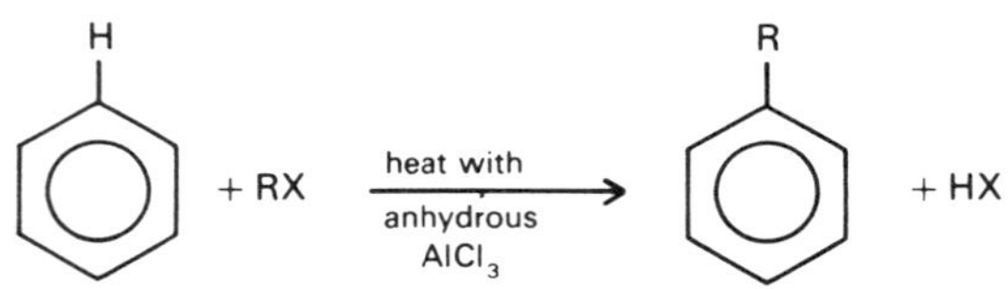 X = halogen; an alkyl benzene	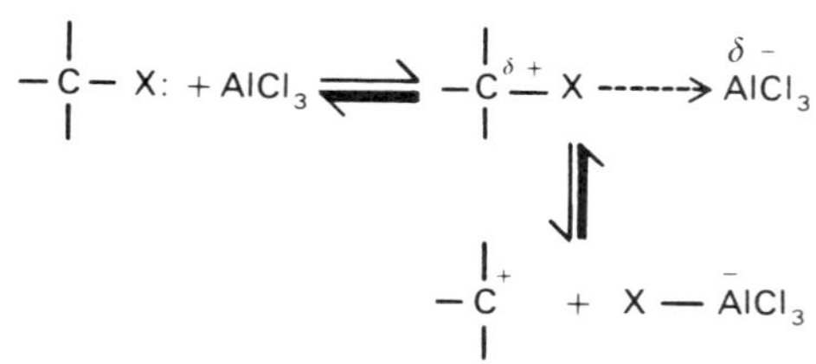	Difficult to control. Primary carbonium ions liable to rearrange

Friedel-Craft acylation

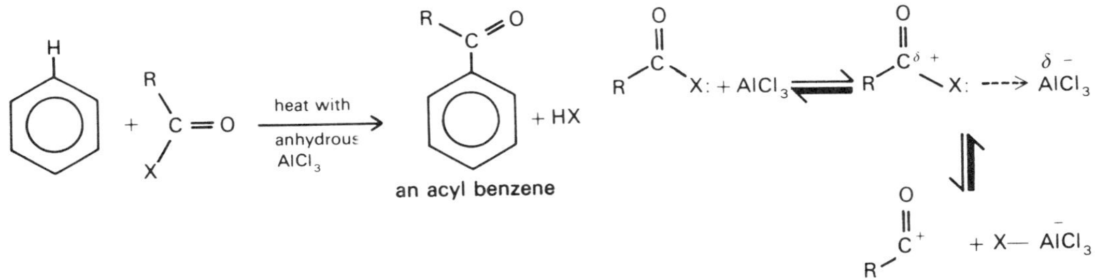

1.1 mole equivalents of catalysts used since it forms a 1:1 complex with product

Halogenation

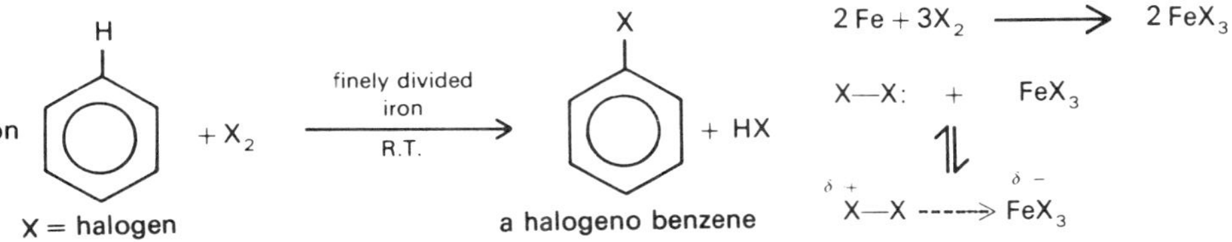

Pyridine or iodine also used as catalysts

Sulphonation

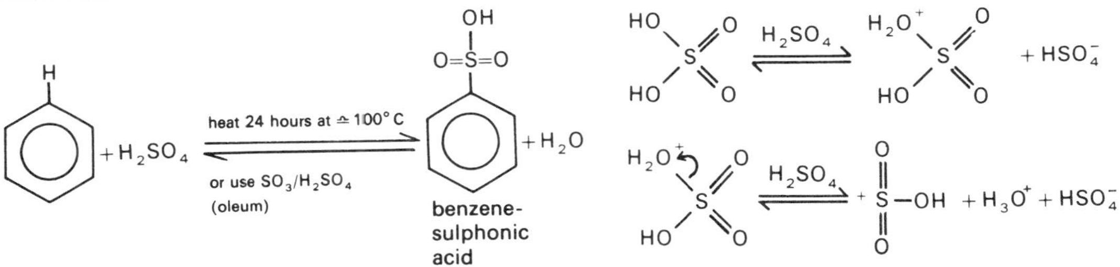

Demonstrably reversible. Slow step involves loss of H^+ from σ complex

immense value both to academic and industrial chemistry and their characteristics have been well authenticated by numerous investigators.

Aromatic substitution *in vivo*

Hydroxylation. Very few aromatic substitution reactions have been found to occur in living cells. At first sight this may seem surprising in view of the fact that many naturally occurring aromatic compounds contain hydroxyl groups or substituents derived from hydroxyl groups. However, direct hydroxylation by substitution would be a very unlikely process *in vivo* or *in vitro* because the only likely reagent, hydroxide ion, is a strong nucleophile. In keeping with this, it is found that hydroxy groups are introduced into naturally occurring aromatic compounds indirectly. A number of reactions are known which formally involve substitution but which in practice involve addition followed by rearrangement (4.58).

$$ArH + O_2 + XH_2 \rightarrow ArOH + X + H_2O \qquad \textbf{(4.58)}$$

They are catalysed by enzymes known as mono-oxygenases or hydroxylases; XH_2 in equation (4.58) is a reduced co-enzyme, usually reduced flavin adenine dinucleotide ($FADH_2$) or a reduced pteridine. The reduced co-enzymes are thought to combine with oxygen in some way (Section 9.2) and it is the product of this combination which attacks the arene. There is some argument about the electronic character of the reagent—it has been suggested to act as a source of free radicals by some investigators or as an electrophile by others. Whatever the truth is in this matter it is known that the first product of the reaction is an epoxide which undergoes rearrangement to give a phenol.

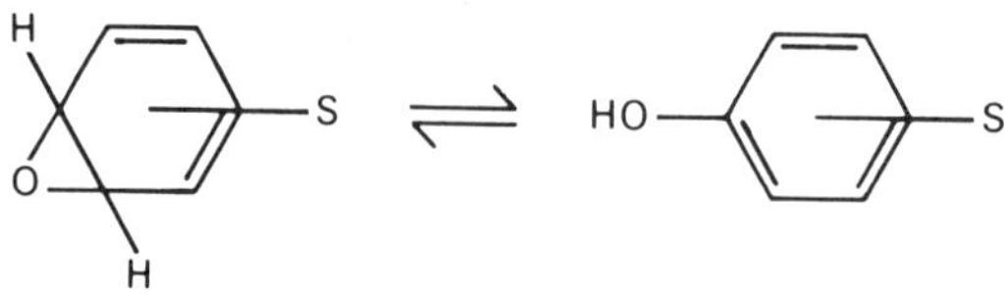

Two representative examples of reactions of the above type involve the conversion of phenylalanine to tyrosine in the liver (4.59) and the production of protocatechuate from 4-hydroxybenzoate in a **Pseudomonas** species (4.60).

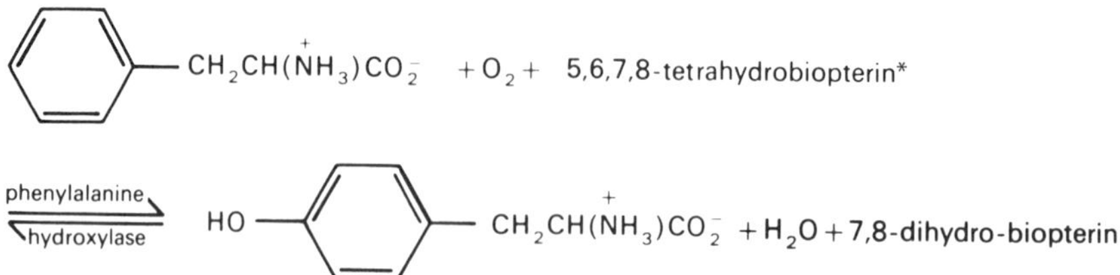

(4.59)

* 2-amino-5,6,7,8-tetrahydro-4-hydoxy-6-(1′,2′-dihydroxypropyl) pteridine

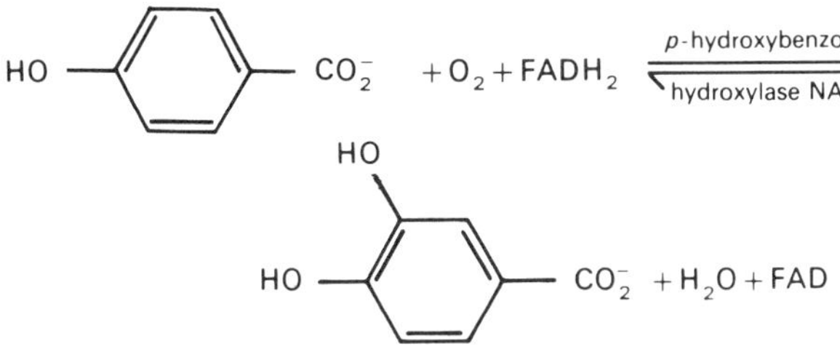

(**4.60**)

Synthesis of thyroxine. The introduction of iodine atoms into tyrosine residues during the synthesis of thyroxine in the thyroid gland may involve direct substitution of iodine for hydrogen. The thyroid gland concentrates iodide ions which are then used to iodinate tyrosine residues in a specific protein, thyroglobulin. The reaction requires the presence of hydrogen peroxide and is catalysed by a membrane bound protein, thyroidal peroxidase. It seems possible that iodide ions are oxidised to iodine which then causes substitution to occur in the benzenoid ring of tyrosine residues (the hydroxyl group of tyrosine would activate the ring towards electrophiles). Subsequent, complex, coupling reactions produce thyroxine residues which are liberated by cleavage of thyroglobulin (4.61). Exact details of the bio-synthetic route remain unknown.

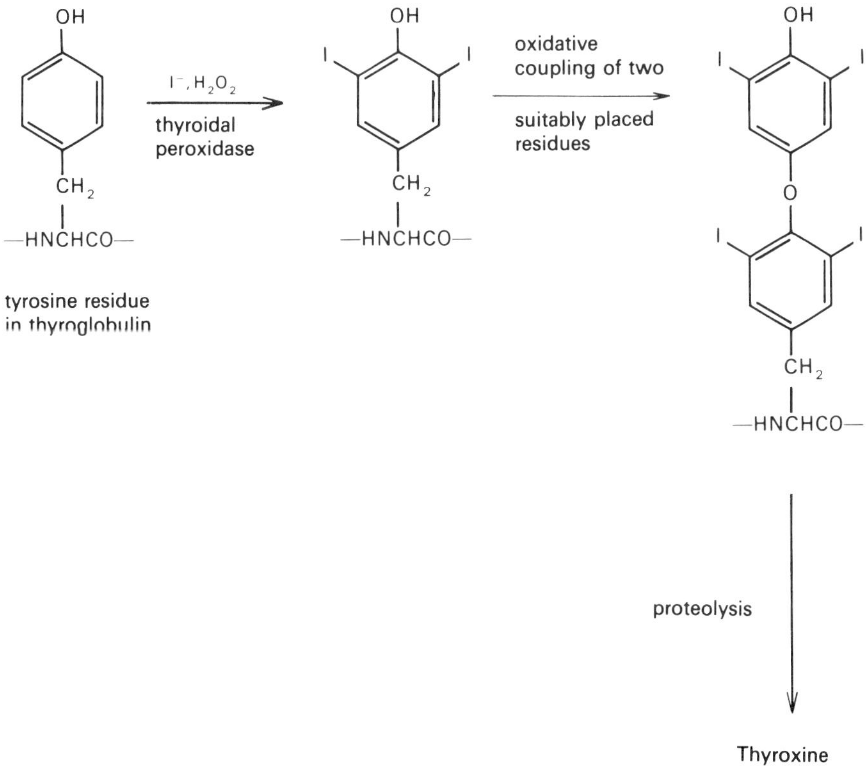

(**4.61**)

Miscellaneous substitutions. The biosynthesis of the 2-amino acid tryptophan is thought to follow the pathway shown by (4.62). Steps (a) and (b) together constitute electrophilic displacement of the carboxyl group.

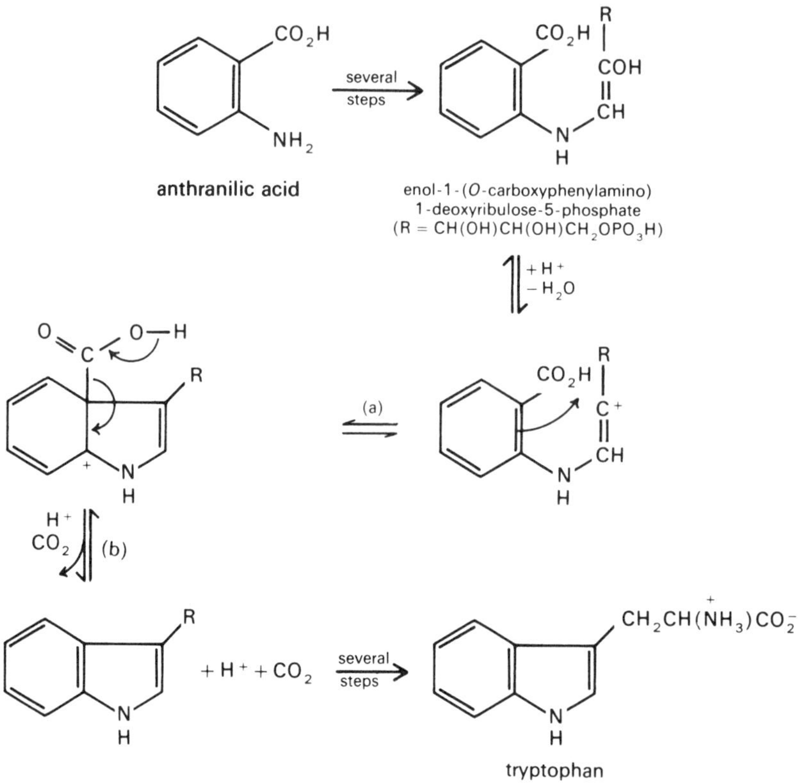

(4.62)

The enzymic decarboxylation of some aromatic acids in plant tissues may also involve electrophilic substitution.

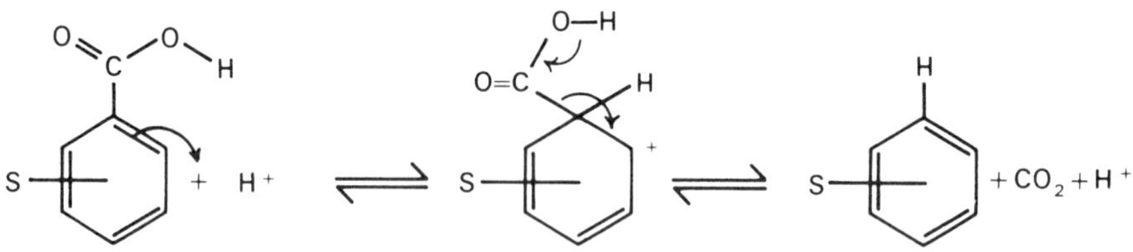

S = an undefined substituent

5

Functional groups of the type $-\overset{|}{\underset{|}{C}}-X$. Halogeno, hydroxy, mercapto and amino compounds and their derivatives

Halogeno, hydroxy, mercapto and amino compounds all contain similar functional groups and exhibit, to varying degrees, the same types of reactivity. It is therefore logical to discuss them collectively.

5.1 Aliphatic compounds

Examples and nomenclature

The suffixes and prefixes which donate the presence of the various functional groups under discussion in a molecule are shown in Table 5.1 and some typical named examples of appropriate compounds in Table 5.2.

Table 5.1 Nomenclature of functional groups of the type $-\overset{|}{\underset{|}{C}}-X$.

Functional group	Suffix	Prefix
X = halogen	–	halogeno
$-OH$	–ol	hydroxy–
$-OR$		alkyloxy– or aryloxy–
$-SH$	–thiol	mercapto–
$-SR$		alkylthio– or arylthio–
$-NH_2$	–amine	amino–
$-NHR$	–	alkylamino or arylamino
$-NRR'$	–	dialkylamino, diarylamino or alkylarylamino

In connection with the nomenclature of the compounds note that:

(i) Compounds which contain a functional group attached to a primary, secondary or tertiary carbon atom are similarly designated as primary, secondary or tertiary compounds. Thus (5.1) and (5.2) are primary halides, while (5.4) is a secondary halide.

(ii) Halogen atoms are always cited using a prefix. All other functional groups are cited using suffixes except when more than one is present. In the latter case one group is designated as the **principle group** and is cited using a suffix. Any other groups present are then cited, in alphabetical order, using prefixes. The order of decreasing priority for designation as the principle group is: 'Onium and similar cations, acids, anhydrides, esters, amides, nitriles, aldehydes, thioaldehydes, ketones, thioketones, alcohols, thiols, phenols, thiophenols, amines and ethers. Functional groups are located by numbering the parent hydrocarbon from the end of the carbon chain which leads to the lowest series of locant numbers.

Table 5.2 Some representative compounds containing functional groups of the type $-\overset{|}{\underset{|}{C}}-X$

CH_3Cl chloromethane (**5.1**)	CH_3CH_2Br bromoethane (**5.2**)	$CH_3CH_2CH_2Cl$ 1-chloropropane (**5.3**)	$CH_3CHClCH_3$ 2-choloropropane (**5.4**)	$CH_3CH_2CHClCH_2Cl$ 1,2-dichlorobutane (**5.5**)
CH_3CH_2OH ethanol (**5.6**)	$CH_3CH_2CH_2CH_2OH$ 1-butanol (n-butanol) (**5.7**)	$CH_3CH_2CH(OH)CH_3$ 2-butanol (sec-butanol) (**5.8**)	$(CH_3)_2CHCH_2OH$ 2-methylpropan-1-ol (isobutanol) (**5.9**)	
$(CH_3)_3COH$ 2-methylpropan-2-ol (tertbutanol) (**5.10**)	$CH_3CH(OH)CH_2CO_2H$ 3-hydroxybutanoic acid (**5.11**)	$HOCH_2CH_2OH$ 1,2-ethanediol (**5.12**)	CH_3CH_2SH ethanethiol (**5.13**)	
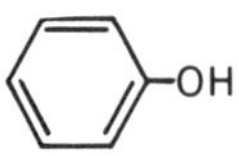 phenol* (**5.14**)	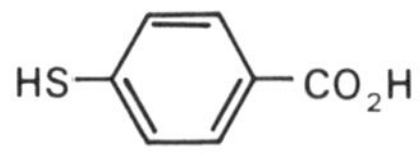 4-mercaptobenzoic acid (**5.15**)	$HOCH_2CH(SH)CH_2SH$ 2,3-dimercaptopropan-1-ol (**5.16**)		
$CH_3O(CH_2)_4CH_3$ methoxypentane (**5.17**)	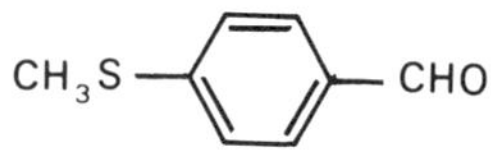 4-methylthiobenzaldehyde (**5.18**)			
$CH_3CH_2NH_2$ ethylamine (**5.19**)	$CH_3CH(NH_2)CH_3$ 2-propylamine (**5.20**)	$H_2NCH_2CH_2OH$ 2-aminoethanol (**5.21**)		
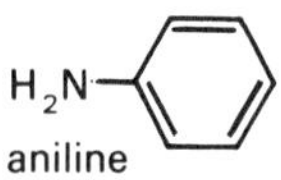 aniline (**5.22**)	$(CH_3CH_2)_2NH$ Diethylamine (**5.23**)	$CH_3CH_2NHCH_3$ *N*-methylethylamine (**5.24**)		
$(CH_3)_3N$ triethylamine (**5.25**)	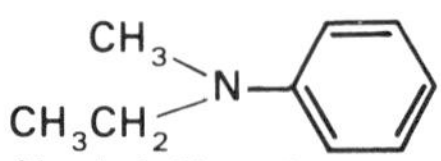 *N*-ethyl-*N*-methylaniline (**5.26**)	CH_3NH $CH_3\overset{|}{C}HCO_2H$ 2-methylamino propionic acid (**5.27**)		
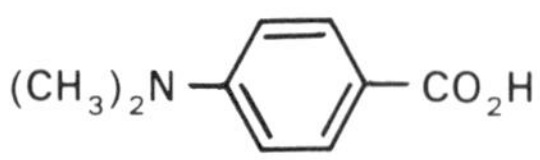 4-dimethylamino benzoic acid (**5.28**)				

(*Trivial names retained by virtue of long usage).

(iii) Amines of the type RNH_2 are designated as primary amines, those of type R^1R^2 NH as secondary amines and compounds of the type $R^1R^2R^3$ N as tertiary amines. Radicofunctional nomenclature is used to name these compounds. The parent radical corresponds to the longest unbranched carbon chain attached to the nitrogen atom. Note that each radical attached to nitrogen must have a locant letter (*N*) unless they are similar, in which case an appropriate multiplying prefix is used. This is necessary in order to distinguish unambiguously between structural isomers such as (5.29) and (5.30).

$(CH_3)_2CHCH_2CH_2NCH_2CH_3$
N-ethyl-3-methylbutylamine
(5.29)

$CH_3CH_2CH_2CH_2\overset{\overset{\displaystyle CH_3}{|}}{N}CH_2CH_3$
N-ethyl-*N*-methylbutylamine
(5.30)

As noted in Section 2.2 oxygen, sulphur and nitrogen can all increase their normal covalency by one by forming a co-ordinate bond. Thus ethers can give rise to tertiary oxonium salts (5.31), thio-ethers to tertiary sulphonium salts (5.32) and tertiary amines to quaternary ammonium salts (5.33).

$$R^1OR^2 + R^3-X \rightleftharpoons R^1\overset{+}{O}R^2R^3X^- \qquad \textbf{(5.31)}$$

$R^1\overset{+}{S}R^2R^3X^-$
(5.32)

$R^1\overset{+}{N}R^2R^3R^4X^-$
(5.33)

All of these 'onium compounds are named as alkyl or aryl substituted salts. For example (5.34) is trimethyloxonium tetrafluoroborate, (5.35) is ethyldimethylsulphonium bromide and (5.36) is trimethylphenylammonium hydroxide.

$(CH_3)_3\overset{+}{O}\,B\overset{-}{F}_4$
(5.34)

$CH_3CH_2\overset{+}{S}(CH_3)_2\,\overset{-}{Br}$
(5.35)

$Ph\overset{+}{N}(CH_3)_3\overset{-}{O}H$
(5.36)

Some compounds of biological interest which contain functional groups of the type $-\overset{|}{\underset{|}{C}}-X$

Some compounds of biological interest which contain functional groups of the type $-\overset{|}{\underset{|}{C}}-X$ are shown in Table 5.3. Very few halogen compounds occur naturally, the most abundant is probably thyroxine (4.43). A number of synthetic chlorinated compounds are of interest, however, since they find extensive use as pesticides and herbicides. Outstanding among these are 2,4-dichlorophenoxyacetic acid (5.37), a systemic herbicide, and the pesticides D.D.T. (2,2-di-4'-chlorophenyl-1,1,1,-trichloroethane, 5.38), gammexane (5.39, often called, incorrectly, hexachlorobenzene) and aldrin (5.40).

A number of naturally occurring hydroxy compounds have already been mentioned, viz: vitamin A (4.24), cholesterol (4.27), tyrosine (4.42), thyroxine (4.43), adrenalin (4.44), eugenol (4.46), methyl salicylate (4.48) and oestradiol (4.49). Carbohydrates are an important class of naturally occurring hydroxy compounds of which α-D-glucose* (5.41) is representative. Fats are mixtures of esters derived from glycerol (1,2,3-propanetriol, 5.42), 3- (or β-) hydroxy acids or esters are involved in many oxidative and biosynthetic pathways, and many hydroxylated compounds, for example the broad spectrum anti-biotic tetracycline (5.43), are produced as metabolites by micro-organisms. Finally, note that 2,4-dichlorobenzyl alcohol (5.44), a fungicide, and the antiseptic 2,4,6-trichlorophenol (T.C.P., 5.45) are representative of a large number of synthetic hydroxy compounds which exhibit physiological activity.

Thiols are of great importance in biochemistry. Cysteine (5.46) is an 2-amino acid which occurs in the active centre of several enzymes, in the ubiquitious tripeptide glutathione (γ-glutamylcysteinylglycine) and, in the oxidised form, in several small, physiologically active polypeptides such as oxytocin and insulin. Lipoic acid (5.47) and co-enzyme A (5.48) are both co-enzymes. The former is required for the function of both the pyruvate and α-ketoglutarate dehydrogenase enzyme complexes while the latter is intimately involved in the enzyme catalysed reactions of acyl groups (Section 9.5).

Ethers do not appear to be of great importance in cellular metabolism. This is particularly so in the case of purely aliphatic ethers. Phenols, however, may be *O*-methylated prior to excretion. For example 3-*O*-methylnoradrenalin (5.49) and 4-hydroxy-3-methoxymandelic acid (5.50) are both produced as degradation products of adrenalin and noradrenalin. Similarly thio-ethers in general are not important in cellular metabolism, although at least one specific thio-ether is. This compound is the essential 2-amino acid methionine (5.51); the adjective essential indicates that, because the amino acid cannot be synthesised by animals it must be present in their diet. All 2-amino acids (5.52), except proline, possess a (protonated) amino group. Lysine (5.53) also possesses an amino group in its side chain and this basic group plays an important role in the mechanism of action of aldolase, acetoacetate decarboxylase and many of the enzymes catalysing the reactions of 2-amino acids. Proteins, known as histones, which are found associated with DNA, contain regularly spaced lysine (and arginine) residues from which they derive their strongly basic character. The polyamines spermidine (5.54) and the closely related spermine are also found in association with nucleic acids. 5-Amino-4-oxopentanoic acid (5.55, δ-aminolaevulinic acid) is an important intermediate in the biosynthesis of porphyrins, which in turn are components of the cytochromes, haemoglobin and myoglobin. The aromatic amino acid 4-aminobenzoic acid (5.56) is an important growth factor for many micro-organisms which incorporate it into folic acid. (Various derivatives of folic acid act as co-enzymes for reactions which involve the transfer of one carbon atom from one reactant to another, Section 9.4). Adrenalin (4.44) which possesses CNS activity has already been cited as an example of both a derivative of an arene and as a hydroxy compound; it is also a secondary amine. A number of related synthetic amines, for example amphetamine (5.57), also possess CNS activity. Tertiary oxonium salts do not occur naturally. However, a tertiary sulphonium salt, *S*-adenosylmethionine (5.58) acts as a biological methylating agent (Section 9.4), while of several naturally occurring quaternary ammonium salts the most important is probably acetylcholine (5.59). This compound acts as a neurotransmitter at the synapses and neuro-muscular junctions in the parasympathetic nervous system of animals.

* The prefex α- defines the stereochemistry of the carbohydrate at C-1 (cf. p. 38).

Table 5.3 Some compounds of biological interest which contain functional groups of the type $-\overset{|}{\underset{|}{C}}-X$.

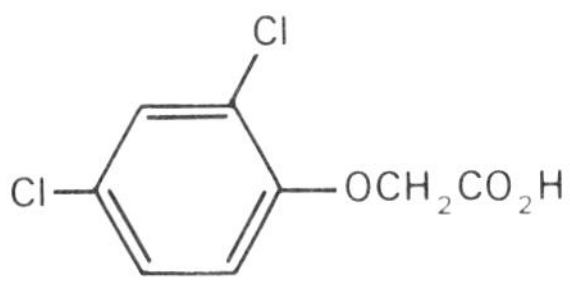

(5.37)

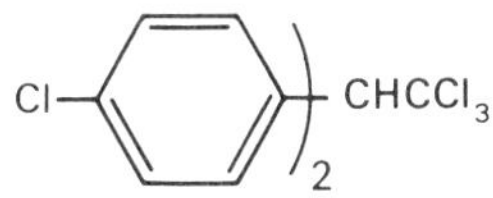

(5.38)

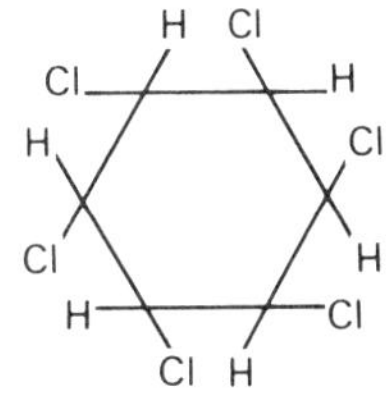

(5.39)

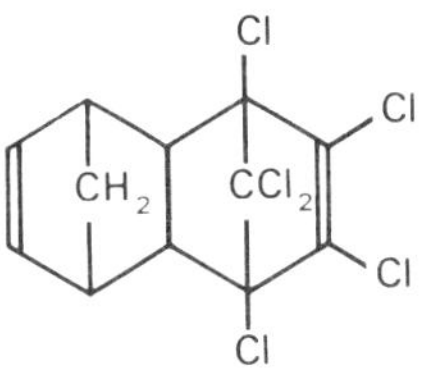

(5.40)

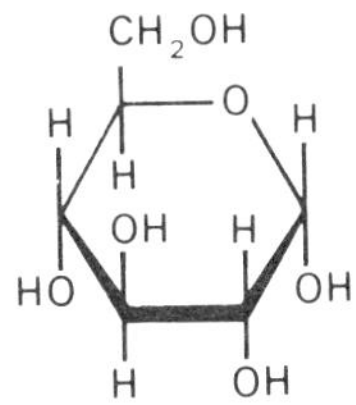

(5.41)

CH_2OH
|
$CHOH$
|
CH_2OH

(5.42)

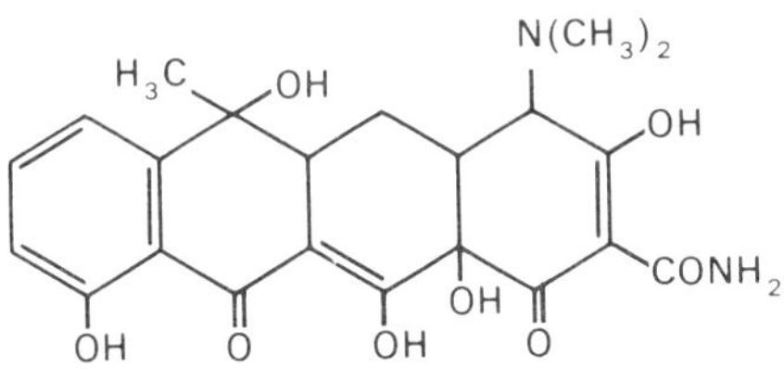

(5.43)

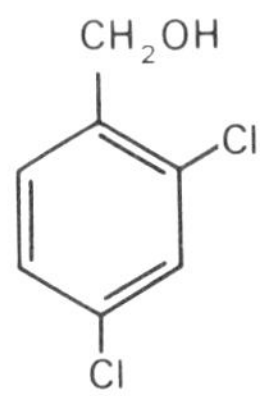

(5.44)

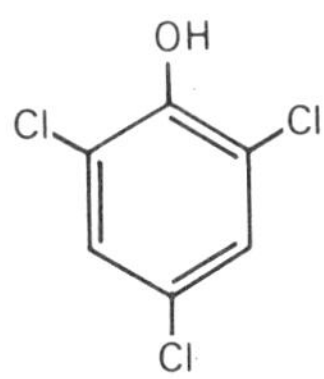

(5.45)

$H_3\overset{+}{N}CH(CH_2SH)CO_2^-$

(5.46)

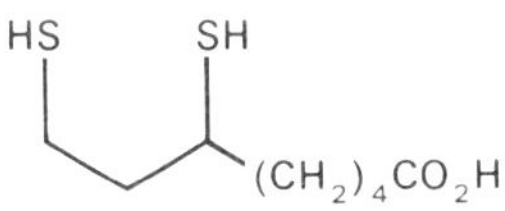

(5.47)

CoA — SH

(5.48)

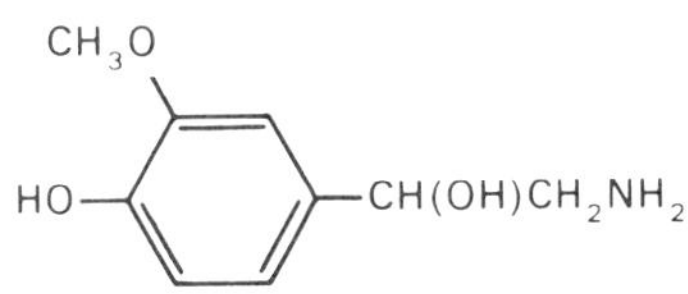

(5.49)

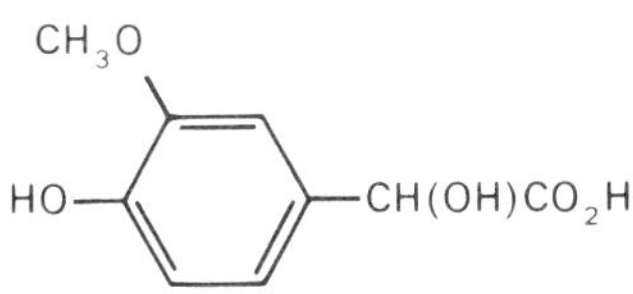

(5.50)

$H_3\overset{+}{N}CH(CH_2CH_2SCH_3)CO_2^-$

(5.51)

$H_3\overset{+}{N}CH(R)CO_2^-$

(5.52)

$H_2N(CH_2)_4CH(\overset{+}{N}H_3)CO_2^-$

(5.53)

$H_2N(CH_2)_3NH(CH_3)_4NH_2$

(5.54)

Table 5.3 (Contd.)

$H_2NCH_2COCH_2CH_2CO_2H$
(5.55)

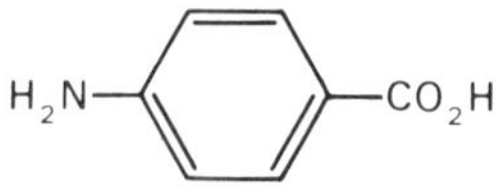

(5.56)

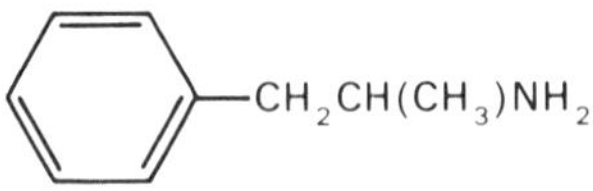

(5.57)

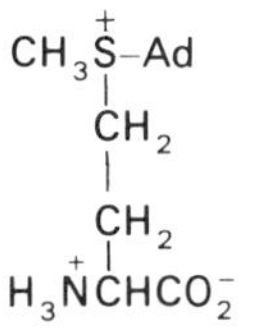

(5.58)

Ad = adenosinyl*

$CH_3CO_2CH_2CH_2\overset{+}{N}(CH_3)_3\overset{-}{O}H$

(5.59)

*See Section 9.1 for the complete structure

Properties of aliphatic halogeno, hydroxy, mercapto and amino compounds and their derivatives

Halogen derivatives of hydrocarbons are volatile, covalent liquids. Polyhalogenated compounds are non-inflammable and have vapour densities greater than that of air; carbon tetrachloride, for example, is used in fire extinguishers. They are excellent solvents for organic materials as evidenced by their use in the dry cleaning industry.

The presence of a hydroxyl group imparts considerable polarity to a molecule. In consequence of this, low molecular mass alcohols are able to hydrogen bond with water and are soluble in it (5.60).

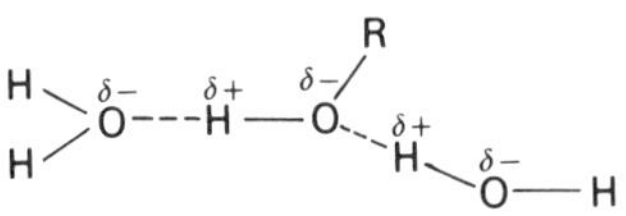

(5.60)

However, as their molecular masses increase the solubility of alcohols in water decreases; the butanols and pentanols are only sparingly soluble while hexanol and higher homologues are essentially insoluble. Liquid alcohols also exhibit inter-molecular hydrogen bonding (5.61).

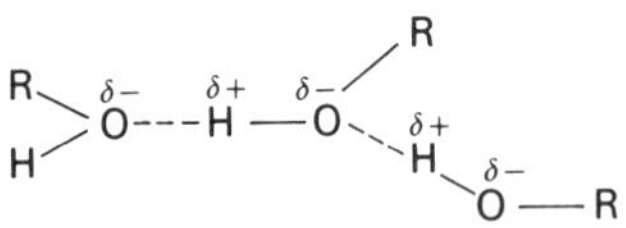

(5.61)

The additional inter-molecular forces introduced by hydrogen bonding lower the volatility and raise the boiling point of an alcohol compared with an isomeric non-hydrogen bonded compound, as noted previously (Section 2.7).

Sulphur is considerably less electronegative than oxygen and does not form hydrogen bonds. Thiols, therefore, are more volatile than alcohols and are typically covalent. Many thiols have characteristic, unpleasant, smells which are detectable at very low concentrations of the thiol. Butanethiol is claimed to be responsible for the odour of the skunk; one part of the thiol can be detected by smell in 10^{10} parts of air!

Ethers and thio-ethers are, for all practical purposes, essentially non-polar. They are usually volatile liquids which are soluble only in non-polar solvents.

Most amines are volatile covalent liquids with characteristic, pungent (fishy), smells. Low molecular mass amines are soluble in water, but solubility decreases rapidly with increasing molecular mass.

'Onium salts are usually nicely crystalline materials which decompose at moderately high temperatures. Those that contain small alkyl or aryl groups are freely soluble in water but insoluble in non-polar solvents. As the size of the substituent groups increases mixtures of water and polar, water miscible solvents such as ethanol or acetone must be used to dissolve the salts.

It is convenient to discuss the chemical properties of aliphatic compounds before those of aromatic compounds. All contain a functional group of type (5.62).

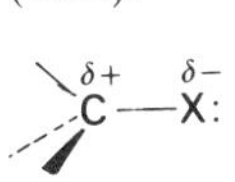

(5.62)

These compounds are potential bases and nucleophiles (5.63) because the atom joined to carbon in each case possesses at least one unshared pair of electrons.

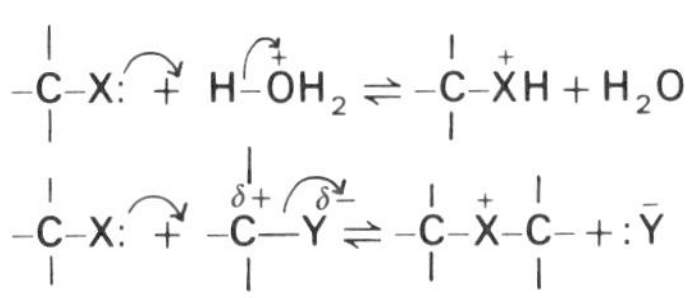

(5.63)

Furthermore, because X is always more electronegative than carbon the C–X bond is polarised as shown and displacement of the functional group by nucleophiles is a second potential type of reactivity (5.64).

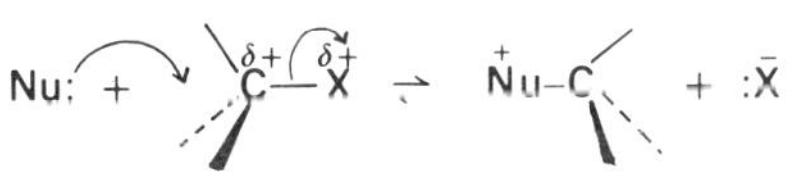

(5.64)

As would be expected, compounds which function as nucleophiles are not attacked readily by nucleophiles; the two activities vary in a complementary manner as illustrated by the reactions discussed below.

Basic Character

Basic character decreases in the group in the order amines > alcohols, ethers > thiols, thioethers > halides. Amines, like their parent compound ammonia, are moderately strong bases. Water soluble amines impart an alkaline reaction to their solution (5.65) and all amines are protonated by strong acids giving salts which are usually crystalline and water soluble (5.66)

$$R^1R^2R^3N\!: + H{-}OH \rightleftharpoons R^1R^2R^3\overset{+}{N}H\,\overset{-}{O}H \qquad \textbf{(5.65)}$$

$$R^1R^2R^3N\!: + H_3\overset{+}{O}\overset{-}{X} \rightleftharpoons R^1R^2R^3\overset{+}{N}H\,\overset{-}{X} + H_2O \qquad \textbf{(5.66)}$$

The basic properties of amines are utilised in various biochemical situations. The association of deoxyribonucleic acids with histones or with spermidine and spermine was mentioned earlier in this section. The protonated amino groups in the histones or polyamines neutralise the negatively charged phosphate groups in the nucleic acid. Amino groups are also involved in forming 'salt bridges' in proteins. These ionic bonds may maintain folding in a single polypeptide chain or may hold two chains in juxtaposition.

Alcohols are also bases, although much weaker than amines. They are protonated only by very strong acids (5.67) and the primary oxonium ions which result cannot usually be isolated.

$$CH_3CH_2\ddot{O}H + H{-}OSO_3H \rightleftharpoons CH_3CH_2\overset{+}{O}H_2\ HS\overset{-}{O}_4 \qquad \textbf{(5.67)}$$

The ions are, however, of great importance in the acid catalysed reactions of alcohols which are discussed below. The situation with ethers is similar.

Despite the fact that sulphur is less electronegative than oxygen, and might therefore be expected to co-ordinate with a proton more easily, thiols and thioethers are weaker bases than their oxygen analogues. Sulphur to hydrogen bonds are less polar, but are weaker, than oxygen to hydrogen bonds and the latter factor appears to be the major one in determining the relative basicities of the two types of compound.

At an elementary level, certainly as far as biochemistry is concerned, it is satisfactory to state that alkyl halides are non-basic.

The ease with which the hetero-atom in the compounds under consideration can utilise an electron pair for binding a proton cannot be correlated precisely with its electronegativity. In fact, no single property explains the observed gradation in basicity satisfactorily. Undoubtedly several factors, including the strength of the X–H bond, the polarisability of X and the effectiveness of stabilisation of the cation by charge delocalisation and solvation are involved in determining basic strengths.

Nucleophilic character

Since basic activity is a specific example of nucleophilic activity a general parallelism between the two would be expected. This is observed when elements in the same period of the periodic table are concerned. Thus both basic and nucleophilic character decrease in the sequence below.

$$\text{Amines} > \text{alcohols} > \text{fluorides}$$

However, the order of basicity and nucleophilicity is often reversed when elements within the same group of the periodic table are considered. Thiols are weaker bases than alcohols but stronger nucleophiles. Polarisability and the efficiency with which the attacking nucleophile can be 'de-solvated' appear to be important factors in determining nucleophilic strength. Thus the approximate order of nucleophilic activity exhibited by the compounds under discussion is

$$\text{amines} > \text{thiols, thioethers} \gg \text{alcohols, ethers} \gg \text{halides}$$

The sequence can be illustrated in terms of reactions involving alkylation, addition to carbonyl groups and acylation.

Alkylation. Primary amines can be alkylated directly using alkyl halides, sulphates or sulphonates under relatively mild conditions, but the reactions are somewhat difficult to control. This is because some of the protonated secondary amine which is the initial product (5.68) will be de-protonated by unreacted primary amine (5.69).

$$R^1{-}X + H_2\ddot{N}R^2 \rightleftharpoons R^1H_2\overset{+}{N}R^2\,\bar{X} \qquad (5.68)$$

$$R^1H_2\overset{+}{N}R^2X^- + H_2\ddot{N}R^2 \rightleftharpoons R^1H\ddot{N}R^2 + H_3\overset{+}{N}R^2X^- \qquad (5.69)$$

The free secondary amine produced is a nucleophile and can be alkylated by any remaining reagent to give a salt of a tertiary amine and this, in the same manner, can give rise to some quaternary ammonium salt. In practice the use of excess alkylating agent usually gives reasonable yields of quaternary ammonium salt while the use of excess amine gives a protonated secondary amine as the major product.

Thiols and alcohols, however, can be alkylated **directly** only by very powerful electrophiles such as carbonium ions illustrating that they are weaker nucleophiles than amines. The carbonium ions themselves are derived by elimination of water from a protonated tertiary alcohol, or by protonation of an alkene. (5.70 and 5.71).

$$\underset{\text{triphenylmethanol}}{Ph_3COH} + H^+ \underset{H_2SO_4}{\overset{\text{conc.}}{\rightleftharpoons}} Ph_3C{-}\overset{+}{O}H_2 \rightleftharpoons Ph_3\overset{+}{C} + H_2O;$$

$$Ph_3\overset{+}{C} + H\ddot{O}CH_3 \rightleftharpoons Ph_3C\overset{+}{\underset{\underset{H}{|}}{O}}CH_3 \rightleftharpoons \underset{\text{triphenylmethoxymethane}}{Ph_3COCH_3} + H^+ \qquad (5.70)$$

$$\underset{\text{isobutene}}{Me_2C{=}CH_2} + H^+ \underset{H_2SO_4}{\overset{\text{conc}}{\rightleftharpoons}} Me_3\overset{+}{C};$$

$$Me_3C^+ + \underset{\text{cysteine}}{H_3\overset{+}{N}CH(CH_2\ddot{S}H)CO_2H} \rightleftharpoons H_3\overset{+}{N}CH(CH_2\overset{+}{S}HCMe_3)CO_2H$$

$$\rightleftharpoons \underset{S\text{-}tert\text{ butylcysteine}}{H_3\overset{+}{N}CH(CH_2SCMe_3)CO_2H} + H^+ \qquad (5.71)$$

These reactions produce ethers containing a tertiary alkyl group. Ethers which do not contain such a group are normally prepared indirectly by alkylating alkoxide or thiolate anions which are stronger nucleophiles than the alcohols or thiols from which they are derived (5.72)

$$CH_3CH_2OH + Na \rightarrow CH_3CH_2\bar{O}{:}\ Na^+ + \tfrac{1}{2}H_2$$

$$CH_3CH_2\bar{O}{:} + \underset{\text{chloroacetate}}{\bar{O}_2CCH_2{-}Cl} \rightleftharpoons \underset{\text{ethoxyacetate}}{CH_3CH_2OCH_2C\bar{O}_2} + {:}\bar{Cl} \qquad (5.72)$$

Ethers and thio-ethers are usually slightly better nucleophiles than alcohols and thiols because alkyl groups increase the electron density on the heteroatom by virtue of their +I effect (Section 3.1). Ethers can be alkylated using powerful reagents such as alkyl fluorosulphonates to give tertiary oxonium salts (5.73).

$$(CH_3)_2O: + CH_3{-}OSO_2F \rightleftharpoons (CH_3)_3\overset{+}{O}:\overset{-}{O}SO_2F \qquad \textbf{(5.73)}$$

methoxymethane — methyl fluorosulphonate — trimethyloxonium fluorosulphonate

As sulphur is more nucleophilic than oxygen, tertiary sulphonium salts can be produced using less powerful reagents (5.74)

$$PhSCH_3 + CH_3Br \rightleftharpoons Ph\overset{+}{S}(CH_3)_2 : \overset{-}{Br} \qquad \textbf{(5.74)}$$

S-methylthiophenol — dimethylphenylsulphonium bromide

Alkyl halides cannot be alkylated indicating that they are very weak nucleophiles.

Addition to carbonyl groups. Amines react readily with aldehydes and ketones producing compounds containing a >C=N– group, known as Schiff bases. Schiff bases derived from aliphatic aldehydes are unstable and usually polymerise (5.75) those derived from aromatic aldehydes are stable, presumably due to conjugation of the carbon to nitrogen bond with the benzenoid ring (5.76).

$$CH_3CHO + H_3NCH_3 \xrightarrow[-H_2O]{\text{warm}} CH_3CH{=}NCH_3 \longrightarrow \text{polymer} \qquad \textbf{(5.75)}$$

aminomethane — *N*-ethylideneaminomethane

$$PhCHO + H_2NCH_2CH_3 \xrightarrow[-H_2O]{\text{warm}} PhCH{=}NCH_2CH_3 \qquad \textbf{(5.76)}$$

aminoethane — *N*-benzylideneaminoethane

Reactions of this type using formaldehyde often produce bis-alkylamino-methylenes as the final product (5.77). The mechanism of Schiff base formation is probably as shown (5.78)

$$H_2CO + H_2NCH_3 \rightleftharpoons H_2C(OH)NHCH_3 \xrightleftharpoons{-H_2O,\ +CH_3NH_2} H_2C(NHCH_3)_2 \qquad \textbf{(5.77)}$$

N,N-dimethylaminomethane

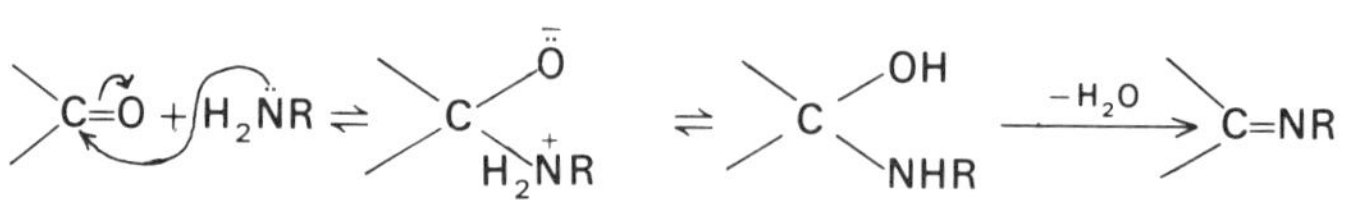

(5.78)

Thiols also react with carbonyl compounds but, because they are weaker nucleophiles than amines, an acid catalyst must be present to increase the polarity in the carbonyl group (5.79).

$$>C{=}O + H^+ \rightleftharpoons [\ >C{=}\overset{+}{O}H \leftrightarrow >\overset{+}{C}{-}OH];$$

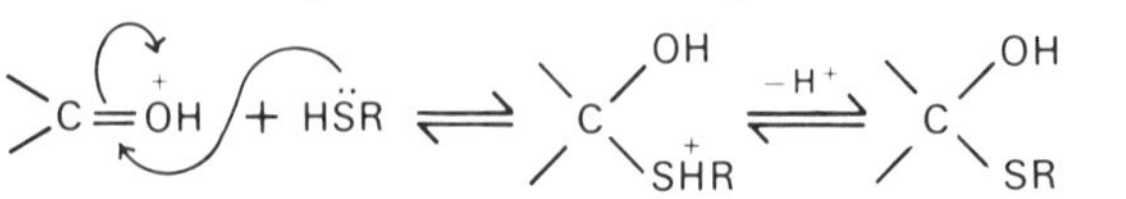

(5.79)

The initial products of the reactions are hemi-thioacetals or hemi-thioketals. However, these compounds usually react with excess reagent to give the corresponding thio-acetal or thioketal (5.80); these compounds are analogous to the bis-alkylaminomethylenes produced in reaction (5.77). In thioacetal and thioketal formation the catalyst facilitates removal of water from the starting material.

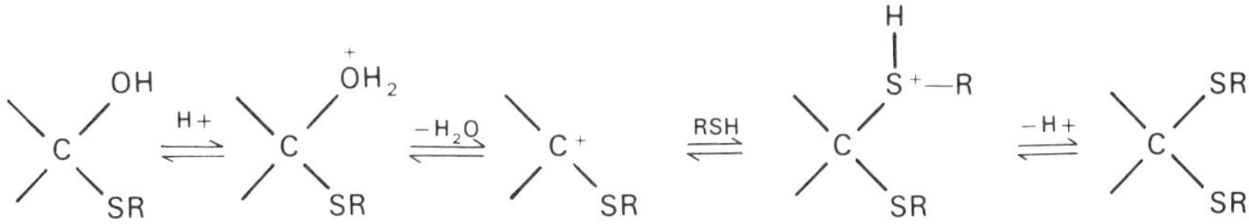

(5.80)

Cyclic thioketals derived from 1,2-ethanedithiol have found extensive use for protecting carbonyl groups in steroids during synthesis and also as intermediates in the reduction of carbonyl to methylene groups (5.81).

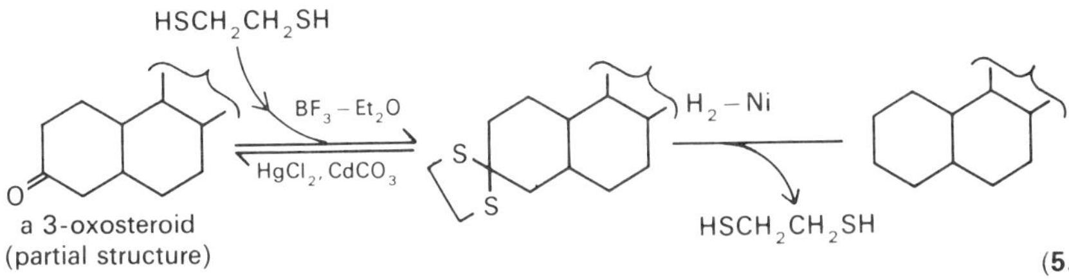

(5.81)

The reactions of alcohols with aldehydes yield hemi-acetals and acetals and are completely analogous to those involving thiols. Acyclic ketals cannot usually be formed by reacting alcohols with ketones in the presence of an acid catalyst. 1, 2-Diols, however, give cyclic ketals which, like their thio analogues, have frequently been used to protect keto functions during synthesis.

Alkyl halides do not act as nucleophiles towards carbonyl groups.

Acylation. The acylation of nucleophiles by carboxylic acids and their derivatives is discussed in Section (6.2); consequently only an outline of the topic will be given here. The reactions involve attack by the nucleophile at the electron deficient carbon atom of the carbonyl group in the acylating agent (5.82).

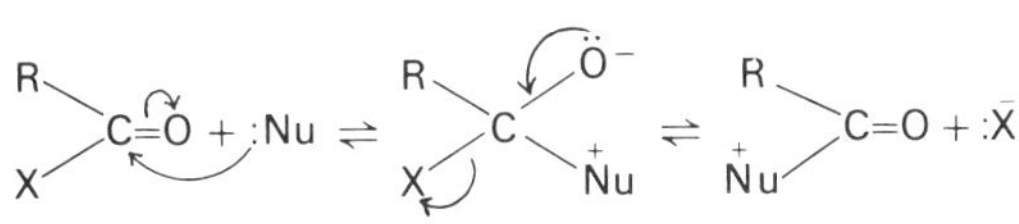

(5.82)

Amines, being moderately strong nucleophiles react smoothly with all of the common acylating agents; the products are named as *N*-acylamines or as amides (5.83).

$$R^1COX + H_2NR^2 \rightleftharpoons R^1CONHR^2 + HX; \quad (5.83)$$

$X = \text{hal}, OCOR^3, OH, OR^3$

The reactions do not require the presence of a catalyst.

Thiols and alcohols will also react readily with powerful acylating agents such as acid halides and anhydrides (5.84 and 5.85) to produce esters.

$$CH_3COCl + CH_3CH_2OH \rightleftharpoons CH_3COOCH_2CH_3 + HCl \qquad (5.84)$$

ethanoyl chloride — ethyl ethanoate

$$(CF_3CO)_2O + HSCH_2CH_3 \rightleftharpoons CF_3COSCH_2CH_3 + CF_3CO_2H \qquad (5.85)$$

trifluoro-acetic anhydride — ethanethiol — *S*-ethyl thio-trifluoroacetate — trifluoro-acetic acid

In contrast to amines, however, they will react with carboxylic acids, again producing esters, only in the presence of an acidic catalyst.

$$CH_3CO_2H + HOCH_2CH_3 \overset{H^+/\text{heat}}{\rightleftharpoons} CH_3CO_2CH_2CH_3 + H_2O$$

This fact reflects the relative reactivities of the reactants in question; thiols and alcohols are weaker nucleophiles than amines, and carboxylic acids are less effective acylating agents than acid halides and acid anhydrides. Acylation using a carboxylic acid is only possible because the catalyst enhances the reactivity of the carboxylic acid, as explained in Section 6.2.

Alcohols and thiols can also be esterified with inorganic acids using methods which are analogous to those used for carboxylic acids. The formation of sulphate (5.86, 5.87) and phosphate esters (5.88, 5.89) is particularly relevant due to the frequent occurrence of this type of compound in living cells. The mechanisms of these reactions are probably very similar to those of analogous acylations (5.82).

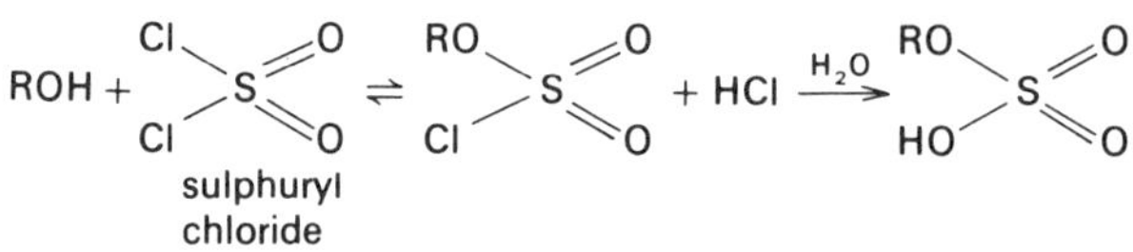

(5.86)

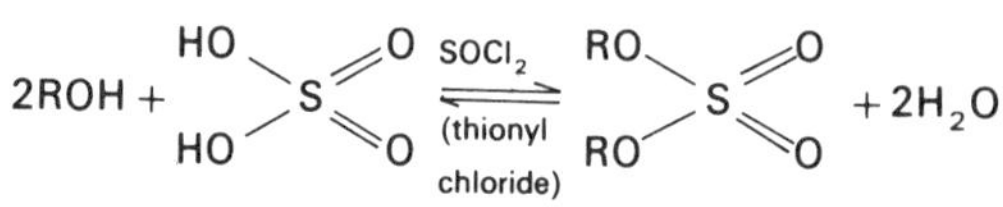

(5.87)

$$3\,ROH + Cl_3P{=}O \overset{\text{pyridine}}{\rightleftharpoons} (RO)_3P{=}O + 3HCl \qquad (5.88)$$

phosphoryl chloride

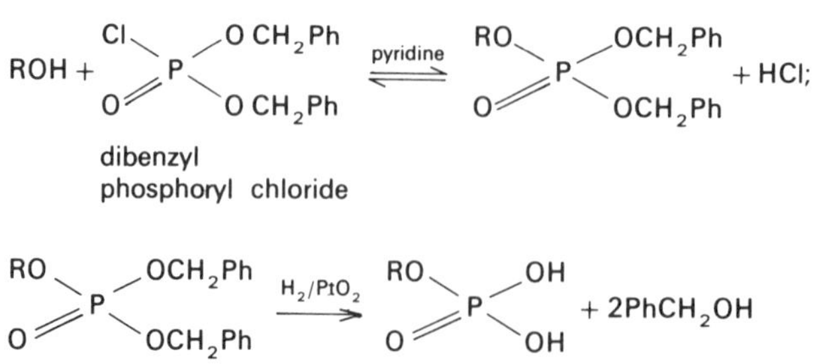

(5.89)

As might be expected from the preceding account of alkylation and addition to carbonyl groups, it is found that alkyl halides cannot be acylated.

Nucleophilic character of alkyl halides. Although alkyl halides are non-nucleophilic as far as the three types of reaction discussed above are concerned, other reactions do provide evidence of weak nucleophilic character. For example in Friedel-Craft alkylation reactions (Table 4.8) polarisation of the halide by the catalyst involves donation of an electron pair from the halogen atom to electron deficient aluminium. Similarly many nucleophilic substitution reactions of alkyl halides are catalysed by the presence of metal ions, particularly Ag^+. A general mechanism which has been suggested for such reactions is shown below. It can be seen that an electron pair from the halogen atom co-ordinates with the metal ion, a process in which the halide is acting as a nucleophile.

$$\mathrm{Rhal{:}} + \mathrm{Ag^+X^-} \rightleftharpoons [\mathrm{Rhal{:}} \dashrightarrow \mathrm{Ag^+X^-}]$$

$$\bar{\mathrm{X}}{:} + [\mathrm{Rhal{:}} \dashrightarrow \mathrm{Ag^+X^-}] \xrightleftharpoons{\text{slow}} [\mathrm{X^-R^+hal^-Ag^+}] + \mathrm{X^-}$$

$$[\mathrm{X^-R^+hal^-Ag^+}] \rightarrow \mathrm{RX} + \mathrm{Aghal} + \text{alkene} + \mathrm{HX}$$

Biochemical reactions in which amines, thiols or alcohols act as nucleophiles.
Alkylation. Numerous examples of biochemical reactions involving alkylation (almost invariably methylation) of an aliphatic amine or thiol are known.

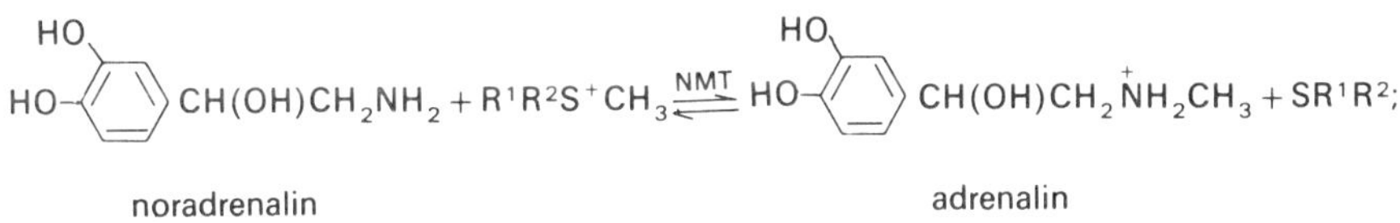

$R^1 = -CH_2CH_2CH(\overset{+}{N}H_3)CO_2^-$; R^2 = adenosinyl; NMT = noradrenalin-*N*-methyltransferase.
(5.90)

Reaction (5.90), which occurs in the adrenal medulla, interconverts noradrenalin, which acts mainly as a neurotransmitter, and adrenalin whose main function is as a hormone. The methylating agent involved is *S*-adenosylmethionine as it is, also, in reaction (5.91) which produces phosphatidyl choline, an important component of biological membranes.

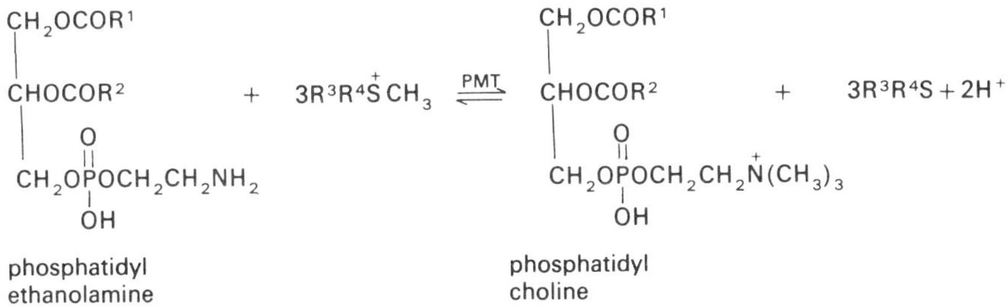

$R^1 = R^2$ = alkyl or alkenyl; $R^3 = -CH_2CH_2CH(\overset{+}{N}H_3)CO_2^-$; R^4 = adenosinyl; PMT = phosphatidylethanolamine methyltransferase
(5.91)

Reaction (5.92) represents the last step in the biosynthesis of the amino-acid methionine in certain bacteria (but not in man).

$$\underset{\text{homocysteine}}{H_3\overset{+}{N}CH(CO_2^-)(CH_2)_2SH} + \underset{\text{dimethylthetin}}{(CH_3)_2\overset{+}{S}CH_2CO_2^-} \underset{}{\overset{\text{DHMT}}{\rightleftharpoons}} \underset{\text{methionine}}{H_3\overset{+}{N}CH(CO_2^-)(CH_2)_2SCH_3} + \underset{\text{methiolacetate}}{CH_3SCH_2CO_2^-} + H^+$$

DHMT = dimethylthetin-homocysteine methyltransferase **(5.92)**

O-Alkylation of alcohols does occur in living cells, but does not seem to be of great importance (*O*-methylated monosaccharide units are known to be a minor component of some plant gums and pectins, and of certain bacterial lipopolysaccharides). In contrast, *O*-methylation of phenols is frequently encountered in metabolic pathways involved with the de-toxification and excretion of aromatic compounds. In the liver, for example, *S*-adenosylmethionine mono-*O*-methylates adrenalin, noradrenalin and 3,4-dihydroxymandelate (a degradation product of noradrenalin), producing the corresponding 3-methoxyderivative in each case (5.93) which is then subject to further degradation prior to excretion.

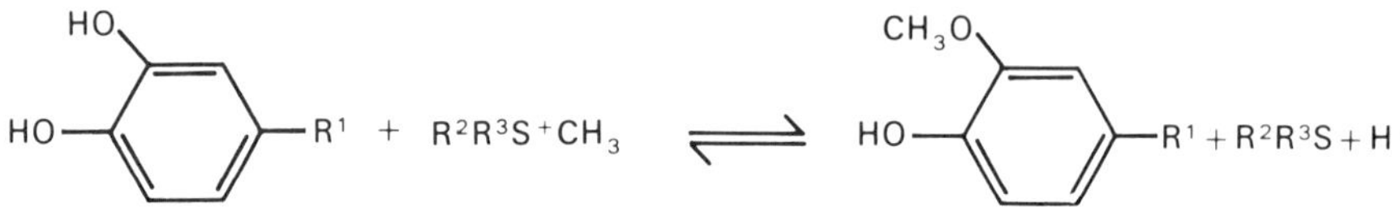

$R^1 = -CH(OH)CH_2NHCH_3$, $-CH(OH)CH_2NH_2$ or $-CH(OH)CO_2^-$;
$R^2 = CH_2CH_2CH(\overset{+}{N}H_3)CO_2^-$; R^3 = adenosinyl **(5.93)**

It should be apparent at this point that many biological methylations employ *S*-adenosylmethionine as the methyl group donor. The action of this compound as a methylating agent will be discussed in Section 9.4.

Addition to carbonyl groups. Schiff base formation is of great importance in enzymology, particularly in the field of amino acid metabolism. Virtually all of the enzymes which catalyse the reactions of 2-amino acids utilise pyridoxal phosphate (5.94) as co-enzyme.

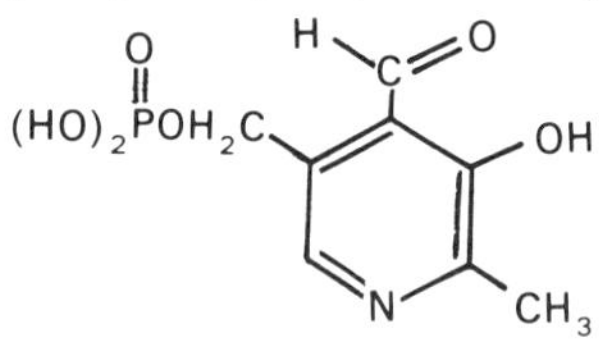

(5.94)

The heterocyclic aldehyde is found bound to the enzyme as a Schiff base involving a lysine residue (5.95); the mechanistic role of the bound co-enzyme is discussed in some detail in Section 9.3.

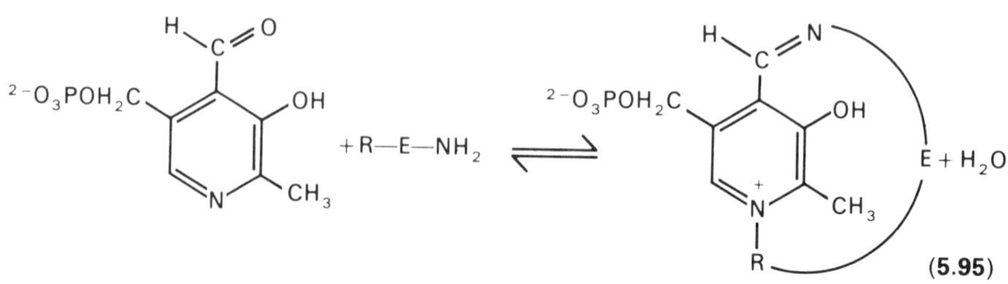

(5.95)

The mechanism of action of acetoacetate decarboxylase and of aldolase and related enzymes also entail Schiff base formation involving the 6-amino group of a lysine residue.

Hemithioacetal formation, from an aldehyde and enzyme thiol group (5.96) appears to be the first step in the oxidation of many aldehydes *in vivo*; this type of reaction is discussed further in Chapters 5 and 9.

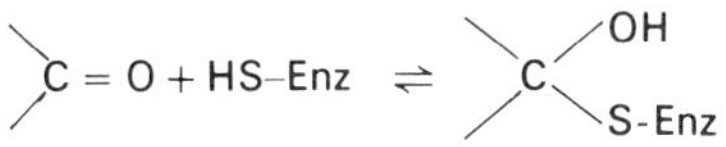

(**5.96**)

Carbohydrates undoubtedly provide the examples of hemi-acetal, hemi-ketal, acetal and ketal formation which are most relevant to biochemistry. Glucose (5.41), which can be taken as being representative of monosaccarides in general, is a cyclic hemi-acetal which arises from attack on the carbonyl group by the C-5 hydroxyl group in the acyclic form of the sugar. Since glucose is a hemi-acetal it can react with other hydroxylic compounds to give acetals, which are known collectively as glucosides (5.97).

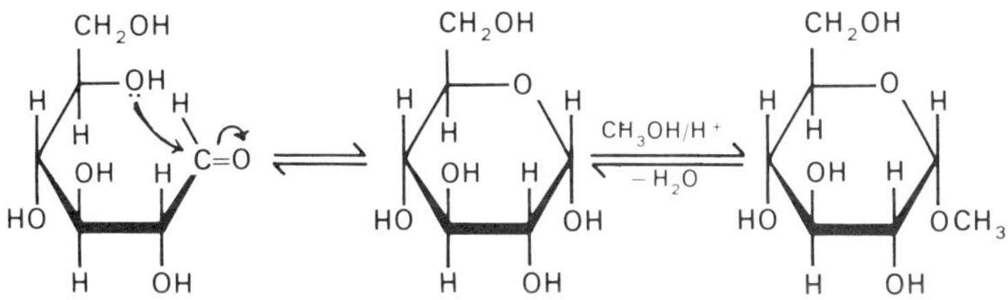

(**5.97**)

acyclic D-glucose α-D-glucose α-D-methyl glucoside

Oligo- and polysaccharides are polymers in which monosaccharide residues are joined together by acetal or ketal linkages.

Acylation. In view of the fact that a variety of acylating agents are available in both the laboratory and in cells it seems desirable to postpone an account of the formation of amides, esters and thiol esters in living cells until acylation has been discussed (Section 6.2)

Displacement of the functional group

The overall order of activity observed for alkyl halides, alcohols, thiols and aliphatic amines acting as bases or nucleophiles is found to be reversed when reactions involving displacement of the functional group are examined. Displacement occurs during two types of reaction, viz: those involving nucleophilic substitution and those involving elimination.

Displacement during Nucleophilic substitution reactions. Alkyl halides are highly reactive towards nucleophiles and undergo many reactions which involve displacement of the functional group.

$$\text{Nu:} + \text{R—X} \rightleftharpoons \text{Nu—R} + \text{:X}$$

These reactions are classified on the basis of the kinetic order of their rate equations, which in turn reflects the temporal order in which bonds are formed and broken. The reactions of most primary and of some secondary halides follow second order rate equations (Rate $\propto$ [halide] [Nu]) and the bond forming and breaking processes appear to be synchronous. The stereochemistry of the products of reactions using optically active halides indicates that the nucleophilic reagent attacks the electrophilic carbon atom on the side opposite to that to which the leaving group is bonded (5.98).

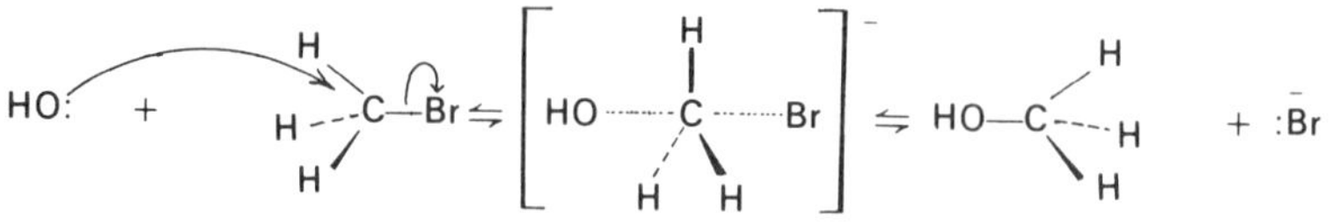

(5.98)

This type of nucleophilic substitution, which occurs by a single, bimolecular, elementary step is designated as an S_N2 reaction—a **bi** molecular, substitution involving a **n**ucleophilic reagent.

In contrast the reactions of tertiary halides, and of other secondary halides, with nucleophiles exhibit kinetics which are first order with respect to the halide and zero order with respect to the nucleophile (Rate $\propto$ [halide]). The stereochemistry of products arising from optically active halides suggests that a planar intermediate, presumably a carbonium ion, is formed during such reactions. The interpretation of these facts is that the carbon to halogen bond is broken before the carbon to nucleophile bond is formed and that the former process is rate determining (5.99). Such reactions are designated as S_N1 reactions.

$$(CH_3)_3CBr \underset{}{\overset{Slow}{\rightleftharpoons}} (CH_3)_3\overset{+}{C}\!:\!\overset{-}{Br}; \quad (CH_3)_3\overset{+}{C} + :\overset{-}{O}H \underset{}{\overset{Fast}{\rightleftharpoons}} (CH_3)_3COH \qquad \mathbf{(5.99)}$$

2 bromo-2-methylpropane (tertbutyl bromide) — 2-methylpropan-2-ol (tertbutanol)

The main factor which determines whether a halide reacts by the S_N2 or by the S_N1 mechanism is the nature of the substituents attached to the α-carbon atom. Electron withdrawing substituents will increase the electrophilic character of the α-carbon atom and make it more susceptible to direct attack by a nucleophile—i.e. they will increase S_N2 reactivity. On the other hand the same substituents will destabilise a carbonium ion and hence decrease S_N1 reactivity. Obviously, the converse is true for electron releasing substituents. This explains why the reactions of most primary halides proceed by the S_N2 mechanism while those of tertiary and some secondary halides proceed by the S_N1 mechanism.

The rates of reaction of a tertiary halide with any nucleophile should be the same since the latter is not involved in the rate determining step. The rates of reaction of a primary halide with a variety of nucleophiles will, however, depend on the nucleophilic strength of the reagent. In general the stronger the nucleophile the easier it will be able to attack the electron deficient α-carbon of the halide. It follows that if the structure of a halide is such that it can react by either mechanism then the use of a strong nucleophile will promote the S_N2 mechanism at the expense of the S_N1 mechanism. It is interesting to note that in the case of alcohol formation, where the use of two alternative nucleophiles (H_2O or $\overline{H}O$:) is possible, primary halides will react at an appreciable rate only with the more powerful nucleophile (5.98) while tertiary halides will react at the same rate with both reagents (5.99 and 5.100).

$$(CH_3)_3CBr \underset{\text{acetone}}{\overset{\text{aqueous}}{\rightleftharpoons}} (CH_3)_3\overset{+}{C}:\overset{-}{Br} \overset{H_2O}{\rightleftharpoons} (CH_3)_3C\overset{+}{O}H_2 \overset{-H^+}{\rightleftharpoons} (CH_3)_3COH \qquad \textbf{(5.100)}$$

Table 5.4. lists some common nucleophilic substitution reactions of alkyl halides.

Table 5.4. Some Nucleophilic Substitution Reactions of Halogeno-alkanes

Nucleophile	Source	Product
H_2O:	—	$-\overset{|}{\underset{|}{C}}-OH$ alcohols (from tertiary halides)
$H\bar{O}$:	Alkali hydroxides in aqueous soln.	$-\overset{|}{\underset{|}{C}}-OH$ alcohols
$R\bar{O}$:	ROH/Na	$-\overset{|}{\underset{|}{C}}-OR$ ethers
$R\bar{S}$:	RSH/NaOH	$-\overset{|}{\underset{|}{C}}-SR$ thio-ethers
$\ddot{N}H_3$	—	$-\overset{|}{\underset{|}{C}}-\overset{+}{N}H_3\,:\overset{-}{X}$ amine hydrohalides (excess NH_3 gives $-\overset{|}{\underset{|}{C}}-NH_2+\overset{+}{N}H_4\overset{-}{X}$)
$\ddot{N}R^1R^2R^3$ (R = H or alkyl etc)	—	$-\overset{|}{\underset{|}{C}}-\overset{R_1}{\underset{R^3}{\overset{+}{N}}}-R^2:\overset{-}{X}$ hydrohalide salts of secondary or tertiary amines or a quaternary amonium salt

Since reaction (5.98) does give reasonable yields of alcohol it is axiomatic that the reverse reaction does not occur to any appreciable extent, i.e. halide ions cannot displace HO– from an alcohol. The explanation of this situation is as follows. The high reactivity of alkyl halides in nucleophilic substitution reactions stems largely from the fact that halide atoms are good leaving groups. Good leaving groups are attached to carbon by a relatively weak, easily polarised, bond; they leave, in possession of the electron pair of the carbon to leaving group bond, as stable ions or neutral molecules. An approximate inverse relationship exists between the nucleophilic activity of a group and its effectiveness as a leaving group. Powerful nucleophiles such as H^-, HO^-, H_2N^- which are small and not easily polarised are very poor leaving groups. However, the easily polarised iodide ion is both a good nucleophile and a good leaving group and a better correlation of effectiveness of a group as a leaving group can be had in terms of the basic strength of the group which is expelled. Thus weakly basic entities such as H_2O, ROH, RSO_3^-, are good leaving groups, whereas strongly basic entities are not. Consequently, halide ions, which are good leaving groups and only weak to moderately strong nucleophiles cannot displace the hydroxyl group of an alcohol as $H\bar{O}$: which is a strong nucleophile and a poor leaving group. Alcohols will, however, undergo substitution if the hydroxyl group is converted into a good leaving group. One way of achieving substitution is to perform the reaction in acidic solution when the first step involves conversion of the alcohol to a primary oxonium ion by protonation. Protonation has two effects. It greatly increases the polarity in the C–O bond by the inductive mechanism, making the α-carbon atom susceptible to attack by relatively weak nucleophiles such as Br^-; it also enables what was the hydroxyl group to leave as a very weakly basic water molecule.

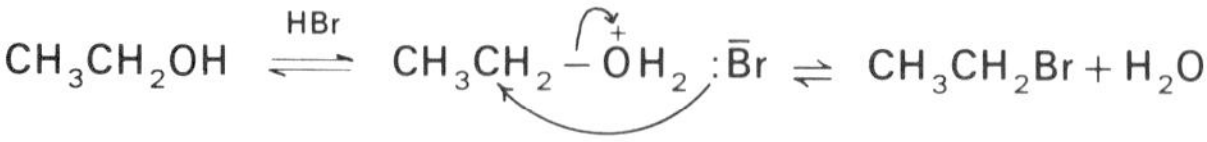

Ethers are alkyl or aryl derivatives of alcohols and, as would be expected, also undergo substitution under acidic conditions. The oxygen function cannot be displaced in neutral solution as it would have to leave as a strongly basic alkoxide or phenoxide ion. However, protonation of the ether forms a secondary oxonium ion from which a molecule of an alcohol can be displaced by the nucleophile. The frequent use of hydrogen iodide to cleave ethers is a reflection of the fact that the iodide ion is the most powerful nucleophile which can be obtained in acidic solution. The most electrophilic α-carbon atom, usually that of a methyl group, if present, is attacked by the nucleophile.

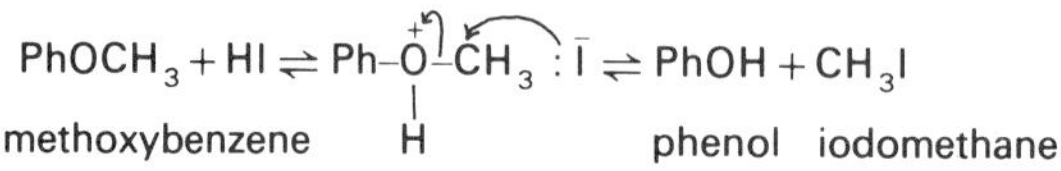

Tertiary oxonium ions already possess a positively charged oxygen atom and hence a good leaving group (an ether molecule) as an integral part of their structure. Consequently they undergo nucleophilic substitution readily and are powerful alkylating agents. Again a methyl group, if present, is transferred to the nucleophile (5.102). Unlike primary and secondary oxonium ions the formation and use of tertiary oxonium ions is not restricted to acidic media.

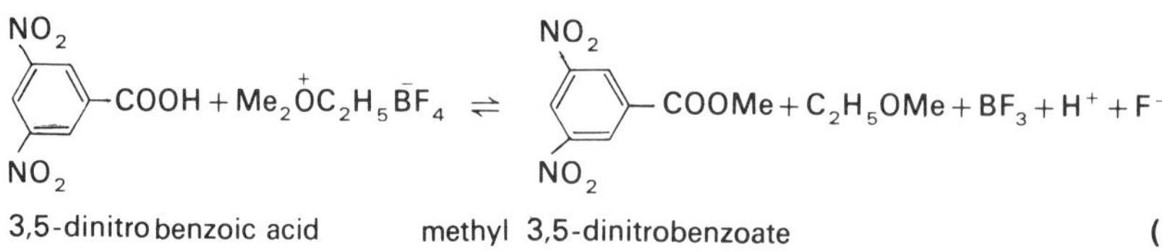

(5.102)

As an alternative to methods which place a positive charge on oxygen the hydroxyl group in an alcohol may be displaced by converting the latter to an ester. For example, alkyl sulphates, toluene-*p*-sulphonates, methanesulphonates and fluorosulphonates are all efficient alkylating agents; the corresponding anions are good, weakly basic leaving groups. (5.103, 5.104)

$$PhO^- + (CH_3)_2SO_4 \rightleftharpoons PhOCH_3 + CH_3SO_4^- \qquad \textbf{(5.103)}$$

phenate anion; dimethyl sulphate; methoxy-benzene; methyl sulphate anion

$$PhS^- + p\text{-}CH_3C_6H_4SO_2OCH_3 \rightleftharpoons PhSCH_3 + p\text{-}CH_3C_6H_4SO_3^- \qquad \textbf{(5.104)}$$

thiophenate anion; methyl toluene 4-sulphonate; methylthio-benzene; toluene-4-sulphonate anion

Similarly, treatment of an alcohol with thionyl chloride produces a chlorosulphite ester from which sulphur dioxide and chloride ion are displaced (5.105); phosphorus halides convert alcohols to chlorides in a similar manner (5.106).

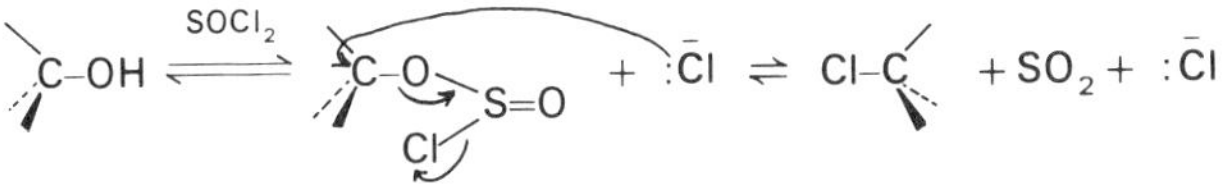

(5.105)

$PCl_5 \equiv \overset{+}{P}Cl_4\overset{-}{Cl}$
phosphorus pentachloride

$$ROH + \overset{+}{P}Cl_4\overset{-}{Cl} \rightleftharpoons RO\overset{+}{P}Cl_3 + HCl + :\overset{-}{Cl}; \qquad RO\overset{+}{P}Cl_3 + :\overset{-}{Cl} \rightleftharpoons RCl + POCl_3 \qquad \textbf{(5.106)}$$

Thiols are weaker bases than alcohols and displacement reactions which occur in acidic solution are of little importance. Displacement reactions of thio-ethers, however, are of some importance. For example, *S*-tritylcysteine has been used as a protected derivative of cysteine for the synthesis of peptides because, despite the low basicity of sulphur, the stability of the trityl carbonion ion allows the trityl group to be removed under mild conditions by treatment with acid. The cysteine portion of the ether acts as leaving group in a typical S_N1 reaction (5.107). It has already been mentioned that tertiary sulphonium ions (usually referred to simply as sulphonium ions) can be formed, by alkylating thio-ethers, more easily than can tertiary oxonium ions. Like their oxygen analogues they are efficient alkylating agents.

$$\underset{S\text{-tritylcysteine}}{H_3\overset{+}{N}CH(CH_2SCPh_3)CO_2H} \underset{\text{in dioxan}}{\overset{HCl}{\rightleftharpoons}} H_3\overset{+}{N}CH(CH_2\overset{+}{S}H CPh_3)CO_2H$$

$$\rightleftharpoons Ph_3\overset{+}{C} + \underset{\text{cysteine}}{H_3\overset{+}{N}CH(CH_2SH)CO_2H}; \qquad \textbf{(5.107)}$$

$$Ph_3\overset{+}{C} + \overset{-}{Cl} \rightleftharpoons Ph_3CCl$$

$$\underset{\text{dimethyl (1-phenylethyl) sulphonium cation}}{Ph(Me)CH-\overset{+}{S}Me_2} + \underset{\text{azide}}{:\overset{-}{N}_3} \rightleftharpoons \underset{\text{1-phenyl azidoethane}}{Ph(Me)CHN_3} + \underset{\text{dimethyl sulphide}}{SMe_2}$$

Direct displacement of the functional group from amines in neutral solution is not possible; the amino group would have to be displaced as an amide ion ($:\overset{-}{N}H_2$) which is a powerful base and nucleophile. As in the case of alcohols, substitution can be brought about by placing a positive charge on the hetero-atom attached to carbon. Protonation is again effective and when amine hydrohalides are heated the anion is alkylated.

$$\underset{\text{1-butylammonium bromide}}{CH_3CH_2CH_2CH_2-\overset{+}{N}H_3\ :\overset{-}{Br}} \underset{300\text{–}360^\circ C}{\overset{\text{Heat at}}{\rightleftharpoons}} \underset{\text{1-bromobutane}}{CH_3CH_2CH_2CH_2Br} + NH_3$$

Cations of secondary and tertiary amines are attacked at the most electrophilic α-carbon atom, that of a methyl group, if present. The Herzig-Meyer method of determining *N*-alkylimide groups is based on this type of reaction. The nitrogenous compound is treated with hydrogen iodide and the hydro-iodide which is formed is pyrollised; the alkyl iodide which results can be determined iodimetrically.

$$>NR + HI \rightleftharpoons >\overset{+}{N}HR\ \bar{I} \rightleftharpoons >NH + RI$$

In a similar manner quaternary ammonium cations can act as alkylating agents.

$$\bar{Br}:\ CH_3\text{—}\overset{+}{N}(CH_3)_3 \underset{300\text{–}360°\,C}{\overset{\text{heat at}}{\rightleftharpoons}} CH_3Br + N(CH_3)_3$$

tetramethylammonium bromide

The substitution reactions of ammonium salts mentioned above are demonstrably reversible and require vigorous conditions. The facts reflect, at least in part, that ammonia and amines although much better than amide anion are still not good leaving groups. This type of reaction has, however, been adapted satisfactorily for synthetic use. For example, gramine methosalts will alkylate cyanide and diethyl acetamidomalonate anions giving products which lead to indolyl-3-acetic acid (heteroauxin) and *N*-acetyltryptophan, respectively (5.108).

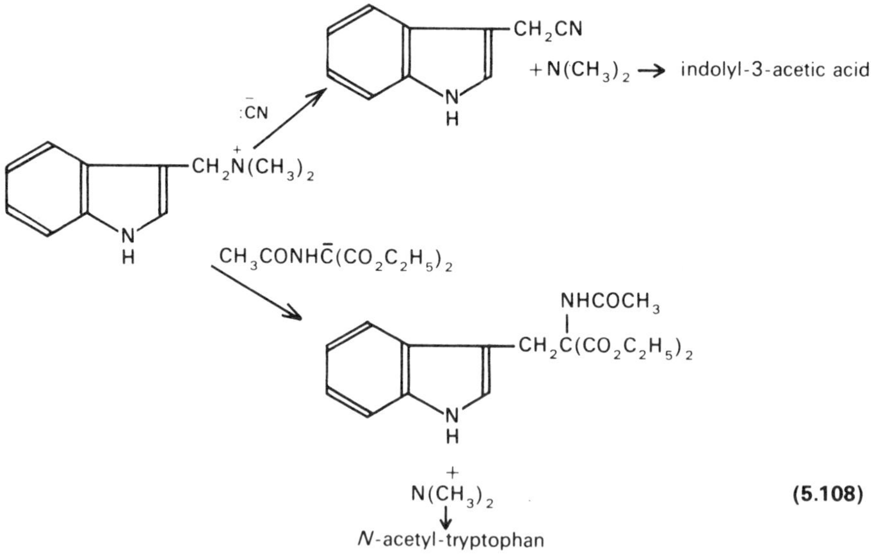

(5.108)

Another way of achieving indirect displacement of the amino group from an amine involves diazotisation. Treatment of an amine with cold nitrous acid gives a diazonium salt containing the grouping $-\overset{|}{\underset{|}{C}}-\overset{+}{N}\equiv N$.

$$-\overset{|}{\underset{|}{C}}-NH_2 + HNO_2 + HCl \rightleftharpoons -\overset{|}{\underset{|}{C}}-\overset{+}{N}\equiv N\ \bar{C}l + 2H_2O$$

Aliphatic diazonium salts in aqueous solution lose nitrogen and the carbonium ion which results is attacked by solvent in an S_N1 type reaction. Although the corresponding alcohol is sometimes the major product the reactions are of little practical value since the carbonium ion can be attacked by any other nucleophile present, and can also lose a proton to produce an alkene (5.109).

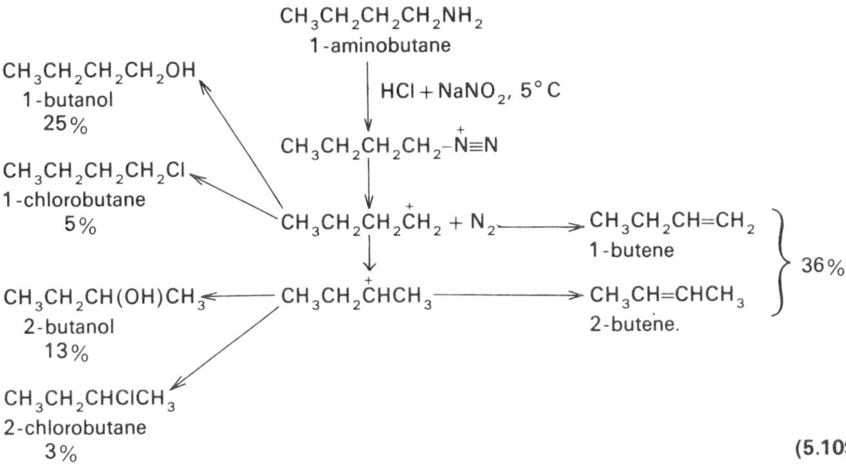

(5.109)

In contrast, aromatic diazonium salts are relatively stable in aqueous solution at 5°C and undergo a number of useful nucleophilic displacements, for example with iodide ions.

$$\text{Ph-}\overset{+}{\text{N}}\equiv\text{N} \rightleftharpoons \overset{+}{\text{Ph}} + \text{N}_2 \overset{:\text{I}^-}{\rightleftharpoons} \text{PhI}$$

phenyldiazonium cation — iodobenzene

Biochemical Alkylating Agents. In reactions such as those described above, in which alkyl halides, alcohols, thiols and alkylamines and their derivatives are attacked by a nucleophile the former act as alkylating agents towards the latter. Most biological alkylations are, in fact, methylations involving the use of *S*-adenosylmethionine (5.58) as the methyl group donor (alkylating agent) This sulphonium salt is usually (but not always) attacked by nucleophiles at the carbon of the methyl group. The thio-ether *S*-adenosylhomocysteine acts as the leaving group in what appear to be typical S_N2 reactions (5.110).

$$\text{Nu:} + \text{CH}_3\text{-}\overset{+}{\text{S}}\text{R}^1\text{R}^2 \rightleftharpoons \overset{+}{\text{Nu}}\text{-CH}_3 + \text{SR}^1\text{R}^2 \qquad \textbf{(5.110)}$$

$$\text{R}^1 = \text{-CH}_2\text{CH}_2\text{CH}(\overset{+}{\text{N}}\text{H}_3)\text{CO}_2^-; \quad \text{R}^2 = \text{adenosinyl}$$

Specific examples of this type of reaction have already been given. (5.90, 5.91, and 5.93). Other 'onium salts such as dimethylthetin (5.92), choline $HOCH_2CH_2\overset{+}{N}(CH_3)_3$ and betaine $(CH_3)_3\overset{+}{N}CH_2CO_2^-$ can act as methylating agents, but only towards homocysteine (5.111); methionine (5.51) is the product in each case.

$$(\text{CH}_3)_3\overset{+}{\text{N}}\text{CH}_2\text{CO}_2^- + \text{HSCH}_2\text{CH}_2\text{CH}(\overset{+}{\text{N}}\text{H}_3)\text{C}\bar{\text{O}}_2 \xrightleftharpoons[]{\text{betaine-homocysteine methyltransferase}} (\text{CH}_3)_2\overset{+}{\text{N}}\text{HCH}_2\text{C}\bar{\text{O}}_2 + \text{CH}_3\text{SCH}_2\text{CH}_2\text{CH}(\overset{+}{\text{N}}\text{H}_3)\text{C}\bar{\text{O}}_2 \qquad \textbf{(5.111)}$$

N,*N*-dimethylglycine

Although *S*-adenosylmethionine is the most common biochemical methylating agent two other compounds, *N*-methyltetrahydrofolate and methylcobalamin, are also known to be involved in methylations. However, it is by no means clear how the methyl group in these compounds is activated and consequently they will not be discussed here.

Displacement during elimination reactions. Substitution reactions are almost invariably complicated, to some extent, by competing eliminations as illustrated in Table 5.5. It is convenient to consider elimination reactions as a type of internal nucleophilic substitution. Nucleophilic reagents, since they possess a pair of electrons which is available for forming a new bond, are also bases. Abstraction of a β-hydrogen atom, as a proton, by the reagent enables the electron pair of the β-H bond to act as an 'internal nucleophile' towards the electron deficient α-carbon atom resulting in the formation of a double bond (5.112).

Eliminations from primary halides usually embody bimolecular rate determining steps and are known as E_2 reactions; those of tertiary and many secondary halides usually embody unimolecular rate determining steps and are designated as E_1 reactions.

Table 5.5. Action of ethoxide ion on alkyl halides

Halide	Products (%)	
CH_3CH_2Br bromoethane	$CH_3CH_2OCH_2CH_3$(99) ethoxyethane	$CH_2{=}CH_2$(1) ethene
$(CH_3)_2CHBr$ 2-bromopropane	$CH_3CH_2OCH(CH_3)_2$(21) 2-ethoxypropane	$CH_3CH{=}CH_2$(79) propene
$(CH_3)_3CBr$ 2-bromo-2-methylpropane	$CH_3CH_2OC(CH_3)_3$(1) 2-ethoxy-2-methylpropane	$(CH_3)_2C{=}CH_2$(99) isobutene

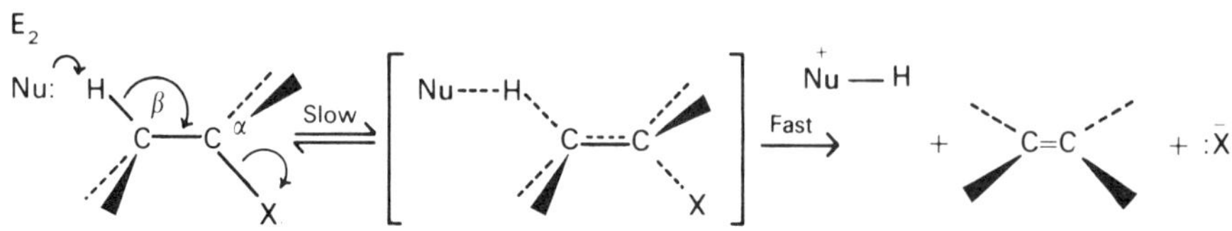

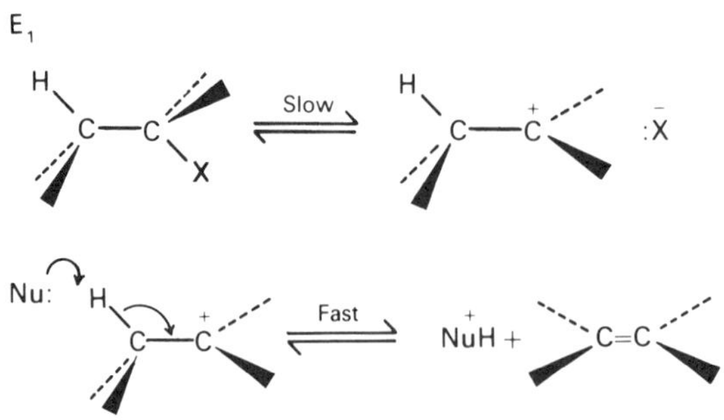

(5.112)

In general, the factors which determine whether a given halide will react by the E_1 or by the E_2 mechanism are the same as those which determine the course of its substitution reactions. Elimination can be promoted at the expense of substitution by employing a reagent which is a more powerful base than it is nucleophile. For example, pyridine, sodium acetate and alcoholic solutions of hydroxides are often used to produce alkenes (5.113, 5.114, 5.115). Due to a combination of electronic and steric factors the best yields of alkenes are usually obtained from secondary and tertiary halides.

$$\underset{\text{3-bromopentane}}{CH_3CH_2CHBrCH_2CH_3} \xrightarrow[115°C]{\text{KOH, alcohol}} \underset{\text{2-pentene}}{CH_3CH_2CH{=}CHCH_3} \qquad \textbf{(5.113)}$$

$$\underset{\text{2-chlorobutanoic acid}}{CH_3CH_2CHClCO_2H} \xrightarrow[\text{pyridine}]{\text{boiling}} \underset{\text{but-2-enoic acid}}{CH_3CH{=}CHCO_2H} \qquad \textbf{(5.114)}$$

$$\underset{\text{3,4-dibromo-4-phenyl butan-2-one}}{PhCHBrCHBrCOCH_3} \xrightarrow[\text{boiling ethanol}]{\text{NaOAc}} \underset{\text{3-bromo-4-phenyl but-3-en-2-one}}{PhCH{=}CBrCOCH_3} \qquad \textbf{(5.115)}$$

Elimination of water from alcohols is undoubtedly the most useful method for obtaining alkenes. As would be expected from the preceding discussion of the substitution reactions of alcohols, the reaction is catalysed by acid; sulphuric or phosphoric acid are preferred to halogen acids for protonating the alcohol because hydrogen sulphate and dihydrogen phosphate anions are only weakly nucleophilic and hence will not cause a large amount of substitution. Surprisingly, at first sight, primary as well as secondary and tertiary alcohols react by the E_1 mechanism. The E_2 mechanism cannot operate in strongly acidic solutions, since it requires a strong base to abstract a β-hydrogen atom. (cf. an S_N2 reaction which requires a powerful nucleophile as reagent). However, because of the more pronounced electron withdrawing effect of the positively charged α-carbon atom, the β-hydrogen atoms of a carbonium ion are much more acidic than those of a protonated primary alcohol and can be removed by very weak bases. Consequently acid catalysed dehydrations of primary alcohols are performed at high temperatures when the protonated alcohol dissociates and the reaction proceeds by the E_1 mechanism. (5.116) As indicated by this equation dehydration of an alcohol is the exact reverse of the addition of water to an alkene (cf. 4.35).

$$CH_3CH_2OH \underset{H_2SO_4}{\overset{\text{conc}}{\rightleftharpoons}} CH_3CH_2{-}\overset{+}{O}H_2 \rightleftharpoons CH_3\overset{+}{C}H_2 + H_2O;$$

$$H_2O{:} + H{-}CH_2{-}\overset{+}{C}H_2 \rightleftharpoons H_3\overset{+}{O} + CH_2{=}CH_2 \qquad \textbf{(5.116)}$$

Understandably, since ethers are usually made from alcohols, elimination of an alcohol from an ether has not been developed as a method for the synthesis of alkenes. However, mechanistic studies show that elimination of methanol from tertbutyl methyl ether at 240°C in the gas phase is catalysed by hydrogen halides. This and similar reactions, show typical E_1 characteristics (5.117).

$$\underset{\text{2-methoxy-2-methyl propane}}{(CH_3)_3COCH_3} \overset{H^+}{\rightleftharpoons} (CH_3)_3C\underset{\substack{|\\H}}{\overset{+}{O}}CH_3 \overset{\text{slow}}{\rightleftharpoons} (CH_3)_3\overset{+}{C} + HOCH_3;\ (CH_3)_3\overset{+}{C} \rightleftharpoons \underset{\text{Isobutene}}{(CH_3)_2C{=}CH_2} + H^+ \qquad \textbf{(5.117)}$$

Little appears to be recorded regarding acid catalysed elimination reactions of thiols. Thio-ethers containing groups which can support a positive charge are potentially capable of undergoing elimination. A few examples are known but they are not of great importance. In contrast, elimination reactions of trialkylsulphonium hydroxides are of some value in preparative chemistry. The salts contain a good leaving group (a thio-ether) and a strong base (hydroxide ion) as part of their structure and undergo elimination when heated.

$$H\bar{O}:\ CH_3CH_2\overset{+}{S}(CH_3)_2 \overset{\text{heat}}{\rightleftharpoons} H_2O + CH_2{=}CH_2 + S(CH_3)_2$$

ethyldimethyl
sulphonium hydroxide

Tetra-alkyl ammonium hydroxides react similarly (5.118). Eliminations of this type from 'onium hydroxides are known collectively as Hoffmann eliminations. Elimination from mono, di- and tri-alkyl ammonium salts does not appear to have been investigated.

$$H\bar{O}\ CH_3CH_2\overset{+}{N}(CH_3)_2CH_2CH_2CH_3 \overset{\text{heat}}{\rightleftharpoons} CH_2{=}CH_2 + (CH_3)_2NCH_2CH_2CH_3 + H_2O \quad \textbf{(5.118)}$$

ethyldimethylpropyl ammonium hydroxide; *N,N*-dimethyl propylamine

Elimination reactions in biochemistry. The reversible elimination of water from an alcohol is frequently encountered in biochemistry. A number of reactions of this type (Section 4.2) have already been discussed, and no further examples will be considered here.

$$HSCH_2CH(\overset{+}{N}H_3)C\bar{O}_2 \underset{\text{desulphydrase, pyridoxal phosphate}}{\overset{\text{cysteine}}{\rightleftharpoons}} H_2S + CH_2{=}C(\overset{+}{N}H_3)C\bar{O}_2 \xrightarrow{H_2O} CH_3COCO_2^- + \overset{+}{N}H_4 \quad \textbf{(5.119)}$$

dehydroalanine; pyruvate

The enzyme cysteine desulphydrase, which is present in several bacteria and also in animal tissues, catalyses the elimination of hydrogen sulphide from cysteine (5.119) and several anaerobic bacteria such as **Clostridium sporogenes** contain a pyridoxal phosphate dependent enzyme which catalyses the elimination of methanethiol from methionine (5.120).

$$CH_3SCH_2CH_2CH(\overset{+}{N}H_3)C\bar{O}_2 + H_2O \rightleftharpoons CH_3SH + \overset{+}{N}H_4 + CH_3CH_2COC\bar{O}_2 \quad \textbf{(5.120)}$$

2-(or α-) oxobutyrate

The elimination of ammonia from amino acids is also frequently encountered both in animal and plant metabolism. The enzymes, known as **ammonia lyases,** which catalyse the reactions, e.g. (5.121), are not dependent on pyridoxal phosphate.

$$PhCH_2CH(\overset{+}{N}H_3)C\bar{O}_2 \underset{\text{ammonia lyase}}{\overset{\text{phenylalanine}}{\rightleftharpoons}} PhCH{=}CHC\bar{O}_2 + \overset{+}{N}H_4 \quad \textbf{(5.121)}$$

phenylalanine; cinnamate

Oxidation

Alcohols can be oxidised in the laboratory using reagents such as an alkaline solution of permanganate, an acidic solution of dichromate, or nitric acid. The product obtained depends on the type of alcohol oxidised and, to some extent, on the type of reagent used. The first product of oxidation of a primary alcohol is the corresponding aldehyde. However, aldehydes are more easily oxidised than alcohols, consequently if a primary alcohol is oxidised using an excess of a non-specific reagent the corresponding carboxylic acid is produced.

$$CH_3(CH_2)_3OH \xrightarrow[100^\circ C]{MnO_4^-/HO^-} CH_3(CH_2)_2CO_2H$$

1-butanol → butanoic acid

Aldehydes can be produced in moderate yields if it is possible to remove them from the reaction mixture by, for example, steam distillation before they can be oxidised. Much better yields can be obtained if reagents which are specific for the oxidation of alcohols to aldehydes, such as *tert*-butyl chromate or a solution of chromic anhydride in pyridine, are used (5.122).

$$(CH_3CH_2)_2C(CH_3)CH_2OH \xrightarrow{CrO_3:N} (CH_3CH_2)_2C(CH_3)CHO \qquad \textbf{(5.122)}$$

2-ethyl-2-methyl butan-1-ol → 2-ethyl-2-methyl butanal

Secondary alcohols are normally converted to ketones by oxidising agents (5.123). However, powerful reagents, which can break carbon to carbon bonds, produce a mixture of carboxylic acids, each acid containing fewer carbon atoms than the alcohol.

$$CH_3CH_2CH_2CH(OH)CH_3 \xrightarrow{H^+/Cr_2O_7^{2-}} CH_3CH_2CH_2\overset{O}{\overset{||}{C}}CH_3 \qquad \textbf{(5.123)}$$

2-pentanol → 2-pentanone

The mechanism of oxidation of primary and secondary alcohols by dichromate is thought to involve an E_2 type elimination from a chromic acid ester (5.124).

$$Cr_2O_7^{2-} + H_2O \rightleftharpoons 2HCrO_4^-$$

$$(CH_3)_2CHOH + HCrO_4^- + H^+ \rightleftharpoons (CH_3)_2CHOCrO_3H + H_2O$$

$$(CH_3)_2\underset{\underset{H}{|}}{C}-O-\overset{+}{Cr}O_3H + H^+ \rightleftharpoons (CH_3)_2C{=}O + H_3\overset{+}{O} + H_2CrO_3 \qquad \textbf{(5.124)}$$

$H_2O:$

$$3H_2CrO_3 + 6H^+ \rightleftharpoons CrO_3 + 2Cr^{3+} + 6H_2O$$

Esters are also thought to be formed during oxidations involving permanganate. These latter reactions, however, are complex and the details of the mechanism are not clear. Tertiary alcohols cannot be oxidised to a carbonyl compound without breaking a carbon to carbon bond. They are stable towards alkaline oxidising agents but may eliminate water under acidic conditions; the alkene which is produced is then oxidised, yielding a mixture of carboxylic acids each containing fewer carbon atoms than the parent alcohol.

Powerful oxidising agents convert thiols to sulphonic acids.

$$CH_3(CH_2)_3SH \xrightarrow{HCO_3H} CH_3(CH_2)_3SO_3H$$

1-butanethiol → butanesulphonic acid

Mild reagents however, produce disulphides (5.125); there is no analogous reaction in the chemistry of alcohols. Disulphides themselves act as oxidising agents towards other thiols (5.126); reactions of this type are usually freely reversible and are said to involve disulphide interchange.

$$2\,H_3\overset{+}{N}CH(CH_2SH)C\bar{O}_2 \xrightarrow[\text{pH 8.5}]{\text{air}} H_3\overset{+}{N}CH(CO_2^-)CH_2\text{–S–S–}CH_2(C\bar{O}_2)CH\overset{+}{N}H_3 \qquad (5.125)$$

cysteine → cystine

$$R^1SSR^1 + 2R^2SH \rightleftharpoons 2R^1SH + R^2SSR^2 \qquad (5.126)$$

All types of amine can be oxidised; the nature of the product(s) obtained depends on the nature of both the amine and the oxidising agent used. The reactions which are most pertinent to biochemistry are those in which amines containing a primary or secondary carbon adjacent to the amino group are oxidised to carbonyl compounds by way of an intermediate imine (5.127).

$$R^1R^2CHNH_2 \underset{}{\overset{-2H,\ MnO_4^-}{\rightleftharpoons}} [R^1R^2C{=}NH] \xrightarrow{H_2O} R^1R^2C{=}O + NH_3 \qquad (5.127)$$

R^2 = H or alkyl.

Another amine oxidation which has some relevance to biochemistry involves the production of *N*-oxides from tertiary amines by the action of Caro's acid (H_2SO_5), ozone, hydrogen peroxide or percarboxylic acids (but not by permanganate)

$$R^1R^2R^3N \xrightarrow{PhCO_3H} R^1R^2R^3N \rightarrow O$$

Biological oxidation of alcohols, thiols and amines. Reactions involving the reversible oxidation of primary or secondary alcohol functions to carbonyl groups are encountered very frequently and are of great importance in biochemistry. The oxidant is almost invariably nicotinamide adenine dinucleotide ($\overset{+}{\text{NAD}}$) or its phosphate derivative ($\overset{+}{\text{NADP}}$) (5.128, see, also Section 9.2).

$$\overset{+}{\text{NAD}} + H\text{–}\overset{|}{\underset{|}{C}}\text{–O–H} \rightleftharpoons \text{NADH} + {>}C{=}O + H^+ \qquad (5.128)$$

This type of reaction involves transfer of hydride ion ($H:^-$) from C-α of the alcohol to the oxidant and hence appears to constitute one of the few examples of a biochemical reaction which has no obvious mechanistic analogue in the laboratory. On the other hand the reverse reaction involving the addition of hydride ion to a carbonyl group has many precedents in the laboratory chemistry of carbonyl compounds. For this reason further discussion of reaction (5.128) is postponed to Section 6.1 which deals with carbonyl compounds and to Section 9.2 which deals with the chemistry of redox co-enzymes. It is worth noting here, however, that the tricarboxylic acid cycle of reactions contains a step which must have evolved solely for the purpose of converting citrate, containing a non-oxidisable tertiary hydroxyl group, to isocitrate, which contains an oxidisable secondary alcoholic group.

$$\bar{O}_2CCH(\overset{+}{N}H_3)CH_2CH_2CONHCH(CH_2SH)CONHCH_2C\bar{O}_2 \qquad (5.129)$$

glutathione (reduced form; GSH)

The interconversion of thiols and disulphides is encountered in several areas of biochemistry. For example, the tripeptide glutathione, L-γ-glutamyl-L-cysteinylglycine (5.129) occurs in all cells in relatively high concentrations. Its function may be to 'protect' cell membranes and enzymes by maintaining their essential thiol groups in the reduced form (5.130).

$$2GSH + RSSR \rightleftharpoons GSSG + 2RSH \qquad (5.130)$$

Although cystine residues, $\begin{array}{c} CH_2-S-S-CH_2 \\ | \qquad\quad | \\ -NHCHCO--NHCHCO- \end{array}$, occur in many proteins the genetic code does not provide for their insertion into polypeptide chains as the latter are formed. Cystine residues appear to be formed by oxidation of spatially adjacent cysteine residues subsequent to protein synthesis. As noted previously lipoic acid (5.47) is a co-enzyme which is essential for the function of both the pyruvate dehydrogenase enzyme complex and the α-ketoglutarate dehydrogenase enzyme complex. In both cases the function of the co-enzyme (Chapter 9) is to transfer an activated acyl group from an enzyme–thiamine complex to Co-enzyme A and in doing so it cycles between the oxidised (disulphide) and reduced (thiol) forms.

The oxidation of amines in living cells appears to be formally similar to reaction (5.127). One of the most important reactions in the field of amino-acid metabolism involves the oxidative deamination of glutamate, catalysed by glutamate dehydrogenase (5.131). This reaction is thought to involve the steps shown in (5.132).

$$\underset{\text{glutamate}}{^-O_2CCH_2CH_2CH(\overset{+}{N}H_3)CO_2^-} + \overset{+}{NAD} + H_2O \overset{GDH}{\rightleftharpoons} \underset{\alpha\text{-ketoglutarate}}{^-O_2CCH_2CH_2\overset{\overset{O}{\|}}{C}CO_2^-} + NADH + \overset{+}{N}H_4 \qquad (5.131)$$

GDH = glutamate dehydrogenase

$$^-O_2CCH_2CH_2CH(\overset{+}{N}H_3)CO_2^- \rightleftharpoons {}^-O_2CCH_2CH_2CH(NH_2)CO_2^- + H^+$$
$$^-O_2CCH_2CH_2CH(NH_2)CO_2^- + \overset{+}{NAD} \rightleftharpoons {}^-O_2CCH_2CH_2C{=}(NH)CO_2^- + NADH + H^+$$
$$^-O_2CCH_2CH_2C{=}(NH)CO_2^- + H_2O \rightleftharpoons {}^-O_2CCH_2CH_2C{=}(O)CO_2^- + NH_3 \qquad (5.132)$$

The oxidative deamination of most other amino acids, catalysed by the FAD dependant L - and D - amino acid oxidases obtained, respectively, from pig's kidney and rattlesnake venom, has also been shown to involve imino acid intermediates (5.133).

$$RCH(\overset{+}{N}H_3)CO_2^- + \text{Enz–FAD} \rightleftharpoons RC{=}(\overset{+}{N}H_2)CO_2^- + \text{Enz–FADH}_2;$$
$$RC{=}(\overset{+}{N}H_2)CO_2^- + H_2O \rightleftharpoons RC = (O)CO_2^- + NH_4^+;$$
$$\text{Enz–FADH}_2 + O_2 \rightleftharpoons \text{Enz–FAD} + H_2O_2 \qquad (5.133)$$

The oxygen consumed in the reactions is involved in re-oxidising the reduced co-enzyme which is produced when the imino acid is formed; the hydrogen peroxide which is produced in the reaction is removed by the action of catalase. The stoichiometry of other amine oxidations is the same as that of the reactions of

(5.133). For example the mono-amine oxidase in mitochondria of the adrenal medulla catalyses the first step in the degradation of noradrenalin, as shown in (5.134).

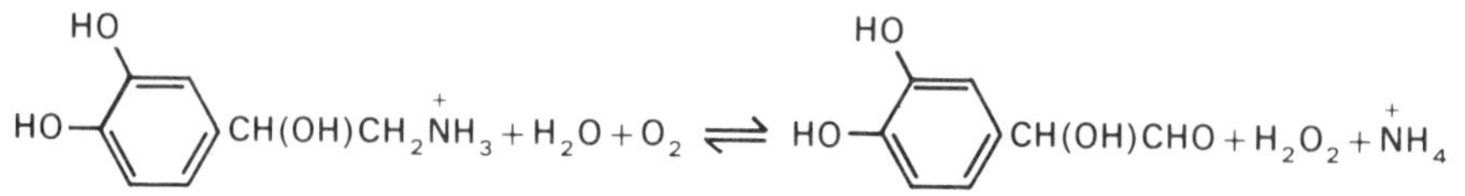

2-hydroxy-2-(3′,4′-dihydroxyphenyl) ethanal

(5.134)

Miscellaneous reactions

Metalation. Alcohols, thiols and amines will all react with strongly electropositive metals.

$$R\bar{X}H + M \rightarrow R\bar{X}:\overset{+}{M} + \tfrac{1}{2}H_2, \quad X = O, S, NH$$

Sodium, potassium or lithium react vigorously with liquid alcohols producing solutions of the corresponding alkoxide in the alcohol.

$$CH_3CH_2OH + Na \rightarrow \underset{\text{sodium ethoxide}}{CH_3CH_2\bar{O}:\overset{+}{Na}} + \tfrac{1}{2}H_2$$

Alkoxide ions are very powerful bases (and nucleophiles) and consequently solutions of them must be protected from moisture to prevent regeneration of the alcohol.

$$R\bar{O}: + H{-}\bar{O}H \rightleftharpoons ROH + :\bar{O}H \qquad \textbf{(5.135)}$$

Since alkoxide ions are very strong bases, (alcohols are very weak acids) the equilibrium in reaction (5.135) lies to the right and no appreciable amount of alkoxide ions can be produced by the action of hydroxide ions on alcohols.

In contrast to alcohols, thiols are appreciably acidic; consequently thiolate ions can be produced from them in aqueous solution by the action of hydroxide ion.

$$RSH + :\bar{O}H \rightleftharpoons R\bar{S}: + H_2O$$

Thiols also react with metals such as copper, lead, mercury and silver. The products, known as mercaptides, contain sulphur to metal bonds which are largely covalent; those derived from mercury are particularly stable.

$$RSH + Hg^{2+} \rightleftharpoons RSHg^{+} + H^{+};$$

$$2RSH + Hg^{2+} \rightleftharpoons (RS)_2Hg + 2H^{+}$$

The thiol can be regenerated from a mercaptide by treating the latter with an excess of a second thiol.

$$(R^1S)_2Hg + 2R^2SH \rightleftharpoons 2R^1SH + (R^2S)_2Hg$$

Thiols also form numerous complexes with transition metals. Those containing iron are particularly important since some of them provide models for the structure of Ferrodoxin and related 'non-haem iron proteins' which contain iron co-ordinated to the thiol groups of cysteine residues and to sulphide ions. These proteins function, in a manner which is still not clear, as electron carriers in redox systems.

Amines can be metalated by lithium, giving derivatives which are useful for various synthetic operations. They are not, however, relevant to biochemistry and will not be considered here.

Aromatic Compounds

Aromatic halogeno-, hydroxy, mercapto and amino compounds, like their aliphatic analogues, possess several common properties and are logically discussed together. In particular, the differences between the properties of any given aromatic compound and its aliphatic analogue can be explained similarly. The most notable differences are:—

(i) Aryl compounds are less reactive in nucleophilic substitutions than their aliphatic analogues.

(ii) Aryl compounds are weaker bases and nucleophiles than their aliphatic analogues. For example ethylamine has p*K* 10.8, aniline p*K* 4.63; phenols cannot be acylated by a protonated carboxylic acid.

(iii) Phenols and thiophenols are demonstrably acidic, whereas their aliphatic analogues are not.

(iv) With some specific exceptions aryl compounds are much less susceptible to oxidation than their aliphatic analogues.

These differences all reflect, at least in part, the distribution of charge within the compounds. Physical measurements, for example of bond lengths and dipole moments, indicate that an electron pair on the exocyclic atom is partially delocalised over the phenyl group, cf, (3.15).

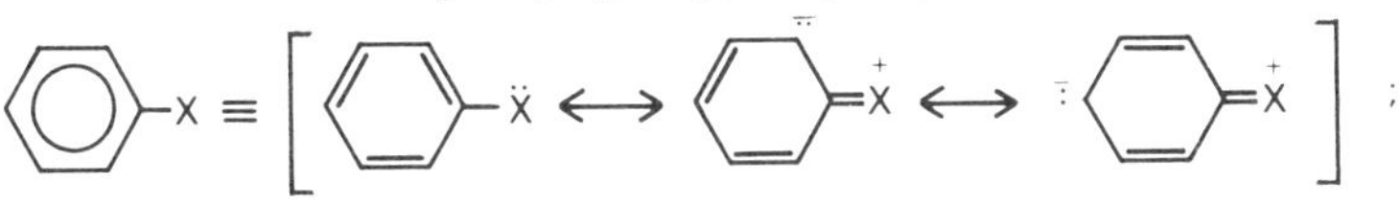

In simple terms this de-localisation of charge will make C-1 of the phenyl group less electrophilic and therefore less susceptible to attack by nucleophiles. Furthermore X will be more difficult to displace because the C–X bond has some double bond character. Since there is some conjugation of charge normally associated with the lone pair of the exocyclic atom with the aromatic ring the compounds are less efficient bases and nucleophiles than their aliphatic analogues.

Phenols and thiophenols are weakly acidic because the anions which they produce are more effectively stabilised by charge delocalisation than are the undissociated molecules (5.136; multiply charged diagrams are always indicative

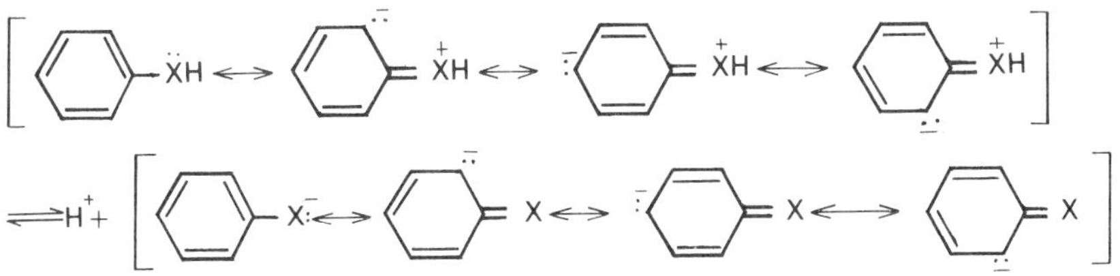

(5.136)

$$PhOH + \bar{O}H \rightleftharpoons Ph\bar{O}: + H_2O \qquad \textbf{(5.137)}$$

phenol

of a higher energy than singly charged diagrams). Unlike alkoxide ions, phenate and thiophenate ions can be produced, and used, in aqueous solution (5.137).

The stability of aromatic compounds towards oxidation can be explained by saying that the process would be energetically unfavourable because it would destroy the low energy, conjugated aromatic system. There are, however, some important exceptions to this generalisation. 1,2- and 1,4- (but **not** 1,3-) dihydroxy, diamino or aminohydroxybenzenes are oxidised readily by reagents such as Fremy's salt, ($(KSO_3)_2NO$), ferric chloride, manganese dioxide in sulphuric acid or acidic dichromate. The product is a 1,2- or 1,4-benzoquinone (5.138 and 5.139).

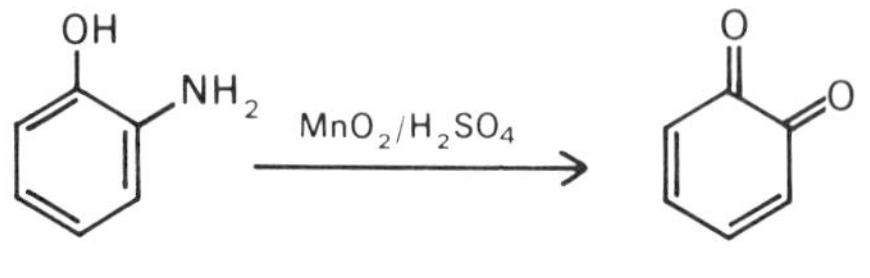

(**5.138**)

2-aminophenol 1,2-benzoquinone

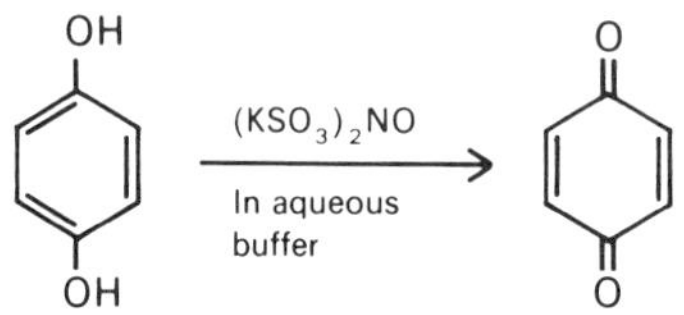

(**5.139**)

1,4-dihydroxybenzene (hydroquinone) 1,4-benzoquinone

Quinones are easily reduced by, for example, molecular hydrogen in the presence of a platinum catalyst. The interconversion of hydroquinone and 1,4-benzoquinone is a step-wise process. Thus, incomplete reduction of a 1,4-benzoquinone, or incomplete oxidation of a hydroquinone can give either a semi-quinone, as the radical anion or as the neutral species, or a quinhydrone. The latter is a 1:1 quinone-hydroquinone complex which results from disproportionation of the neutral semi-quinone (5.140).

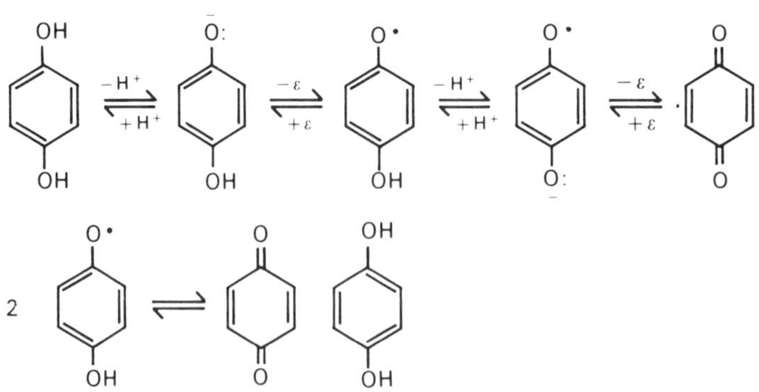

(**5.140**)

Finally the description of aromatic halogeno, hydroxy, mercapto and amino compounds as resonance hybrids can be used to explain the fact that, in all cases, electrophilic substitution occurs in the *ortho* and *para* positions. It can be seen from (5.141) that it is when the substituent enters one of these positions that the intermediate cation (and hence, presumably, the transition state also) can be most effectively stabilised.

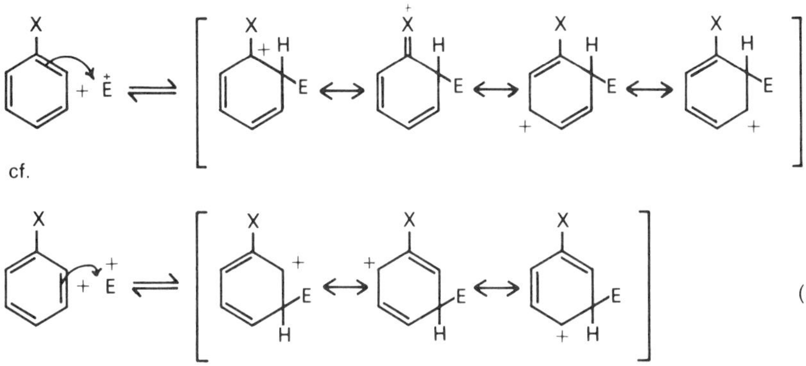

(5.141)

Biochemical reactions of aromatic halogeno, hydroxy mercapto and amino compounds

Although very few aryl halides occur naturally their properties have an important bearing on their use (and misuse!) as pesticides and herbicides. Chloromethane, in solution in aqueous acetone, will undergo very slow hydrolysis at room temperature. Chlorobenzene, on the other hand, is extremely difficult to convert to phenol and would be essentially stable under the same conditions.

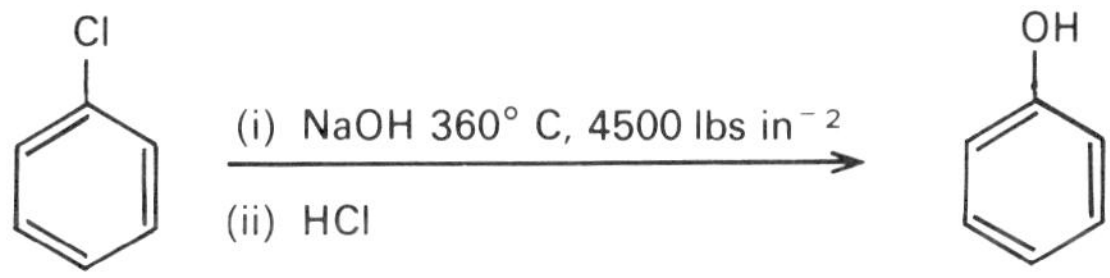

Similarly aryl halides which are used as pesticides are not degraded in the environment by hydrolysis or by oxidation at a significant rate and accumulate. The problems of toxicity which result are well known.

The oxidation of 1,4-diphenols to quinones is of particular relevance to biochemistry because one component of the electron transport chain in mitochondria is a 1,4-benzoquinone. This compound is ubiquinone or co-enzyme Q (5.142a), which is thought to 'cycle' between the oxidised quinone and the reduced hydroquinone forms, accepting two electrons from NADH oxido-reductase (and two protons from the medium) and then transferring them to cytochrome b. Another quinone, plastoquinone (5.142b), is involved in a similar manner in photosynthetic electron transport in green plants.

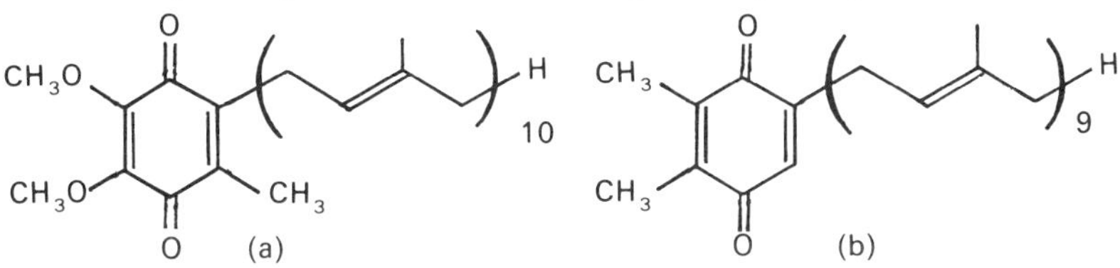

(5.142)

A number of other quinones also occur naturally. Vitamin K (5.143) is a naphthaquinone derivative which is essential for the synthesis of blood clotting factors in mammals. Other naphthaquinones and some anthraquinones (5.144) are produced as secondary metabolites in plants, marine organisms and in bacteria.

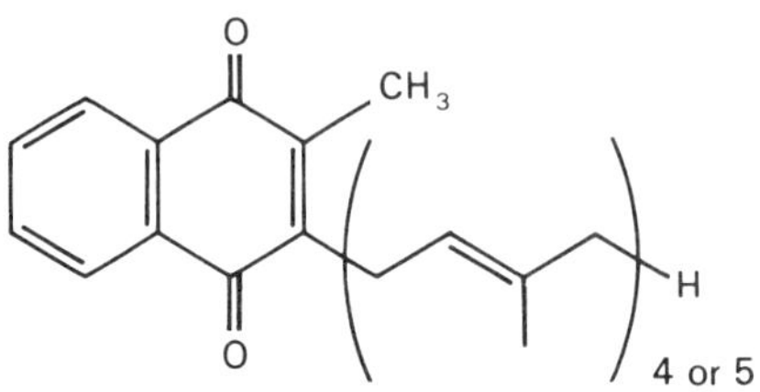

(**5.143**)

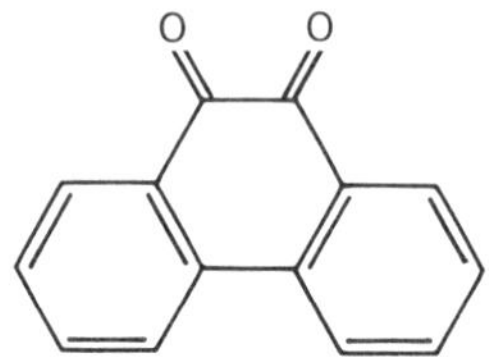

(**5.144**)

6

Functional groups of the type $_{Y}\!>\!C=O$, Carbonyl compounds, Carboxylic acids and derivatives

A carbonyl group ($>C=O$) contains a carbon atom joined to an oxygen atom by a double bond and as such is characteristic of aldehydes and ketones. It also forms part of a number of more complex functional groups which are characteristic of carboxylic acids and their derivatives; the nomenclature of these groups is shown in Table 1.1.

Typical reactivities of the carbonyl group

All the available evidence indicates that the carbonyl group, as would be expected on the basis of the relative electronegativities of carbon and oxygen, is polarised as shown in (6.1) and that reactions of the group are almost invariably initiated by attack by a nucleophile at the electron deficient carbon atom (6.2).

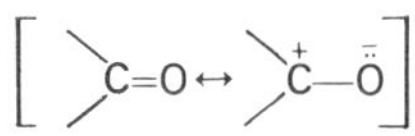

(6.1)

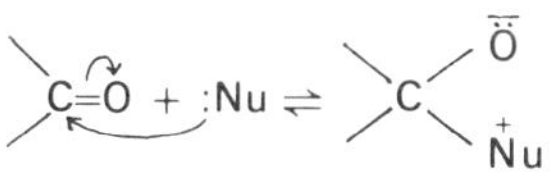

(6.2)

In the case of aldehydes and ketones the nucleophile is usually one of two types, viz.

(i) :X–H. In this case the initial step in the reaction is followed by proton migration.

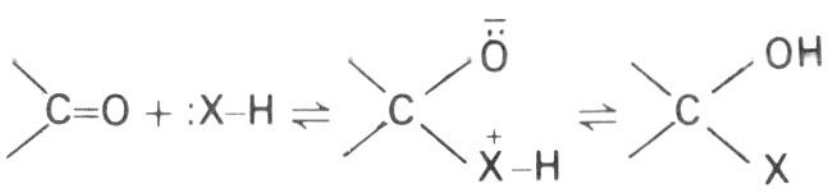

(ii) X–H. Here the nucleophile is generated by removal of a proton from the reagent by a base.

$$X\text{–}H + :B \rightleftharpoons \bar{X}: + H\overset{+}{B}$$

The oxygen anion which results from nucleophilic attack at the carbonyl group then abstracts a proton from the protonated base. Since the base is regenerated in this step the reactions are **base catalysed**.

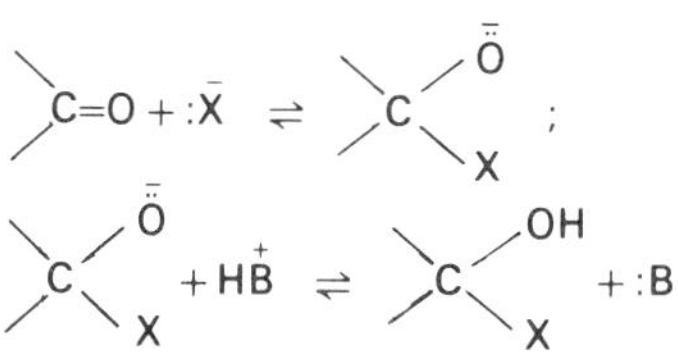

Both types of reaction lead, overall, to addition of the nucleophile across the double bond. In summary, carbonyl groups in aldehydes and ketones undergo reactions involving **nucleophilic addition**.

In the case of carboxylic acids and their derivatives the situation is different. The same type of intermediate is formed by nucleophilic attack at the carbonyl group, but the negatively charged oxygen in the intermediate is not subsequently protonated. Instead, it is energetically more favourable for a double bond to be re-formed with concomitant expulsion of Y as :$\bar{Y}$ (6.3).

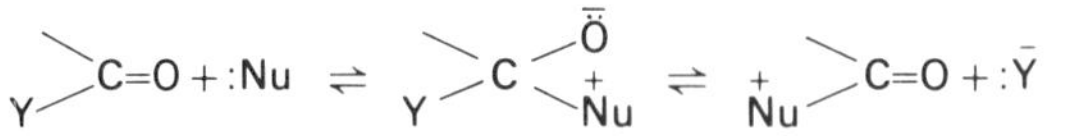

(6.3)

It can be seen that such reactions involve **nucleophilic substitution** (or displacement) **at unsaturated carbon,** which is the typical reactivity of carboxylic acid and their derivatives.

The difference between the type of reactivity exhibited by aldehydes and ketones on the one hand and carboxylic acids and their derivatives on the other can be explained in general terms by comparing the groups which are, or would be, expelled during substitution. The Y groups (6.3) of carboxylic acids and derivatives are all better leaving groups than the corresponding substituents of aldehydes and ketones. The latter would have to be expelled as $\bar{H}$: or as a carbanion; both anions are powerful bases and nucleophiles and are very poor leaving groups (cf. page 71 Section 5.1).

Because of the difference in reactivity between the two types of compound, more detailed discussions of their chemistry will be presented separately in the remainder of this chapter.

6.1. Aldehydes and Ketones

Aldehydes and ketones are often referred to collectively as **carbonyl compounds;** carboxylic acids and their derivatives are excluded from this designation despite the fact that they contain a carbonyl group and this potential source of confusion should be borne in mind.

Nomenclature and Examples

The class suffix which denotes the presence of an aldehyde group is **-al**, the class prefix is **formyl**. An aldehyde group can occur **only** at the end of a carbon chain; locant numbers are therefore superfluous to the nomenclature of aliphatic compounds. The class suffix which denotes the presence of a keto group is **-one**, the prefix is **oxo**;* the carbonyl group is located within the parent hydrocarbon using a locant number. Radical nomenclature must be used for ketones containing aryl substituents. The structures and names of some representative aldehydes and ketones are given in Table 6.1.

(*) Biochemical literature, unfortunately, still uses the prefix *keto*.

Table 6.1. The structures and nomenclature of some representative aldehydes and ketones

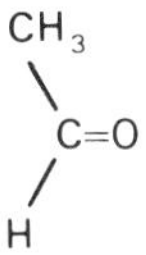

ethanal
(acetaldehyde)

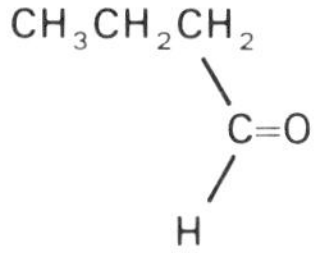

butanal

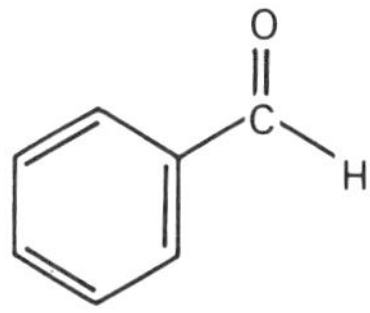

benzaldehyde
(formylbenzene)

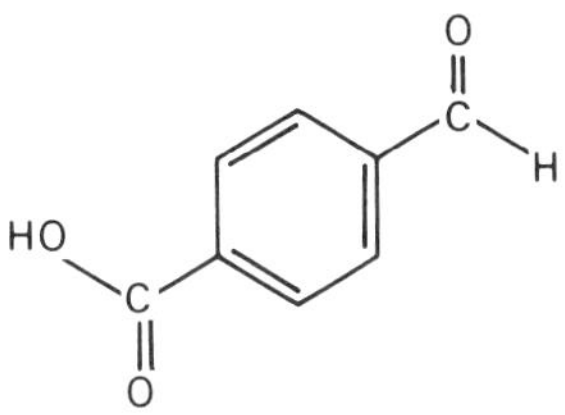

4-formyl benzoic acid

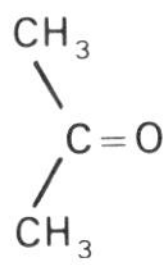

propanone
(acetone)

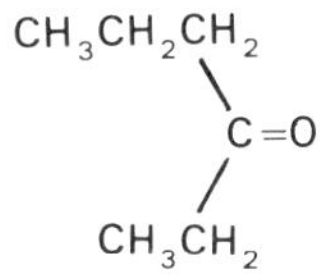

hexan-3-one

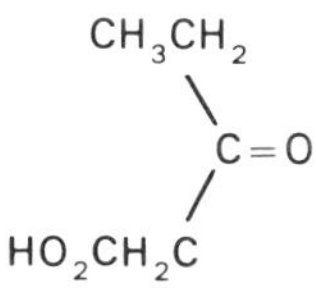

3-oxopentanoic
acid

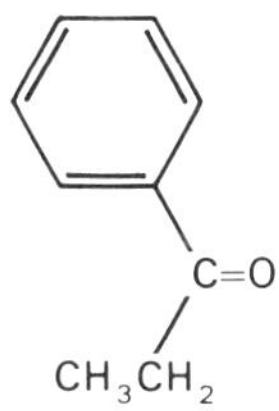

ethyl phenyl ketone

Carbonyl Compounds of Biological Importance

A variety of naturally occurring compounds contain carbonyl functions; some representative examples are shown in Table 6.2.

Solutions of simple sugars such as glucose (6.4) and fructose (6.5) contain small amounts of acyclic aldehyde or keto forms. Pyruvic acid (6.6) is produced in living cells during the oxidation of glucose while 2-oxopentanedioic acid (α-ketoglutaric acid, 6.7) is an intermediate in the tricarboxylic acid cycle of reactions and 3-oxo-acids, such as acetoacetic acid (6.8), are intermediates in the conversion of fatty acids to acetylco-enzyme A. Testosterone (6.9) and aldosterone (6.10) are hormones; the first is a sex hormone, the second is concerned with salt balance in animals. Pyridoxal phosophate (6.11), as has already been mentioned in Section (5.1), is a co-enzyme which is involved in the catalysis of many reactions in amino-acid metabolism. Citral (6.12), camphor (6.13) and vetivone (6.14) are examples of the wide variety of carbonyl compounds which occur in the 'essential oils' of many plants.

Table 6.2. Some naturally occurring carbonyl compounds

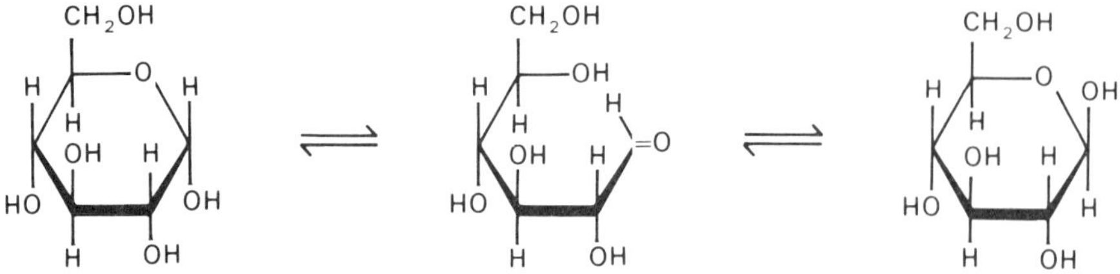

α-D-glucose (**6.4a**) ⇌ acyclic aldehydo form ⇌ β-D-glucose (**6.4b**)

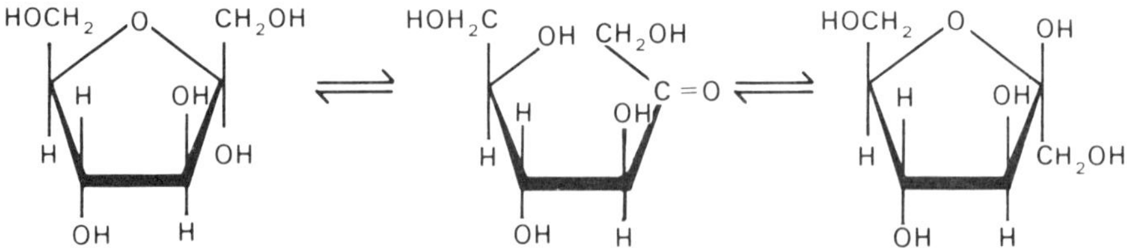

α-D-fructose (**6.5a**) ⇌ acyclic keto form ⇌ β-D-fructose (**6.5b**)

CH_3COCO_2H
pyruvic acid
(**6.6**)

$HO_2CCH_2CH_2COCO_2H$
2-oxopentanedioic acid
(α-ketoglutaric acid)
(**6.7**)

$CH_3COCH_2CO_2H$
3-oxobutanoic acid
(acetoacetic acid)
(**6.8**)

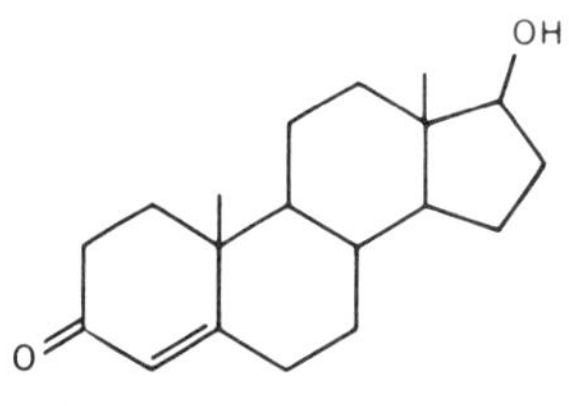

testosterone
(**6.9**)

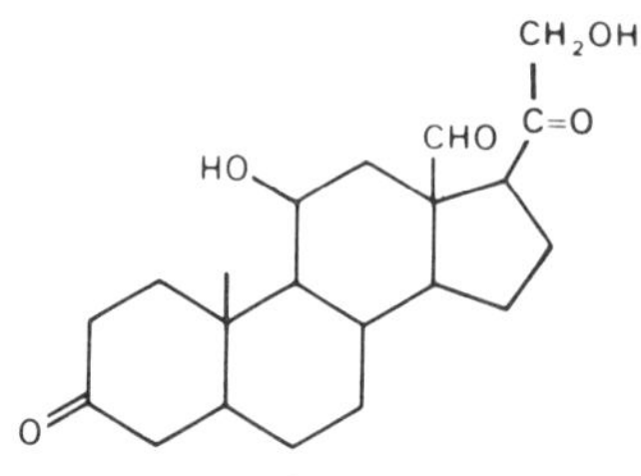

aldosterone
(**6.10**)

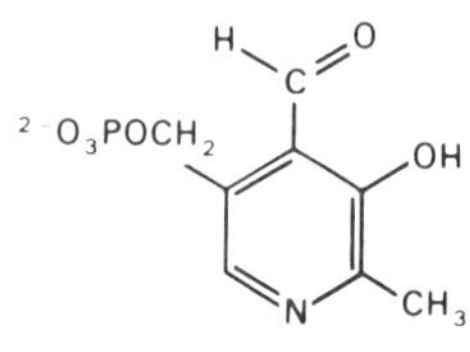

pyridoxal phosphate
(**6.11**)

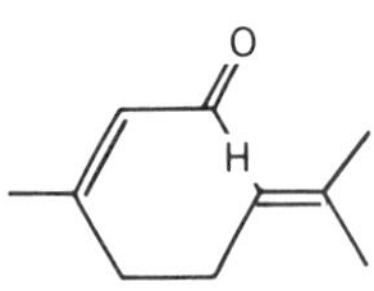

citral
(**6.12**)

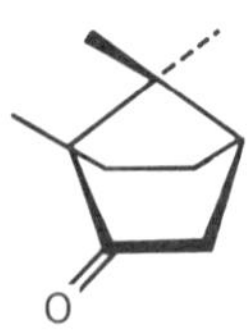

camphor
(**6.13**)

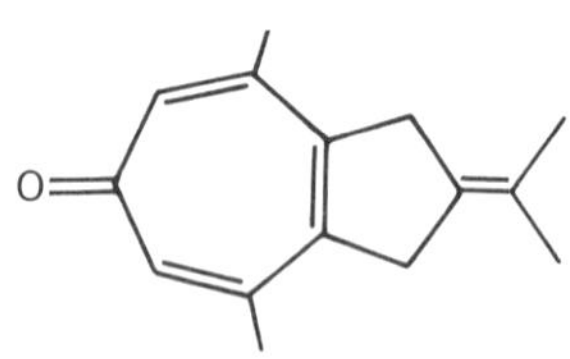

vetivone
(**6.14**)

Properties of Carbonyl compounds

Only low molecular mass compounds such as ethanal and propanone are soluble in water. Most simple carbonyl compounds are volatile, relatively non-polar liquids which are miscible with most non-polar solvents.

Before discussing specific reactions two general points relating to nucleophilic addition to carbonyl groups are worth noting. Firstly, ketones are usually less reactive than aldehydes towards nucleophiles and secondly both aromatic aldehydes and ketones are less reactive than analogous aliphatic compounds. In both cases the decrease in reactivity is due to decreased electrophilic character of the carbonyl carbon atom, both alkyl groups (+I) and phenyl groups (+M) being electron donating compared with hydrogen.

Addition of water and its analogues

The hydration of carbonyl compounds in aqueous solution is described by equation (6.15).

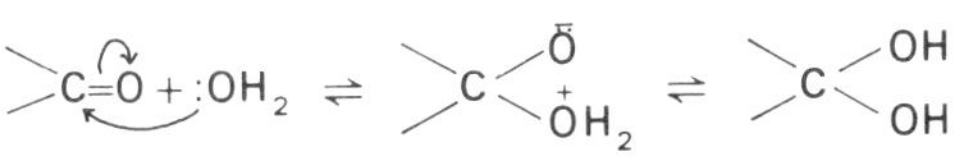

(6.15)

Water is only a weak nucleophile and the equilibrium lies to the right only for carbonyl compounds in which the carbon atom of the carbonyl group is strongly electrophilic. Thus, while methanal is 99.99 % hydrated in aqueous solution, and 2,2,2-trichloroethanal (chloral) and indane-1,2,3-trione give crystalline hydrates, the degree of hydration of propanone in aqueous solution is negligible. Analogues of water, such as alcohols, thiols and amines also add to carbonyl groups. Reactions involving these nucleophiles have been discussed in Chapter 5 to which the reader is referred.

Addition of carbanions—the aldol reaction

In Section 4.2 (p. 42) it was explained that carbonyl compounds which possess an α-hydrogen atom are very weakly acidic. Consequently treatment of such a compound with a strong base will produce significant quantities of the corresponding anion which can then act as a nucleophile towards the same, or a different carbonyl compound (6.16).

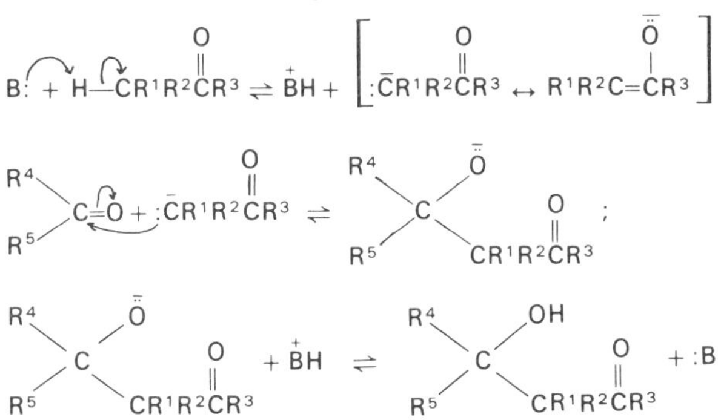

(6.16)

Note that the base is regenerated and that the overall reaction is therefore **base catalysed**. Note also that the anion could, theoretically, act as an oxygen nucleophile towards the carbonyl group. It does not do so because it is energetically more favourable for a strong carbon to carbon bond to be formed rather than a weaker carbon to oxygen bond. Reactions of the type described are known as **aldol reactions** after the simplest example.

$$\underset{\text{ethanal}}{2CH_3CHO} \underset{\text{quantity of NaOH or } Na_2CO_3}{\overset{\text{catalytic}}{\rightleftharpoons}} \underset{\text{3-hydroxybutanal or aldol}}{CH_3CH(OH)CH_2CHO} \underset{-H_2O}{\overset{\text{warm}}{\longrightarrow}} \underset{\text{2-butenal}}{CH_3CH{=}CHCHO}$$

Aldol reactions in vivo. Several aldol reactions occur *in vivo* in connection with the metabolism of carbohydrates. The outstanding example is undoubtedly the reaction which equilibrates fructose-1,6-diphosphate with glyceraldehyde-3-phosphate and dihydroxyacetone phosphate. This reaction is catalysed by the enzyme **aldolase** and is common to both glycolysis and gluconeogenesis. In the direction of gluconeogenesis the reaction involves addition of dihydroxyacetone phosphate (the anion donor, or nucleophile) to the carbonyl group of glyceraldehyde-3-phosphate (6.17). Other examples of aldol reactions can be found in the pentose phosphate pathway of carbohydrate metabolism and in the Calvin cycle of photosynthesis.

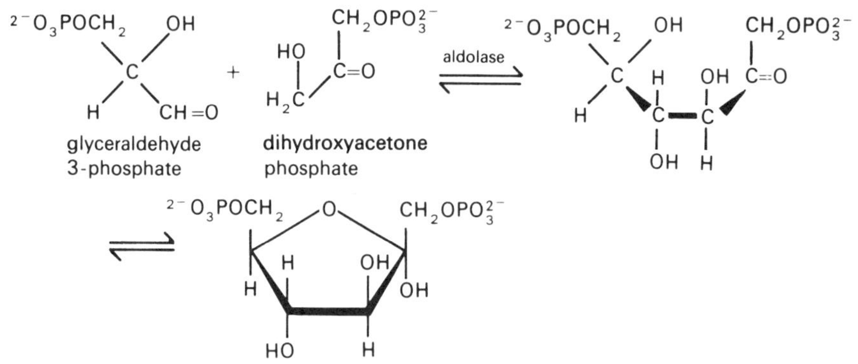

D-fructose-1,6-diphosphate **(6.17)**

Addition of hydrogen cyanide

In the presence of a catalytic quantity of base, hydrogen cyanide will add to a carbonyl group producing a 2- (or α-) hydroxynitrile (6.18) 2-Hydroxypropionitrile (lactonitrile, (6.19)) results from ethanal, while benzaldehyde provides 2-hydroxy-2-phenylethanonitrile (mandelonitrile, 6.20).

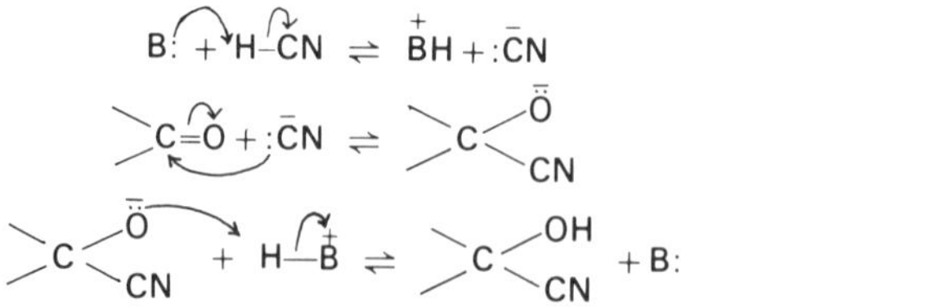

(6.18)

$$CH_3CHO + HCN \xrightarrow{\text{base}} CH_3CH(OH)CN \qquad \textbf{(6.19)}$$

$$PhCHO + HCN \xrightarrow{\text{base}} PhCH(OH)CN \qquad \textbf{(6.20)}$$

This type of reaction forms the basis of Kiliani's method for adding a carbon atom to a sugar, and was used extensively in the elucidation of the stereochemistry of these compounds. Note that cyanide ions, in contrast to hydrogen cyanide together with catalytic quantity of a base, catalyse the formation of 2-hydroxyketones, or **acyloins**, from aromatic aldehydes (6.21).

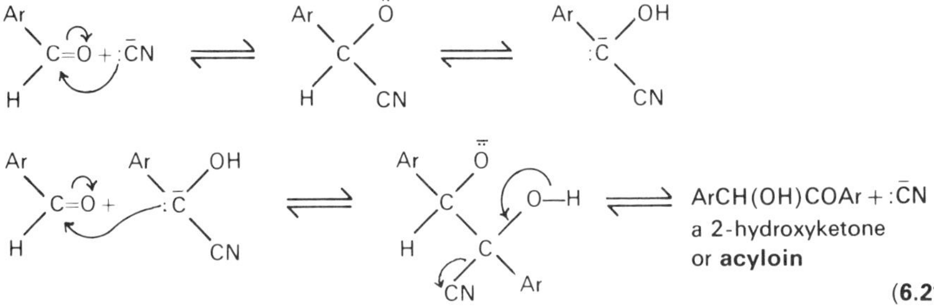

(**6.21**)

When benzaldehyde is used in this reaction the product is benzoin, PhCH(OH)COPh, leading to the description of similar acyloin forming reactions as 'benzoin condensations'. The efficiency of cyanide as a catalyst for benzoin condensation stems from the combination of good nucleophilic acitvity, ability to labilise the aldehyde hydrogen by virtue of the polarisation in the group (6.22) and good leaving properties.

$$[—C\equiv N: \leftrightarrow —\overset{+}{C}\equiv\overset{-}{\ddot{N}}:] \qquad (\mathbf{6.22})$$

Note that acyloins can also be synthesised by treating esters with metallic sodium (6.23).

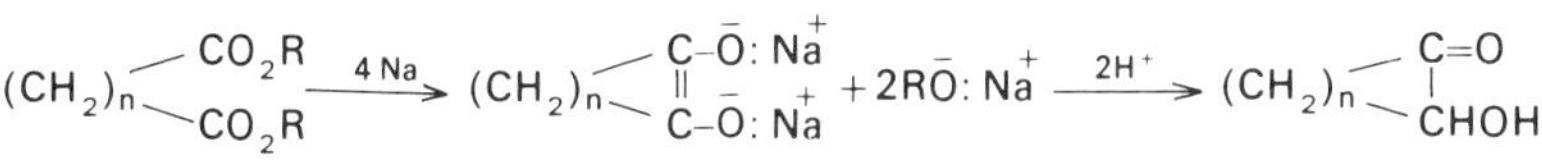

(**6.23**)

This last type of reaction is known as an **acyloin condensation** and must not be confused with a benzoin condensation.

Acyloin reactions in vivo. In vivo acyloin condensations are not restricted to aromatic aldehydes, as are laboratory reactions. For example, transketolase catalyses a number of benzoin condensations of which (6.24), an important reaction of the pentose phosphate pathway, is typical.

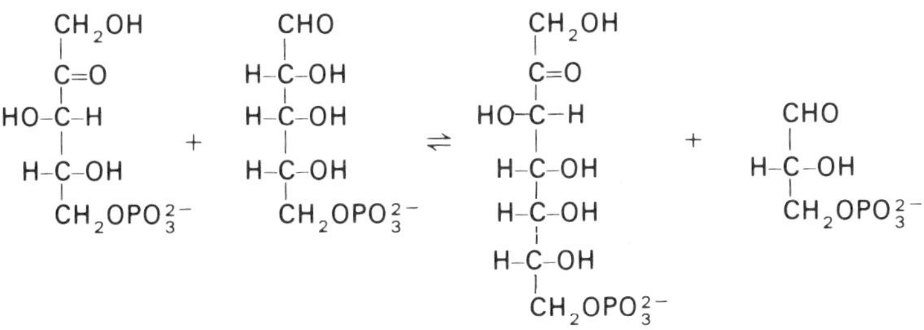

(**6.24**)

xylulose-5-phosphate ribose-5-phosphate sedoheptulose-7-phosphate glyceraldehyde-3-phosphate (G-3-P)

Investigation of this reaction has shown that the first step is a reverse acyloin condensation in which a two carbon unit, 2-hydroxyethanal (glycoaldehyde) is removed from xylulose-5-phosphate (6.25).

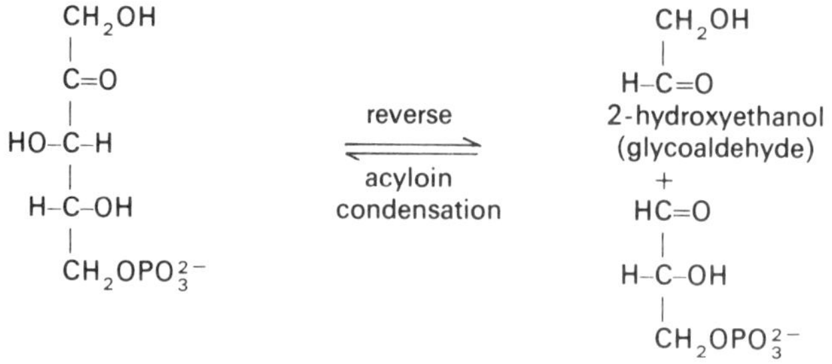

(6.25)

The glycoaldehyde is then condensed with ribose-5-phosphate to form a second acyloin, sedoheptulose-7-phosphate. Transketolase requires thiamine pyrophosphate (TPP) as co-enzyme; the mechanistic role of the co-enzyme is discussed in Section 9.3.

Reduction and oxidation of carbonyl compounds

Reduction. Reduction of an aldehyde or ketone produces, respectively, a primary or a secondary alcohol. A variety of reagents can be used, such as sodium-amalgam and ethanol, molecular hydrogen in the presence of Raney Nickel, aluminium isopropoxide in isopropanol or sodium borohydride or lithium aluminium hydride. Of these reagents the last two are of most interest because they are selective (6.26 and 6.27), and because their mode of action is closely analogous to the mode of action of NADH which reduces many carbonyl compounds *in vivo* (Chapter 8).

$$CH_3COCH_2CO_2CH_3 + NaBH_4 \xrightarrow{\text{isopropanol}} CH_3CH(OH)CH_2CO_2CH_3 \qquad (6.26)$$

methyl 3-oxobutanoate → methyl 3-hydroxybutanoate

$$PhCH{=}CHCHO + LiAlH_4 \xrightarrow{\text{ether}} PhCH{=}CHCH_2OH \qquad (6.27)$$

cinnamaldehyde → cinnamyl alcohol

Both borohydride and aluminohydride anions act as donors of hydride ion, $\bar{H}$:, which attacks the carbonyl group of the carbonyl compound producing an alkoxide which is converted to an alcohol when the reaction mixture is acidified. The overall result is addition of hydrogen across the carbonyl group (6.28).

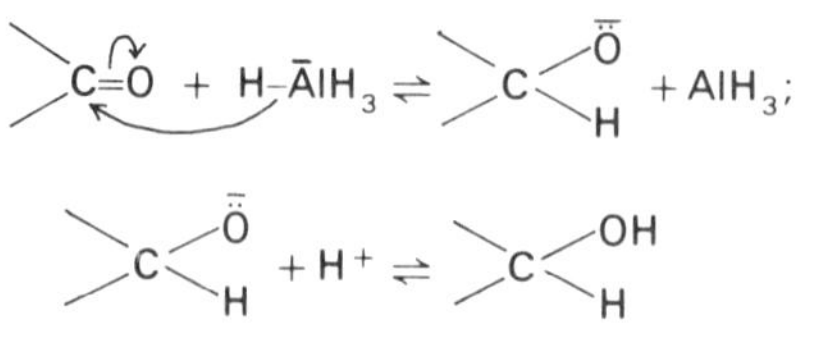

(6.28)

In living cells it is found that carbonyl groups are almost invariably reduced by

NADH or NADPH in the presence of a proton (6.29 to 6.31).

$$\underset{\text{pyruvate}}{CH_3COCO_2^-} + NADH + H^+ \xrightleftharpoons[\text{dehydrogenase}]{\text{lactate}} \underset{\text{lactate}}{CH_3CH(OH)CO_2^-} + \overset{+}{NAD} \qquad \textbf{(6.29)}$$

$$\underset{\text{oxaloacetate}}{^-O_2CCOCH_2CO_2^-} + NADH + H^+ \xrightleftharpoons[\text{dehydrogenase}]{\text{malate}} \underset{\text{malate}}{^-O_2CCH(OH)CH_2CO_2^-} + \overset{+}{NAD} \qquad \textbf{(6.30)}$$

$$\underset{\text{a 3-oxoacylACP}}{RCOCH_2COSACP} + NADPH + H^+ \xrightleftharpoons[\text{reductase}]{\text{3-oxoacylACP}} \underset{\text{3-hydroxyacylACP}}{RCH(OH)CH_2COSACP} + \overset{+}{NADP} \qquad \textbf{(6.31)}$$

ACP = acyl carrier protein

The first two reactions are components of the tricarboxylic acid cycle of reactions, the third is involved in the biosynthesis of fatty acids. In these, and similar reactions, NAD(P)H, like the metal hydrides (above), is thought to act as a donor of hydride ion to the carbonyl group (Section 9.2).

Oxidation. In the laboratory aldehydes can be oxidised to the corresponding carboxylic acid by a variety of reagents, using relatively mild conditions (6.32).

$$\underset{\text{heptanal}}{CH_3(CH_2)_5CHO} \xrightarrow[H_2SO_4]{CrO_3} \underset{\text{heptanoic acid}}{CH_3(CH_2)_5CO_2H} \qquad \textbf{(6.32)}$$

The classical reagents for detecting the presence of an aldehyde group, Fehling's reagent (an alkaline solution of copper tartrate) and Tollen's reagent (an ammoniacal solution of silver nitrate) are both oxidising agents. The former is reduced to cuprous oxide which precipitates (6.33), the second is reduced to metallic silver which is deposited as a film on the walls of the containing vessel (6.34).

$$2Cu^{2+} + RCHO + 5\bar{O}H \rightarrow Cu_2O + RC\bar{O}_2 + 3H_2O \qquad \textbf{(6.33)}$$

$$2Ag^+ + RCHO + 3\bar{O}H \rightarrow 2Ag^\circ + RC\bar{O}_2 + 2H_2O \qquad \textbf{(6.34)}$$

In contrast, simple unsubstituted ketones can be oxidised, using extreme conditions, only by reagents which can break a carbon–carbon bond; a mixture of carboxylic acids, each containing fewer carbon atoms than the ketone, results. It is important to note, however, that 2-hydroxyketones **are** susceptible to oxidation (6.35) in alkaline solution. It seems likely that it is the tautomer of the carbonyl compound, an **ene-diol**, that is actually oxidised (6.36). The easy oxidation of ketose sugars, such as fructose, is explained in a similar manner.

$$\underset{\text{hydroxypropanone}}{HOCH_2\overset{\overset{\displaystyle O}{\|}}{C}CH_3} \xrightarrow[100^\circ C]{\text{Fehling's solutions}} HCO_2H + CH_3CO_2H \qquad \textbf{(6.35)}$$

$$HOCH_2\overset{\overset{\displaystyle O}{\|}}{C}CH_3 \rightleftharpoons \underset{\text{enediol}}{\overset{\overset{\displaystyle OH}{|}}{C}H{=}\overset{\overset{\displaystyle OH}{|}}{C}CH_3} \rightarrow [CH(OH)_3 + CH_3C(OH)_3] \xrightarrow{-2H_2O} HCO_2H + CH_3CO_2H \qquad \textbf{(6.36)}$$

In vivo, aldehydes appear to be oxidised to carboxylic acids by removal of hydride ion from an intermediate hemi-thioacetal. The oxidation which is best understood occurs during glycolysis. It involves the conversion of glyceraldehyde-3-phosphate to 1,3-di-phosphoglyceric acid in the presence of NAD^+ and phosphate under the catalytic influence of glyceraldehyde-3-phosphate dehydrogenase (GPDH, 6.37).

$$^{2-}O_3POCH_2CH(OH)CHO + \overset{+}{NAD} + HPO_4^{2-} \xrightarrow{GPDH} {}^{2-}O_3POCH_2CH(OH)COOPO_3^{2-} + NADH + H^+$$

glyceraldyhyde -3-phosphate → 1,3-diphospho-glyceric acid **(6.37)**

GPDH is an NAD^+ dependent enzyme and contains a thiol group, which is essential for activity, in its active centre. The course of the oxidation which it catalyses is summarised in (6.38).

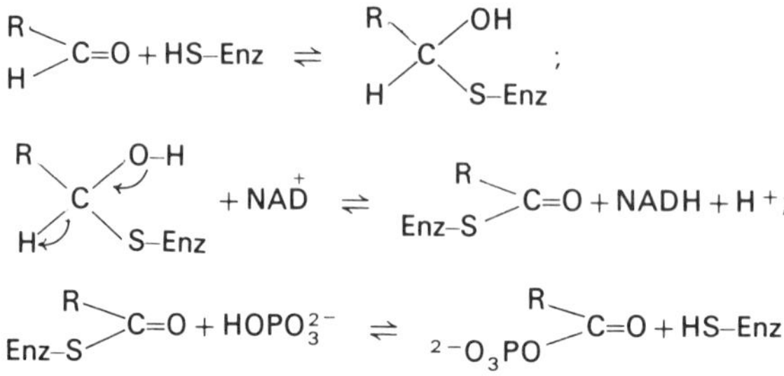

$R = {}^{2-}O_3POCH_2CH(OH)-$ **(6.38)**

As already implied, the first step in the overall reaction involves formation of an enzyme bound hemi-thioacetal (cf. Section 5.1). This intermediate is then oxidised by transfer of hydride ion to NAD^+ (Section 9.2) giving rise to an enzyme bound **thiolester**. Compounds of this type are known to be efficient acylating agents (Section 6.2) and in keeping with this the evidence suggests that the final step in the reaction involves acylation of phosphoric acid by the enzyme thiolester.

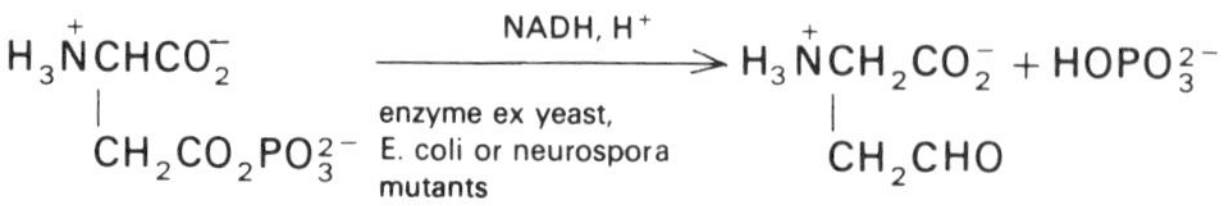

(6.39)

aspartyl-β-phosphate aspartic-β-semialdehyde

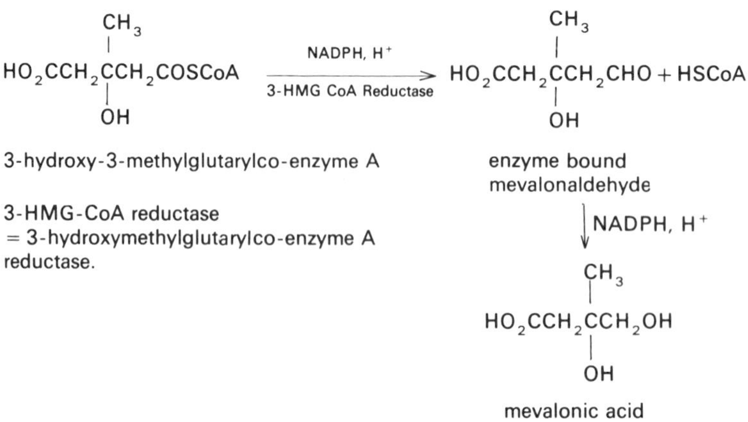

(6.40)

A number of other aldehyde dehydrogenases are known. For example, an enzyme isolated from beef liver will catalyse the oxidation of formaldehyde, acetaldehyde, glyoxal, propanal, butanal, isovaleraldehyde and some aromatic aldehydes. An aldehyde dehydrogenase present in **Clostridium kluyveri** catalyses the formation of a number of esters (RCOSCoA) from the corresponding aliphatic aldehydes and co-enzyme A. Reactions of this type are reversible; two examples where aldehydes are produced from acids are shown in (6.39 and 6.40). Although only one aldehyde dehydrogenase, GPDH, has been purified sufficiently to allow its mechanism of action to be studied in detail the evidence which is available regarding the other enzymes suggests that the sequence hemithioacetal formation, oxidation to a thiol ester followed by acylation of water, phosphate or another acyl acceptor such as co-enzyme A may be a general mechanism for oxidation of aldehydes *in vivo*.

6.2. Carboxylic acids and their derivatives

The structure of the carboxyl group, the functional group which is characteristic of carboxylic acids, has been given in Table 1.1. Acid anhydrides, esters, and amides are formally derived by elimination of water between a carboxylic acid and, respectively, a second molecule of the same or a different acid (organic or inorganic), an alcohol or thiol, or an amine (6.41).

$$\begin{array}{l}R^1\diagdown\\ \quad\ \ C{=}O\\ HO\diagup\end{array} + HX \rightleftharpoons \begin{array}{l}R^1\diagdown\\ \quad\ \ C{=}O\\ X\diagup\end{array} + H_2O \qquad \textbf{(6.41)}$$

$X = R^2CO_2$, $(HO)_2\overset{O}{\overset{\|}{P}}O-$, halogen, OR^2, SR^2, NH_2, NHR^2, NR^2R^3; R^1 = alkyl or aryl.

In mechanistic terms, however, they are substitution products of carboxylic acids in which the hydroxyl group of the acid has been replaced by an acyloxy group or inorganic moiety, an alkoxy, aryloxy, alkylthio or arylthio group, or an amino or substituted amino group (6.3).

Examples and Nomenclature

The structures and names of some representative carboxylic acids, anhydrides, esters, amides and derived acyl groups are shown in Table 6.3. The preferred, systematic, way of naming carboxylic acids is to consider them as substituted hydrocarbons where the substituent ($\begin{array}{l}/\!\!/O\\ \setminus H\end{array}$) is denoted by the suffix **oic acid**; the terminal 'e' in the name of the parent hydrocarbon is elided (examples 6.42, 6.43). However, a number of situations arise where the preferred system of nomenclature cannot be used because no name is available for the parent hydrocarbon to which the functional group is attached. In such cases the complete carboxyl group is considered to be the functional group and the structure to which it is attached is the parent hydrocarbon. Cyclohexane carboxylic acid (6.45) illustrates such a case. Carboxyl groups take precedence over all other functional groups, except 'onium' ions, in order of citation; the carbon atom to which the functional group is attached is always allocated locant number one, irrespective of the positions of other functional groups present. Compound (6.47) is named as a 4-substituted acid. Note that an older method of locating substituents in such compounds is still in use in which the carbon atoms **adjacent** to the carboxyl group are labelled

Table 6.3 Structures and nomenclature of some representative carboxylic acids and derivatives

Carboxylic acids	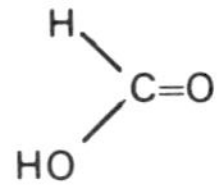	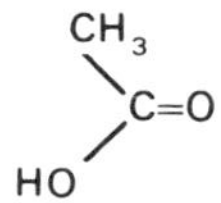	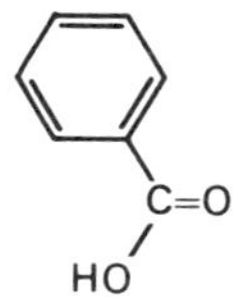	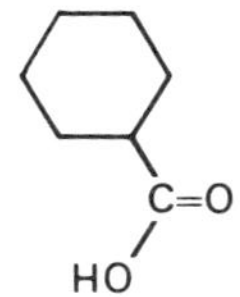	$HO_2C(CH_2)_3CO_2H$	$HOCH_2(CH_2)_2CO_2H$
	methanoic acid (formic acid)	ethanoic acid (acetic acid)	benzoic acid	cyclohexane carboxylic acid	pentanedioic acid (glutaric acid)	4-hydroxybutanoic acid (γ-hydroxybutyric acid)
	(**6.42**)	(**6.43**)	(**6.44**)	(**6.45**)	(**6.46**)	(**6.47**)
Acyl groups	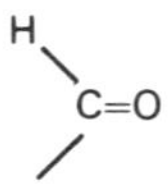	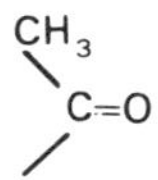	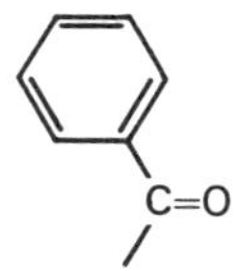	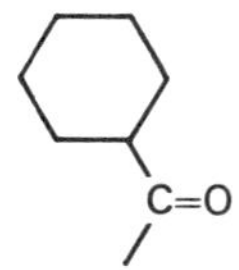	$HO_2C(CH_2)_3CO-$	$HOCH_2(CH_2)_2CO-$
	methanoyl (formyl)	ethanoyl (acetyl)	benzoyl	cyclohexylcarbonyl	glutaryl	4-hydroxybutanoyl (γ-hydroxybutyryl)
	(**6.48**)	(**6.49**)	(**6.50**)	(**6.51**)	(**6.52**)	(**6.53**)
Anhydrides		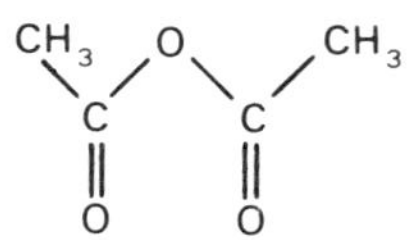	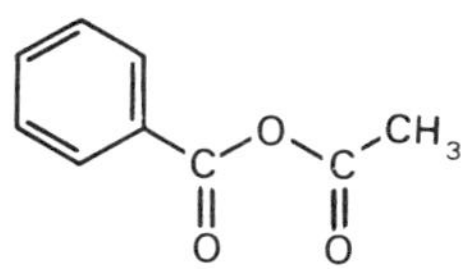		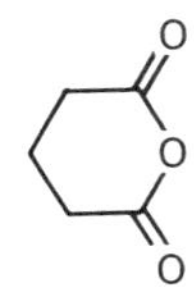	
		ethanoic anhydride (acetic anhydride)	*O*-acetyl benzoic acid (acetic benzoic acid anhydride)		pentanedioic acid anhydride (glutaric anhydride)	
		(**6.54**)	(**6.55**)		(**6.56**)	

Esters	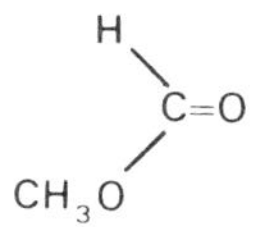	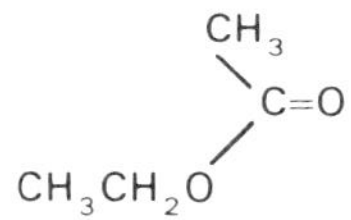	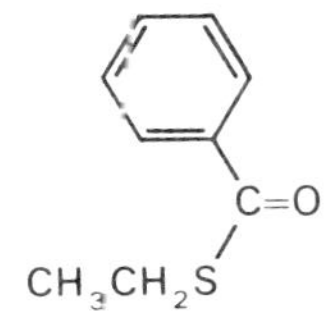	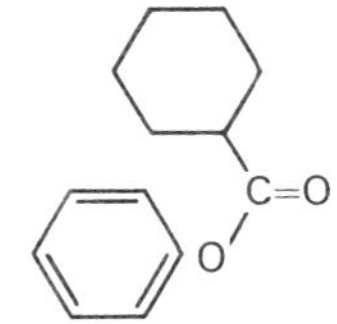	$CH_3O_2C(CH_2)_3CO_2CH_3$	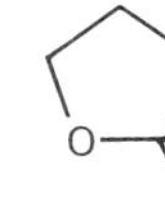
	methyl methanoate (methyl formate) (**6.57**)	ethyl ethanoate (ethyl acetate) (**6.58**)	*S*-etnyl thiobenzoate (**6.59**)	phenyl cyclohexane--carboxylate (**6.60**)	dimethyl pentanedioate (dimethyl glutarate) (**6.61**)	4(or γ)-butyrolactone (**6.62**)
Amides	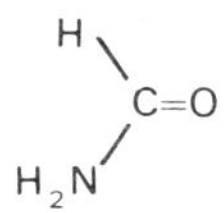	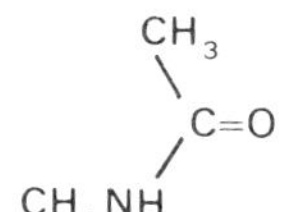	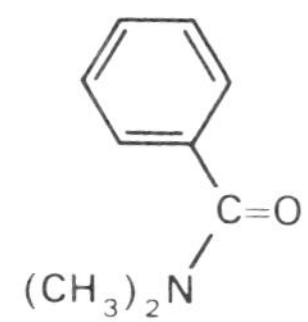	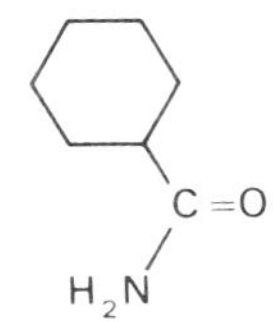	$H_2NOC(CH_2)_3CONH_2$.	
	methanamide (formamide) (**6.63**)	*N*-methylethanamide (*N*-methylacetamide) (**6.64**)	*N,N*-dimethyl--benzamide (**6.65**)	cyclohexane-carboxamide (**6.66**)	pentanediamide (glutaramide) (**6.67**)	

sequentially as α, β, γ, etc.*. If the preferred system of naming carboxylic acids is used then derived acyl groups are named by replacing **oic acid** in the name of the parent acid by **oyl**. Unfortunately, trivial or semi-systematic names are still retained for many acids and acyl groups.

Symmetrical anhydrides are named simply as acid anhydrides (6.54 and 6.56) and unsymmetrical anhydrides (6.55) in terms of the component acids or as acyl derivatives of acids.

Esters are named in a manner analogous to that used for neutral salts. The name of the alkyl or aryl group which replaces the cation is cited first followed by the name of the appropriate anion (cf. sodium acetate, $CH_3CO_2^- Na^+$) (6.57 to 6.61). Note that internal esters such as example (6.62) are called lactones.

Finally, amides are named by replacing the terminal 'e' in the name of the parent hydrocarbon by **amide**; **each** substituent on nitrogen, if present, must be located by a locant letter.

Some carboxylic acids and derivatives of biochemical interest

Some carboxylic acids and derivatives of biochemical interest are shown in Table 6.4.

Table 6.4 Some carboxylic acids and derivatives of biochemical interest

Carboxylic acids

(a) Simple acids

HO_2CCO_2H
(**6.68**)
oxalic acid

$HO_2CCH_2CH_2CO_2H$
(**6.69**)
succinic acid

(b) Substituted acids

$CH_3CH(OH)CO_2H$
(**6.70**)
lactic acid

$HO_2CCH_2C(OH)(CO_2H)CH_2CO_2H$
(**6.71**)
citric acid

$HO_2CCH(OH)CH_2CO_2H$
(**6.72**)
malic acid

$HO_2CC(=O)CH_2CO_2H$
(**6.73**)
oxaloacetic acid

Anhydrides

$CH_3COOP(=O)(OH)_2$
(**6.74**)
acetyl phosphate

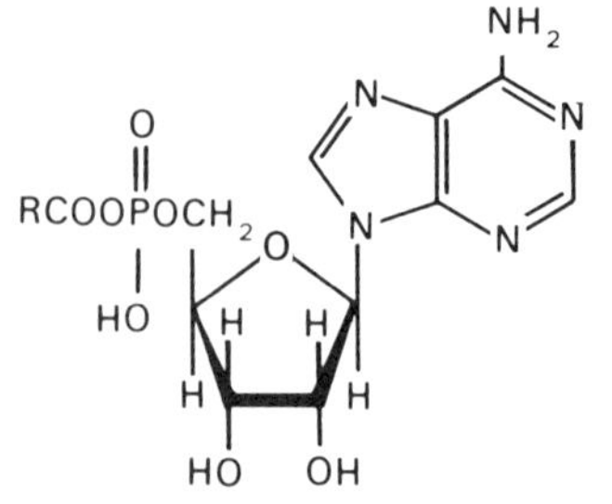

(**6.75**)
an acyl adenylate

* This system is still in common use in biochemical literature, particularly in the field of enzymology. For example, the α-ketoglutarate dehydrogenase enzyme complex is never referred to as the 2-oxopentanedioate dehydrogenase complex!

Table 6. 4 (Contd.)

Esters

(6.76)
a triacylglycerol

$CH_3COSCoA$

(6.77)
acetyl coenzyme A

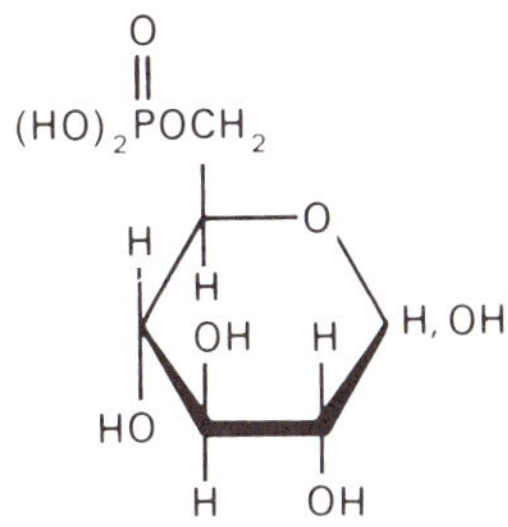

(6.78)
glucose-6-phosphate

Amides

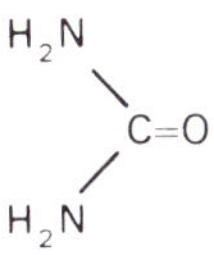

(6.79)
urea

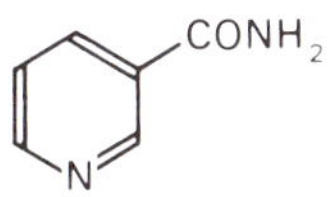

(6.80)
nicotinamide

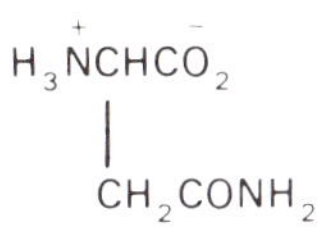

(6.81)
asparagine

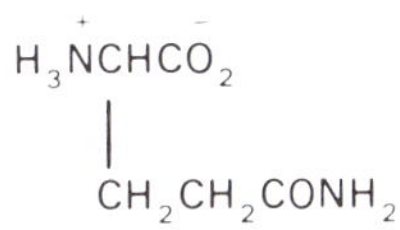

(6.82)
glutamine

$H_2NCH(R^1)CO{-}(NHCH(R)CO)_n{-}NHCH(R^{n+2})CO_2H$

(6.83)
a peptide or protein

Dicarboxylic acids occur in nature more commonly than monocarboxylic acids. Note, however, that acetic acid is produced, in the form of malt vinegar, by the oxidation of ethanol mediated by the bacterium **Mycoderma aceti**. Of the dicarboxylic acids listed in Table 6.4 oxalic acid (6.68) is produced as the potassium hydrogen or calcium salt by members of the **Oxalis** group such as rhubarb, dock and sorrel. Both succinic (6.69) and fumaric acids (4.28) are produced during the oxidation of acetate in living cells. Of the substituted acids listed in the table lactic acid (6.70), together with pyruvic acid (6.6) are products of glycolysis, while citric acid (6.71), malic acid (6.72) and oxaloacetic acid (6.73) are all, like succinic acid, involved in the oxidation of acetate. 2-Amino acids (5.52) are the monomers from which peptides and proteins are constructed and 4-aminobenzoic acid (5.56) is an essential growth factor for many bacteria, as mentioned previously.

Many hydroxyl, sulphydryl and amino groups are acylated by acid anhydrides in living cells. Acetylphosphate (6.74), is produced by a number of bacteria as an intermediate in the biosynthesis of acetylcoenzyme A and acyl adenylates (anhydrides formed from a carboxylic acid and adenylic acid, 6.75) act as acylating agents in processes ranging from the oxidation of fatty acids to protein synthesis.

Esters are encountered frequently in living systems. Fats, which are one of the most important storage substances of plant and animal cells, are mixtures of esters (6.76) derived from the trihydric alcohol glycerol and three molecules of the same, or different, long chain carboxylic acids (fatty acids). Acetyl coenzyme A (6.77) is another very important ester (more precisely a thiol ester) encountered in the metabolism of living cells. It is the vehicle by which acetate is oxidised, and it also enters into a number of other important metabolic reactions. Glucose-6-phosphate (6.78) is representative of the many sugar-phosphate esters which are produced during the metabolism of carbohydrates and acetyl choline (5.59) is yet another ester of importance in living cells. Finally, ripe fruits and the essential oils of plants are particularly rich in esters; methyl salicylate (4.48), the principal constituent of 'oil of wintergreen' is an example which has already been given in Table 4.7.

Urea, (6.79) the compound by which most mammals excrete nitrogen, is the di-amide of carbonic acid (or the mono-amide of carbamic acid, H_2NCO_2H). Nicotinamide (6.80) is a component of the important co-enzyme nicotinamide adenine dinucleotide, while asparagine (6.81) and glutamine (6.82) are, respectively, the β- and γ-amides of aspartic and glutamic acid. Finally, it should be noted that peptides and proteins (6.83) contain substituted amide bonds, –CONH– which are known as **peptide** bonds.

Properties of carboxylic acids and their derivatives

Low molecular mass aliphatic carboxylic acids are volatile, pungent liquids which are freely soluble in water and in organic solvents. High molecular mass aliphatic acids, and most aromatic acids exist as crystalline solids at room temperature and are only sparingly soluble in water.

Acylation

It has already been explained (page 88) that reactions of carboxylic acids and their derivatives typically involve nucleophilic substitution at the carbonyl group. These reactions introduce an acyl group into the nucleophile; carboxylic acids and their derivatives are **acylating agents**, (6.84). Scheme (6.85) summarises the effectiveness of the compounds under discussion as acylating agents. The decrease in the activity of the acylating agents as the table is descended reflects decreasing inductive polarisation in the C–X bond and increasing ability of X to release charge towards the carbonyl group by the mesomeric effect. Both factors serve to decrease the electrophilic character of the carboxyl carbon atom and make it less susceptible to nucleophilic attack. Additionally, the order reflects decreasing stability of the leaving groups $\bar{Cl}$, $RC\bar{O}_2$, $H\bar{O}$, $R\bar{O}$, $H^2\bar{N}$ and O^{2-}.

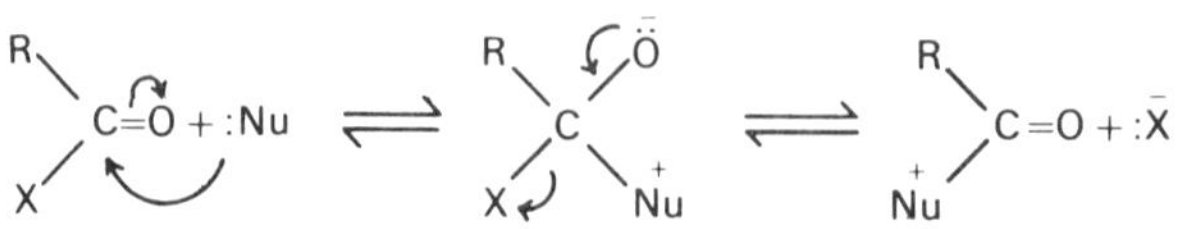

(6.84)

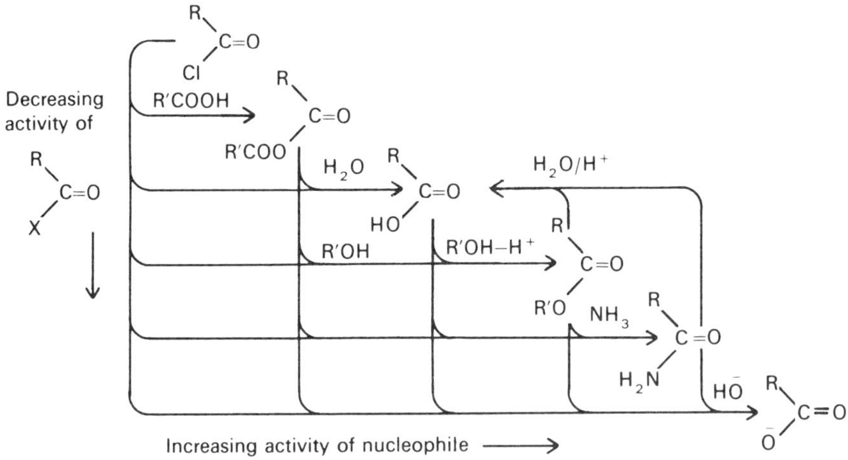

(**6.85**) Carboxylic Acids and their derivatives as acylating agents

Reproduced with permission. Hendrickson, Cram and Hammond: Organic Chemistry, Third Edition, 1970, McGraw-Hill

Acid halides and anhydrides are very effective acylating agents and are commonly used in the laboratory to acylate both weak and strong nucleophiles (5.84, 5.85, 6.86, 6.87).

$$\underset{\text{benzoyl chloride}}{PhCOCl} + \underset{\text{aniline}}{H_2NPh} \xrightarrow[HO^-]{\text{shake with}} \underset{N\text{-phenylbenzamide}}{PhCONHPh} \qquad (\mathbf{6.86})$$

$$\underset{\text{ethanoic acid anhydride}}{CH_3COOCOCH_3} + \underset{\text{excess}}{CH_3CH_2OH} \xrightarrow{\text{warm}} CH_3COOCH_2CH_3 + CH_3CO_2H \qquad (\mathbf{6.87})$$

Carboxylic acids can acylate strong nucleophiles, such as amines directly (6.88) but will react with weak nucleophiles, such as alcohols, only in the presence of an acid catalyst. (6.89).

$$\underset{\text{methanoic acid}}{HCO_2H} + \underset{\text{dimethyl aniline}}{HN(CH_3)_2} \xrightarrow{\text{heat}} \underset{N,N\text{-dimethylmethanamide}}{HCON(CH_3)_2} + H_2O \qquad (\mathbf{6.88})$$

$$\underset{\text{butanedioic acid}}{HO_2CCH_2CH_2CO_2H} + 2CH_3OH \xrightarrow{H^+} \underset{\text{dimethyl butanedioate}}{CH_3O_2CCH_2CH_2CO_2CH_3} + 2H_2O \qquad (\mathbf{6.89})$$

The catalyst increases the polarity in the carbonyl group making the carbon atom considerably more electrophilic (6.90).

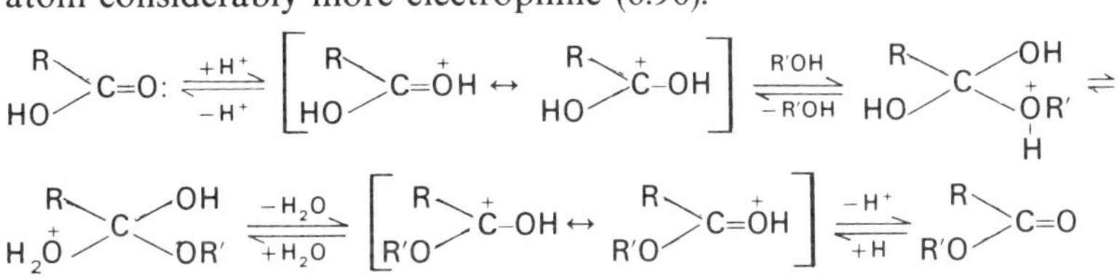

(**6.90**)

It is, perhaps, not generally appreciated that esters are exceptionally useful acylating agents. They will acylate ammonia and amines using quite mild conditions (6.91).

$$\begin{array}{l} CH_2CO_2CH_3 \\ | \\ CH_2CO_2CH_3 \end{array} + 2NH_3 \underset{\text{in aq. } CH_3OH}{\overset{\text{12 hr at R.T.}}{\rightleftharpoons}} \begin{array}{l} CH_2CONH_2 \\ | \\ CH_2CONH_2 \end{array} + 2CH_3OH \qquad \textbf{(6.91)}$$

dimethyl butanedioate — butanediamide

This type of reaction has been developed and used both for amino group protection and for coupling protected amino acids during the synthesis of peptides (6.92 and 6.93).

$$CF_3COSCH_2CH_3 + H_2NCH(R)CO_2^-\overset{+}{Na} \rightleftharpoons CF_3CONHCH(R)CO_2^-\overset{+}{Na} + CH_3CH_2SH$$

S-ethyl thiotrifluoroacetate — 2-amino acid (sodium salt) — *N*-trifluoroacetyl-amino acid (sodium salt) — ethanethiol

(6.92)

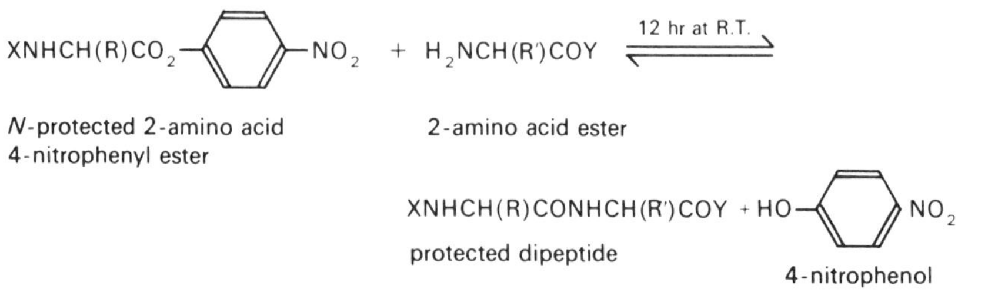

(6.93)

4-Nitrophenyl and similar esters which contain an electron withdrawing group in their alkyloxy or aryloxy moiety are particularly effective acylating agents and are known as 'activated esters'. They owe their increased activity as acylating agents, as compared to normal esters, to the fact that the electron withdrawing group stabilises the leaving group (6.94) and also activates the carboxyl group towards nucleophilic attack to some extent.

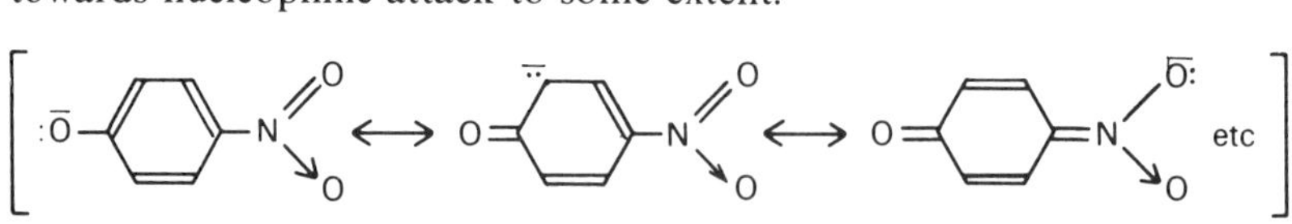

(6.94)

Scheme (6.90) shows that acid catalysed esterification is reversible. Bearing in mind that alcohols can be considered as alkyl substitution products of water it should not be surprising to find that esters undergo acid catalysed transesterification in which an ester acylates an alcohol other than one from which it is derived (6.95).

$$\begin{array}{l} R^1 \\ \quad \diagdown \\ \quad\;\; C{=}O \\ \quad \diagup \\ R^2O \end{array} + HOR^3 \overset{H^+}{\rightleftharpoons} \begin{array}{l} R^1 \\ \quad \diagdown \\ \quad\;\; C{=}O \\ \quad \diagup \\ R^3O \end{array} + HOR^2 \qquad \textbf{(6.95)}$$

This reaction is also subject to base catalysis (6.96) and has been used industrially to manufacture terylene (6.97).

$$B{:} + HOR^3 \rightleftharpoons \overset{+}{B}H + {:}\bar{O}R^3;$$
$$R^1COOR^2 + {:}\bar{O}R^3 \rightleftharpoons R^1COOR^3 + {:}\bar{O}R^2;$$
$${:}\bar{O}R^2 + \overset{+}{B}H \rightleftharpoons HOR^2 + {:}B \quad \textbf{(6.96)}$$

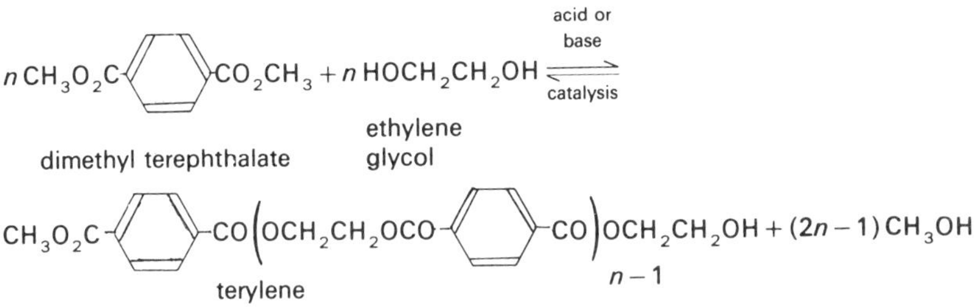

(6.97)

The hydrolysis of esters, catalysed by acid, involves the acylation of water by the ester; the mechanism of the reaction is the reverse of that shown in (6.90) for acid catalysed esterification. Saponification by heating with alkali, followed by conversion of the carboxylate anion which is produced to the free acid also effectively causes hydrolysis of an ester (6.98).

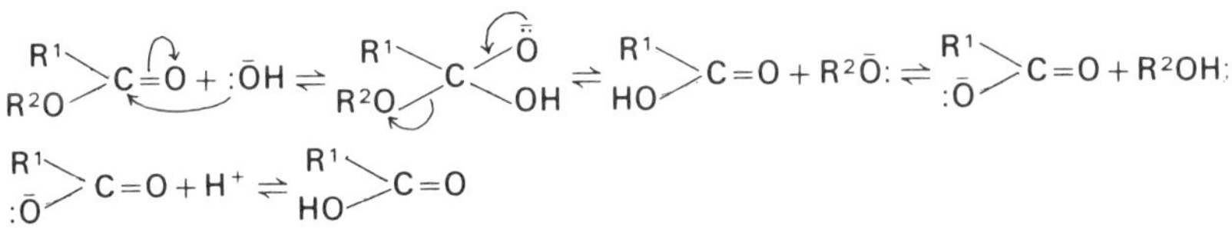
(6.98)

It is interesting to compare the mechanism of saponification with the reaction pathway used by chymotrypsin to catalyse the same overall reaction (Section 8.3).

The only reactions of importance in which amides act as acylating agents are those in which they are hydrolysed, directly by water in the presence of an acid catalyst, or indirectly by the successive action of hot alkali and strong inorganic acid. The mechanisms of the reactions are essentially the same as those of the corresponding reactions of esters. (6.90 and 6.98).

Acylation and related reactions in living cells. Several types of acid anhydride are produced in living cells for use as acylating, sulphating or phosphorylating agents. For example, fatty acids are activated prior to oxidation by conversion to the corresponding esters of coenzyme A; a mixed anhydride of the fatty acid and adenylic acid acylates the thiol (6.99).

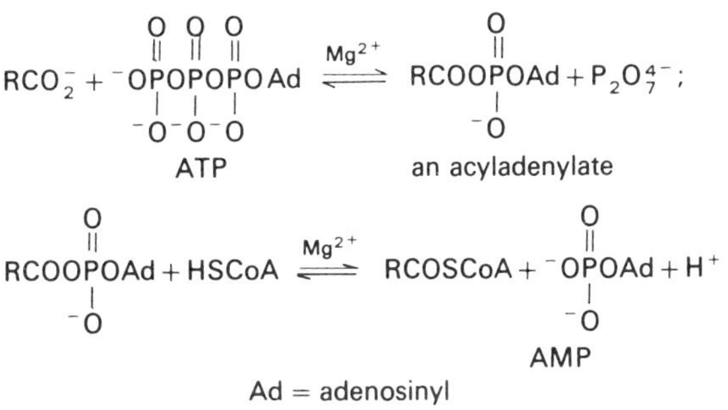

(6.99)

Both steps in the reaction are catalysed by a fatty acid thiokinase. Amino acids are also activated prior to incorporation into peptide chains by conversion to an ester, in this case an ester of the appropriate transfer RNA. The esters are again formed via a mixed anhydride involving adenylic acid (6.100). Both steps in the overall reaction are catalysed by an aminoacyl-tRNA synthetase.

$$H_3\overset{+}{N}CH(R)C\overset{-}{O}_2 + ATP^{4-} \overset{Mg^{2+}}{\rightleftharpoons} H_2NCH(R)COOP\overset{-}{O}_2Ad + P_2O_7^{4-} + H^+$$

an amino-acyladenylate

$$H_2NCH(R)COOPO_2^-Ad + tRNA\text{–}OH \rightleftharpoons H_2NCH(R)COO\text{-}tRNA + AMP^{2-} + H^+$$

Ad = adenosinyl — an amino-acyl tRNA

(6.100)

Similarly many micro-organisms are able to synthesise acetylcoenzyme A from acetate and co-enzyme A using acetyl phosphate as the acylating agent (6.101).

$$CH_3CO_2^- + ATP^{4-} \overset{\text{acetate kinase}}{\rightleftharpoons} CH_3COOPO_3^{2-} + ADP^{3-}$$

$$CH_3COOPO_3^{2-} + HSCoA \overset{\text{phosphotransacetylase}}{\rightleftharpoons} CH_3COSCoA + HOPO_3^{2-} \qquad \textbf{(6.101)}$$

Sulphate esters are produced in living cells as intermediates in the detoxification of several types of compound, particularly phenols and sterols. Sulphate is converted to a mixed sulphuric–phosphoric acid anhydride by reaction with ATP. The anhydride, 3′-phospho-adenosine-5′-phosphosulphate (PAPS) transfers a sulphuryl group (SO_3^{2-}) to a hydroxy group (6.102).

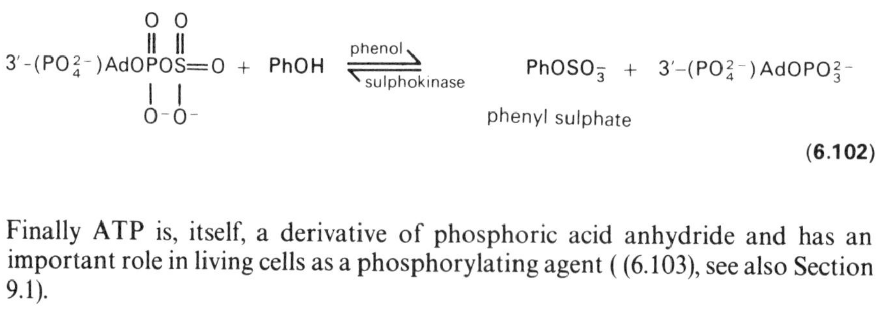

(6.102)

Finally ATP is, itself, a derivative of phosphoric acid anhydride and has an important role in living cells as a phosphorylating agent ((6.103), see also Section 9.1).

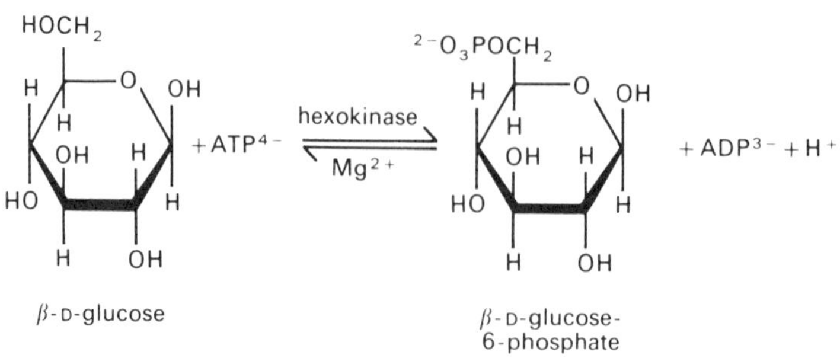

(6.103)

Carboxylic acids do not appear to be used as acylating agents *in vivo*. The logical explanation of this is that, unless they are located in an atypical, specialised environment carboxyl groups will be ionised at physiological pH and that carboxylate anions are not effective acylating agents (6.85).

In addition to anhydrides, esters also function as acylating agents in living cells just as they can in the laboratory. The outstanding example of their use in this capacity is the formation of peptide bonds. Aminoacyl-tRNAs, as explained above, are esters. The synthesis of a protein chain in prokaryotic cells always commences with *N*-formylmethionine-tRNA binding to a ribosome. The aminoacyl-tRNA corresponding to the next amino acid in the protein chain (specified by the appropriate messenger RNA) then binds to the ribosome in a site adjacent to that occupied by *N*-formylmethionine-tRNA. The latter then acylates the amino group of the former under the catalytic action of peptidyl transferase (6.104), producing a ribosome bound *N*-formyldipeptidyl-tRNA.

$$\underset{\displaystyle HCONHCHCOO\text{-}tRNA_{N\text{-}f\text{-}Met}}{\overset{(CH_2)_2SCH_3}{|}} + H_2NCH(R^1)COO\text{-}tRNA_{a.a.^1} \underset{\text{transferase}}{\overset{\text{peptidyl}}{\rightleftharpoons}}$$

$$\underset{\displaystyle HCONHCHCONHCH(R')COO\text{-}tRNA_{a.a.^1}}{\overset{(CH_2)_2SCH_3}{|}} + HO\text{-}tRNA_{N\text{-}f\text{-}Met} \qquad \mathbf{(6.104)}$$

The protein chain is then extended by the *N*-protected dipeptidyl-tRNA and so on. When the synthesis of the protein chain has been completed the *N*-formylmethionine residue is removed and other modifications of the chain may then be made.

In living organisms transesterification appears to be the prerogative of acetylcoenzyme A. For example, acetyl groups are transferred from acetylcoenzyme A to acyl carrier protein (ACP) to form the primer necessary for the biosynthesis of fatty acids (6.105). Esters of co-enzyme A also acylate carnitine during the transfer of fatty acids from the cytoplasm to the mitochondria, where they are oxidised.

$$CH_3COSCoA + HSACP \overset{\text{acetyltransacylase}}{\rightleftharpoons} CH_3COSACP + HSCoA \qquad \mathbf{(6.105)}$$

It is interesting that evolution has produced esters of a thiol, co-enzyme A, as acylating agents in so many biochemical reactions because laboratory studies show that thiol esters are more active in this context than oxygen esters. Evidence is available which shows that a thiol ester is less well stabilised by π electron delocalisation than the corresponding oxygen ester. More specifically, it seems that despite the fact that sulphur is less electronegative than oxygen, diagram (c) in (6.106) has a lower weighting in the description of a thiol ester than the analogous diagram has in the description of an oxygen ester. (Period III elements such as Si, P and S are known to form double bonds less easily than their period 2 cogeners C, N and O, apparently because 3p orbitals cannot overlap efficiently with 2p orbitals.) The carbonyl carbon in a thiol ester is, therefore, more electrophilic than that in an oxygen ester. It has been claimed that an additional factor in determining the charge distribution in a thiol ester is the existence of pπ–dπ

bonding as represented by diagram (d) in (6.106).

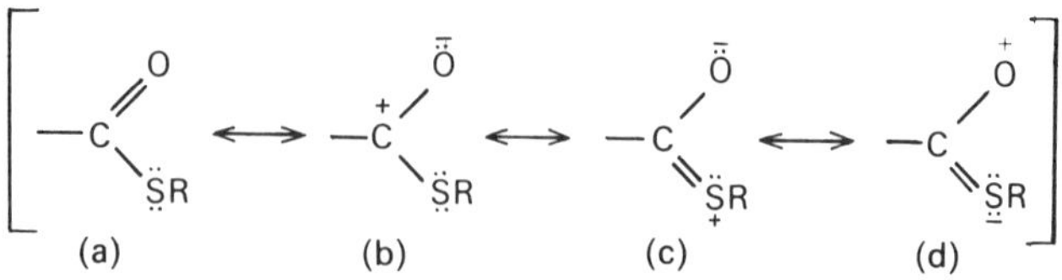

(6.106)

It is claimed that molecular orbital calculations show that this type of bonding will reduce the charge density of the carbon atom. However, the contribution of pπ–dπ bonding to the structure is probably small.

The hydrolysis of esters and amides is the last important biochemical acylation reaction of these compounds which will be considered here. Most cells contain a number of 'non-specific' esterases which catalyse the hydrolysis of a wide range of esters. Other esterases, however, exhibit a well defined specificity. A good example here is nerve cell acetylcholinesterase which catalyses the hydrolysis of the transmitter substance acetylcholine (6.107). The reaction pathway used by this enzyme is probably very similar to that used by chymotrypsin (Section 8.3).

$$CH_3COOCH_2CH_2\overset{+}{N}(CH_3)_3 + H_2O \underset{\text{cholinesterase}}{\overset{\text{acetyl-}}{\rightleftharpoons}} CH_3CO_2H + \underset{\text{choline}}{HOCH_2CH_2\overset{+}{N}(CH_3)_3} \quad \textbf{(6.107)}$$

Living cells also contain a number of phosphatases and sulphatases which catalyse the hydrolysis of phosphate (6.108) and sulphate (6.109) esters respectively.

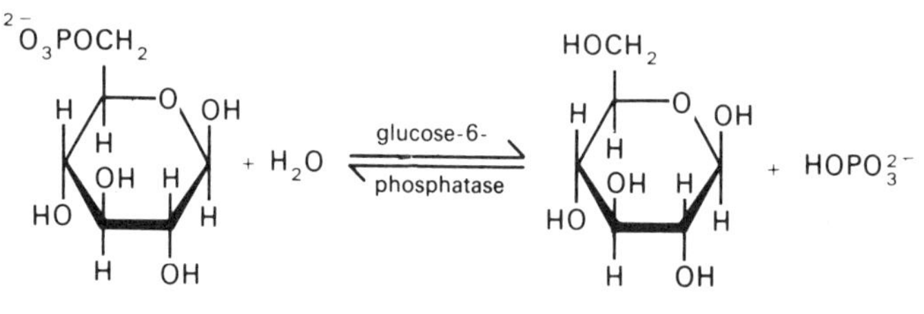

β-D-glucose-6-phosphate β-D-glucose **(6.108)**

$$\underset{\text{phenyl sulphate}}{PhOSO_3^-} + H_2O \overset{\text{m.a.s.}}{\rightleftharpoons} \underset{\text{phenol}}{PhOH} + HSO_4^- \quad \textbf{(6.109)}$$

m.a.s. = microsomal arylsulphatase

Examples of the hydrolysis of both simple and complex amides in living cells are known. Glutamine is thought to be one form in which ammonia (which is toxic to living cells) is stored. When ammonia is required for metabolic purposes it can be liberated by hydrolysis of the γ-amide catalysed by glutaminase (6.110). In man, cells present in the stomach and pancreas produce a number of proteases—pepsin, trypsin, chymotrypsin and carboxypeptidase—which catalyse the hydrolysis of specific peptide bonds in ingested dietary proteins. The mechanisms of all of these enzyme catalysed reactions *in vivo* are undoubtedly closely analogous to those of the corresponding *in vitro* laboratory reactions.

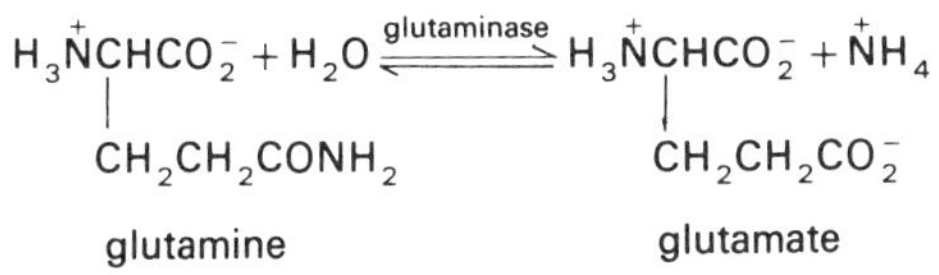

(**6.110**)

Specific properties and reactions of individual types of compound

Carboxylic acids

Acidity. Soluble carboxylic acids dissociate to a small extent in aqueous solution producing relatively stable carboxylate anions in which the negative charge is delocalised equally between the two oxygen atoms (6.111).

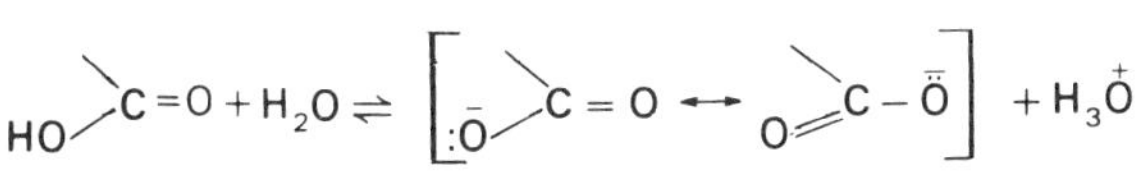

(**6.111**)

Table 6.5 Dissociation constants and pK_a values of some organic acids

Acid	K_a	pK_a
2,2-dimethylpropanoic (pivalic)	9.4×10^{-6}	5.03
acetic	1.76×10^{-5}	4.75
formic	1.77×10^{-1}	3.75
chloroacetic	1.40×10^{-3}	2.85
trichloroacetic	2.0×10^{-1}	0.70
p-hydroxybenzoic	3.3×10^{-5}	4.48
benzoic	6.46×10^{-5}	4.19
p-nitrobenzoic	3.93×10^{-4}	3.41

Substituents in the side chain which assist in the delocalisation of charge promote ionisation. Thus the chloroacetic acids are stronger acids than acetic acid due to the −I effect of chlorine and 2- and 4-nitrobenzoic acids are stronger than benzoic acid due to the −I and −M effects of the nitro group. Conversely, electron donating substituents will decrease the stability of the anion and depress acidity (Table 6.5).

Decarboxylation. Many carboxylic acids and/or carboxylate ions undergo decarboxylation when heated. For example, one of the first reactions detailed in many elementary texts on carbon chemistry describes the formation of methane by heating sodium acetate in the presence of soda lime (6.112).

$$CH_3C\bar{O}_2 + \bar{O}H \rightleftharpoons CH_4 + CO_3^{2-} \qquad (\mathbf{6.112})$$

Decarboxylation reactions are thought to proceed via a carbanion or a carbanion-like intermediate (6.113); thus they are the reverse of carboxylation reactions in which a nucleophile adds to the carbon atom of carbon dioxide.

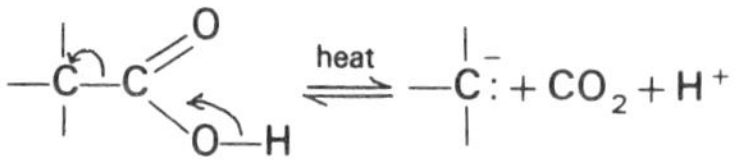

(6.113)

Logically, any structural feature in the side chain of an acid which can stabilise a carbanion by delocalisation of its charge will facilitate decarboxylation. In keeping with this it is found that malonic acid and its derivatives, and 3-oxoacids such as 3-oxobutanoic acid (acetoacetic acid) and its derivatives, can be decarboxylated particularly easily (6.114 and 6.115 respectively).

$$HO_2CCH(R)CO_2H \xrightarrow{heat} RCH_2CO_2H + CO_2 \qquad \textbf{(6.114)}$$

$$CH_3COCH(R)CO_2H \xrightarrow{heat} CH_3COCH_2R + CO_2 \qquad \textbf{(6.115)}$$

In such reactions, it is probable that the 3- (or β-) carbonyl group acts as an electron 'sink' and that the bond forming and breaking processes are facilitated by the favourable geometry of a six membered transition complex (6.116).

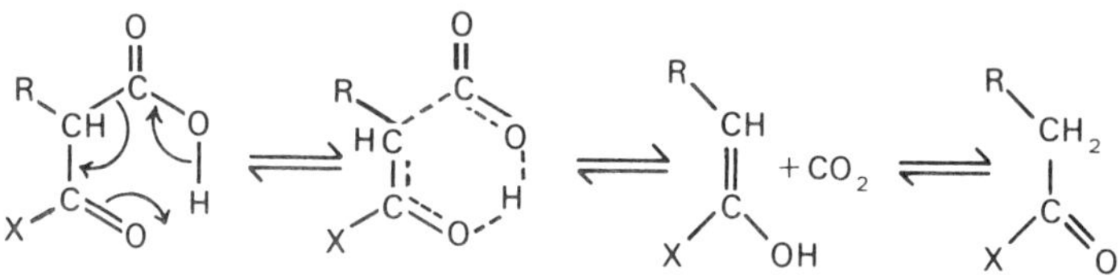

(6.116)

The initial product of the reaction is an enol which immediately rearranges to the keto tautomer. 2- or α-oxoacids can also be decarboxylated; the mechanism of the reactions is probably similar to that described above for 3-oxo acids except that a five membered transition complex is produced. Suitable nitrogen heterocyles can also de-localise negative charge as evidenced by the easy decarboxylation of quinolinic acid (6.117).

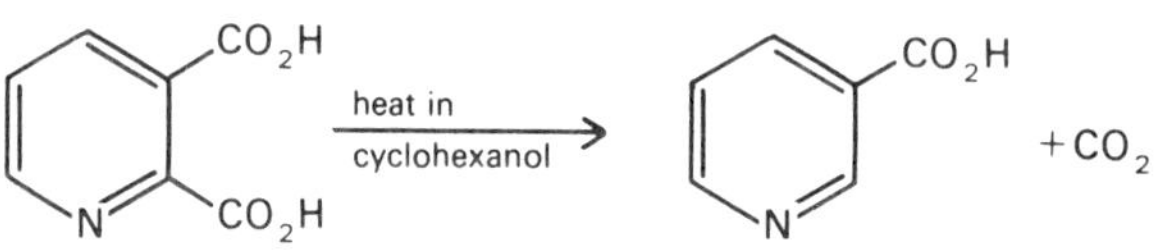

(6.117)

quinolinic acid nicotinic acid

Decarboxylation is the most important reaction by which carbon dioxide is returned from the biosphere to the geosphere. For example when glucose is oxidised in cells, the overall stoichiometry of the reaction demands that six molecules of carbon dioxide should be produced per molecule of glucose (6.118).

$$C_6H_{12}O_6 + 6O_2 \rightarrow 6CO_2 + 6H_2O \qquad \textbf{(6.118)}$$

One molecule of glucose is converted to two molecules of pyruvate by the reactions of glycolysis. One molecule of carbon dioxide is then produced when each molecule of pyruvate (a 2-oxo acid) is converted to acetylcoenzyme A in the presence of CoASH, NAD^+ and the pyruvate dehydrogenase enzyme complex (6.119).

$$CH_3COCO_2^- + \overset{+}{NAD} + HSCoA \underset{}{\overset{PDH}{\rightleftharpoons}} CH_3COSCoA + NADH + CO_2 \qquad (6.119)$$

PDH = pyruvate dehydrogenase complex

A further two molecules of carbon dioxide per molecule of acetylCoA are produced when the acetyl group is oxidised by the reactions of the tricarboxylic acid cycle giving a total of six molecules per molecule of glucose. Specifically, carbon dioxide is liberated when 2-oxoglutarate is produced from isocitrate (6.120) and also when the 2-oxoglutarate is converted to succinylcoenzyme A (6.121).

$$\underset{\text{isocitrate}}{{}^-O_2CCH(OH)\underset{\displaystyle CH_2CO_2^-}{\underset{|}{C}}HCO_2^-} + \overset{+}{NAD} \rightleftharpoons \underset{\text{oxalosuccinate}}{{}^-O_2C\overset{\displaystyle O}{\overset{\|}{C}}\underset{\displaystyle CH_2CO_2}{\underset{|}{C}}HCO_2^-} + NADH + H^+$$

$${}^-O_2C\overset{\displaystyle O}{\overset{\|}{C}}\underset{\displaystyle CH_2CO_2^-}{\underset{|}{C}}HCO_2^- \overset{+H^+}{\rightleftharpoons} \underset{\text{2-oxoglutarate}}{{}^-O_2C\overset{\displaystyle O}{\overset{\|}{C}}CH_2CH_2CO_2^-} + CO_2 \qquad (6.120)$$

$${}^-O_2CCH_2CH_2\overset{\displaystyle O}{\overset{\|}{C}}CO_2^- + \overset{+}{NAD} + HSCoA \overset{TPP}{\rightleftharpoons} \qquad (6.121)$$

$${}^-O_2CCH_2CH_2COSCoA + NADH + CO_2$$

TPP = thiamine pyrophosphate

Both reactions involve 'oxidative decarboxylation'. In reaction (6.120), which is catalysed by isocitrate dehydrogenase, the substrate is first oxidised. The oxidation product, which remains bound to the enzyme, is a 3- (or β-) oxo-acid which then undergoes de-carboxylation. Reaction (6.121), which is catalysed by the 2-oxoglutarate dehydrogenase complex, differs from reaction (6.120) in that de-carboxylation of the substrate, which is a 2- (or α-) oxo-acid, precedes oxidation. The mechanism of this reaction, and that of reaction (6.119) which is essentially the same, is discussed in the chapter dealing with co-enzymes (Section 9.3).

Amino acid metabolism is another field in which many enzyme-catalysed decarboxylations occur. For example a number of important amines, including dopamine (6.122), serotonine (6.123), 4-aminobutyrate (6.124), histamine (6.125)

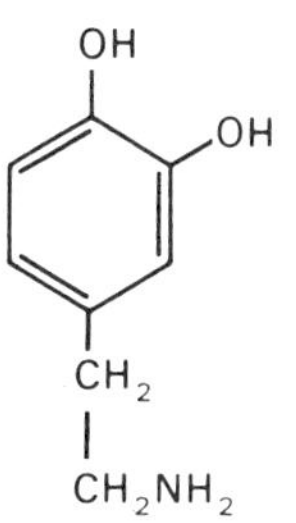

(6.122)

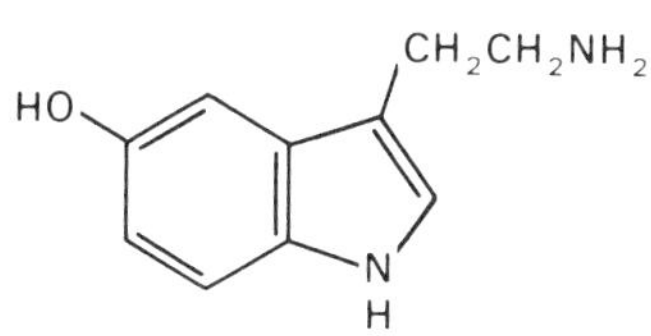

(6.123)

$${}^-O_2CCH_2CH_2CH_2\overset{+}{N}H_3$$

(6.124)

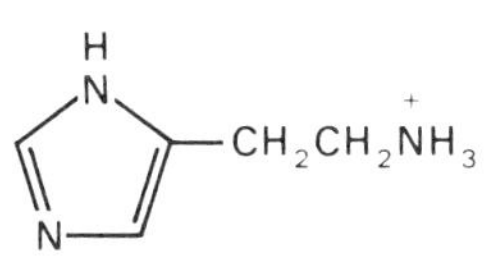

(6.125)

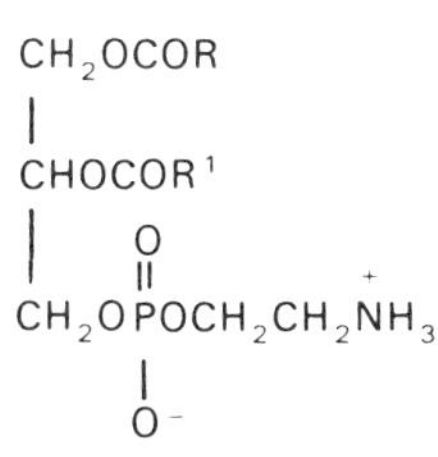

(6.126)

and phosphatidylethanolamine (6.126) all arise by decarboxylation of the corresponding 2-amino acids.

Dopamine, serotonin and 4-aminobutyrate have neurological functions, histamine is associated with allergic reactions and phosphatidylethanolamine is an important component of phospholipids. The enzymes which catalyse decarboxylation of amino acids invariably require the presence of pyridoxal phosphate as co-enzyme, the function of which is outlined in Section 9.3.

Esters

Acidity of α-hydrogen atoms—the Claisen ester condensation. The α-hydrogen atoms of esters are very weakly acidic, one factor leading to dissociation being production of an anion in which the charge can be delocalised effectively (6.127).

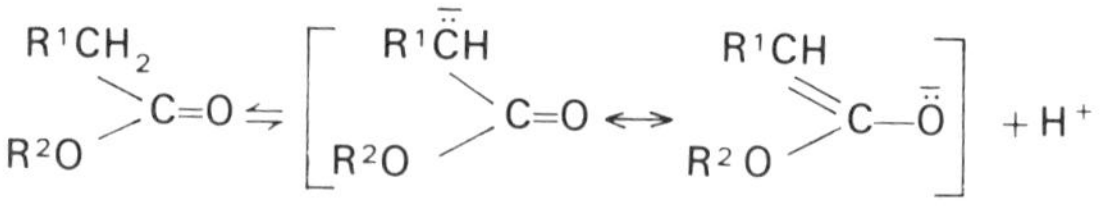

(**6.127**)

Consequently, in the presence of a catalytic quantity of a strong base a significant quantity of the anion will be produced which can act as a nucleophile towards undissociated ester and undergo acylation (6.128 cf. aldol reactions of carbonyl compounds, page 91).

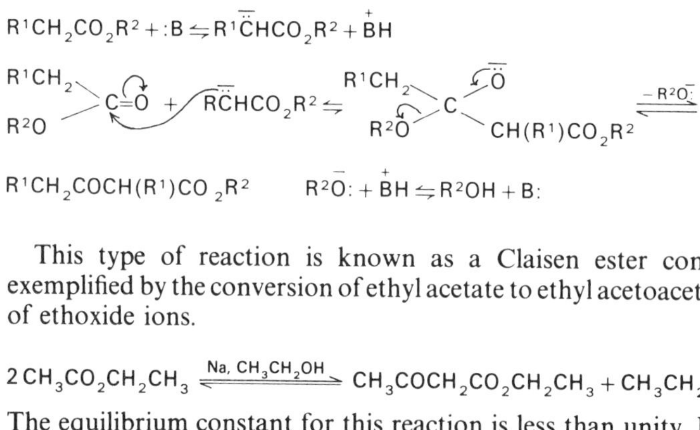

(**6.128**)

This type of reaction is known as a Claisen ester condensation and is exemplified by the conversion of ethyl acetate to ethyl acetoacetate in the presence of ethoxide ions.

$$2\,CH_3CO_2CH_2CH_3 \xrightleftharpoons{Na,\ CH_3CH_2OH} CH_3COCH_2CO_2CH_2CH_3 + CH_3CH_2OH$$

The equilibrium constant for this reaction is less than unity. However, product formation can be promoted by using a molar equivalent of sodium rather than a catalytic quantity. The small amount of ethanol which is added serves to promote the reaction by allowing the formation of a catalytic quantity of ethoxide ion. However, since the reaction produces ethanol, the use of a molar equivalent of sodium ensures that the same quantity of ethoxide ion is effectively present. The product of the condensation, being a 2-oxo-ester, is appreciably acidic; ethoxide ion converts it to the corresponding anion and thus shifts the equilibrium for the overall reaction to the right.

$$CH_3COCH_2CO_2CH_2CH_3 + :\bar{\ddot{O}}CH_2CH_3 \rightleftharpoons CH_3CO\bar{\ddot{C}}HCO_2CH_2CH_3 + HOCH_2CH_3$$

Not only are thiol esters stronger electrophiles than oxygen esters but their α-hydrogen atoms are also more strongly acidic. It appears that because an alkylthio group is less effective than an alkoxy group in electron release towards the carbonyl group (page 107), the latter, when present in a thiol ester, is better able to stabilise a carbanion

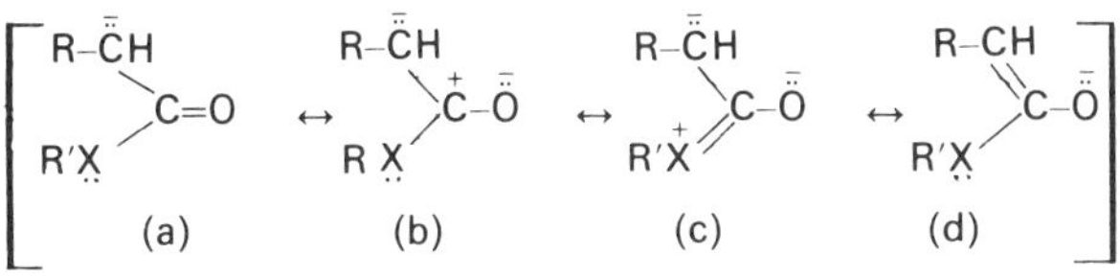

(6.129)

In terms of resonance theory the description of a thiol ester anion would not include a term such as (6.129c) which would de-stabilise the anion. In view of this it appears to be significant that Claisen ester condensations which occur *in vivo* involve esters of a thiol, co-enzyme A. At least one reaction of acetylcoenzyme A appears to be a Claisen ester reaction. This is the last reaction of each cycle of fatty acid oxidation (β-oxidation) in which a molecule of acetylcoenzyme A is formed from a 3-oxoacylcoenzyme A and co-enzyme A itself (6.130).

$$RCH_2COCH_2COSCoA + HSCoA \xrightleftharpoons{\beta\text{-ketothiolase*}} RCH_2COSCoA + CH_3COSCoA \qquad \textbf{(6.130)}$$

As noted previously in relation to laboratory reactions of this type, the equilibrium favours 3-oxoester cleavage. A number of other reactions of acetyl co-enzyme A, while not Claisen ester condensations are similar in that, formally, they could proceed by formation of an anion which then attacks a carbonyl group. One of the most important is undoubtedly that in which citrate is formed from the thiol ester and oxaloacetate under the catalytic influence of citrate synthetase (6.131).

$$^{-}O_2CCOCH_2CO_2^{-} + CH_3COSCoA + H_2O \underset{\text{synthetase}}{\overset{\text{citrate}}{\rightleftharpoons}} {}^{-}O_2CCH_2\underset{\underset{CO_2^{-}}{|}}{\overset{\overset{OH}{|}}{C}}CH_2CO_2^{-} + HSCoA + H^{+} \qquad \textbf{(6.131)}$$

This reaction 'feeds' acetate, derived from carbohydrate or from fat, into the cycle of oxidative reactions known as the tricarboxylic acid cycle. Similarly malonylcoenzyme A, the compound which provides two carbon units for the biosynthesis of fatty acids arises from the reaction of acetylcoenzyme A with carbon dioxide (6.132). This reaction, which is catalysed by acetylcoenzyme A carboxylase in the presence of biotin, possibly involves production of the thiol ester anion which attacks carbon dioxide bound to biotin.

$$CO_2 + CH_3COSCoA \underset{\text{carbozelase biotin}}{\overset{\text{acetyl Co A}}{\rightleftharpoons}} {}^{-}O_2CCH_2COSCoA + H^{+} \qquad \textbf{(6.132)}$$

Amides

Characteristics of the amide group. Amides contain an amino ($-NH_2$) group but, in practice, are not basic as amines are. This is because the 'lone pair' of electrons on the amide nitrogen interacts with the electron deficient carbonyl

* There are, in fact, several forms of β-ketothiolase (often referred to simply as thiolase), each being specific for a different range of acyl group chain lengths.

carbon and is therefore not available for co-ordinating with a proton. The amide group is planar because of the double bond character of the C–N bond, a fact which has important implications for the structure of proteins.

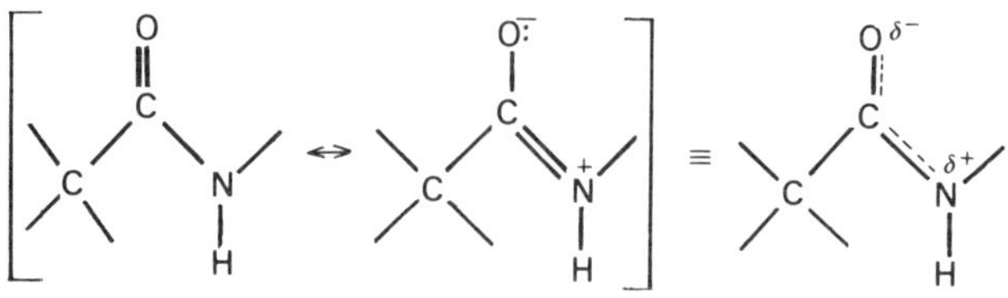

7
Stereochemistry

The branch of chemistry which deals with the properties of molecules in relation to their three-dimensional structure is known as **stereochemistry**. It is a topic of great importance to biochemistry in that most cellular constituents, and the enzymes which catalyse their reactions, possess specific stereochemical properties.

7.1. Optical Isomersion

The area of stereochemistry which is most pertinent to biochemistry is undoubtedly that which deals with optical isomerism. Many organic compounds exist in two or more isomeric forms which can be differentiated by examining the interaction of their solutions with plane polarised light. Before proceeding further it is useful to detail how such interactions are investigated in the laboratory.

Polarimetry

A normal beam of light can be described as an electromagnetic radiation in which the electric vector has every of the possible orientations between 0° and 360° (7.1a).

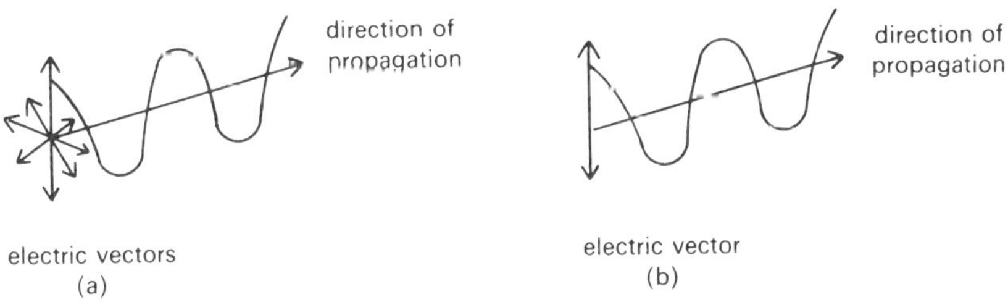

(**7.1**) (a) un-polarised and (b) plane polarised light

If the beam is passed through a polariser, consisting of a suitably cut crystal of a doubly refracting material such as calcite, then the electric vector of the emergent ray is confined to one plane only (7.1b—the polariser can be thought of as acting as a very narrow slit) and is said to be plane polarised. A polarimeter (7.2) consists of a light source, usually a sodium lamp, which illuminates a polariser. The polarised light which is produced passes through a sample tube containing a solution of the material under investigation and then through a second polariser. This second polariser, known as the analyser, is attached to a circular disc calibrated in degrees and can be rotated about the optical axis of the polarimeter. The light emerging from the analyser is viewed through an eyepiece or allowed to impinge on a photo-electric cell.

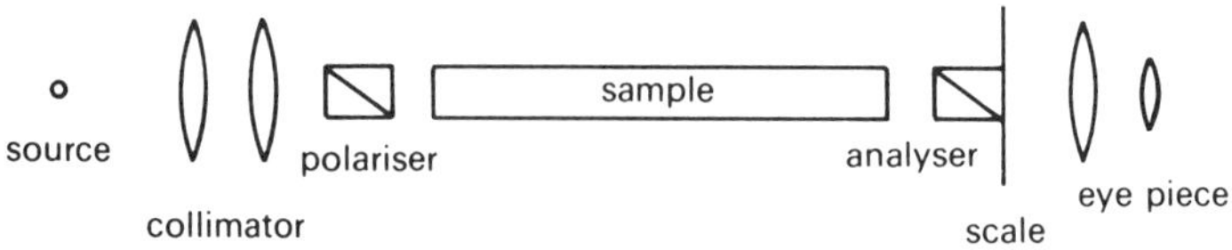

(**7.2**) Diagrammatic representation of a polarimeter

If the sample tube contains only water then the transmitted light will have maximum intensity when the plane of polarisation of both polariser and analyser are co-incident. (i.e. when the hypothetical slits are aligned) and minimum intensity when the plane of polarisation of the analyser is at 90° to that of the polariser. (Intensity maxima are detected electronically or by using a half shadow device.) Solutions of many organic compounds, however, are found to rotate the plane of polarisation of the incident beam clockwise or anticlockwise; such compounds are said to be **optically active**. In such cases the analyser will have to be rotated by the same amount (until its plane of polarisation is co-incident with the plane of polarisation of the light beam impinging on it) for the emergent beam to have maximum intensity. If the analyser has to be rotated clockwise the sample is said to produce a positive (+) rotation and to be dextrorotatory, if it has to be rotated anticlockwise the sample has a negative (−) rotation and is laevorotatory.

Using a specified light source and solvent at constant temperature the angle through which the plane of polarisation is rotated, and hence the angle α through which the analyser must be rotated in order to produce maximum intensity of transmitted light, is proportional to the concentration of solute ($g\,cm^{-3}$) and to the path length (l decimetres) through the solution (7.3).

$$\alpha = \kappa l c \qquad (7.3)$$

The proportionality constant, κ, in equation (7.4) is thus defined as the rotation produced by a solution containing $1\,g\,cm^{-3}$ when contained in a 1 decimetre tube. It is referred to as the specific optical rotation of the solute and is given the symbol $[\alpha]$. Hence (7.3) becomes (7.4) from which (7.5) follows.

$$\alpha = [\alpha]^t_\lambda lc \qquad (7.4)$$

$$[\alpha]^t_\lambda = \frac{\alpha}{cl} \qquad (7.5)$$

In both equations the subscript λ indicates the wavelength of the light source used and t indicates the temperature at which the measurement of α was made. Specific rotations are usually reported as illustrated by (7.6) and are characteristic of the solute when the **specified solvent** is used.

$$[\alpha]^t_\lambda = \pm x^0 (c = y, \text{solvent}) \qquad (7.6)$$

The dependence of $[\alpha]$ on the nature of the solvent used cannot be over-emphasised. Not only can a change in the nature of the solvent change the numerical value of [α] but it can also **change the sign of rotation** in certain cases. The most outstanding examples of this phenomenon involve the amino acids listed in Table 7.1. Additionally a few neutral compounds are known to exhibit different signs of rotation in different solvents. In the case of the amino acids the sign of rotation changes when acid or alkali is added to the solution because a new species is produced (7.7)

Table 7.1. Specific optical rotations of some 2-amino acids

	[α]			
L-amino acid	Solvent 5 M HCl	1 M HCl	Water	1 M NaOH
Asp	33.8		6.7	−2.3
Asn		37.8	−7.4	−2.4
Cys	7.9		−20.0	
His	18.3		−59.8	
Leu	21.0	7.6 (3M HCl)	−14.4	
Met	34.6		−14.9	
Ser		15.9	−7.9	
Trp		5.7	−68.8	6.17 (0.5 M NaOH)

$$H_2NCH(R)CO_2^- \overset{H^+}{\rightleftharpoons} H_3\overset{+}{N}CH(R)CO_2^- \overset{H^+}{\rightleftharpoons} H_3\overset{+}{N}CH(R)CO_2H \qquad (7.7)$$

In the case of neutral compounds a change in the solvent changes the nature of the solute by altering its solvation and hence the rotation also changes. This situation illustrates the necessity for precision in this and other scientific areas. Clearly it is not satisfactory to classify a compound as (+) or (−) unless the solvent in which the rotation was measured is specified. Unfortunately, for reasons of convenience, or neglect, solvents used for measurements of specific rotations are not always quoted. In this text rotations are quoted for aqueous solutions unless stated otherwise.

Compounds containing one chiral centre

Three isomeric forms of lactic acid (2-hydroxypropanoic acid, $CH_3CH(OH)CO_2H$) are known. Dextrorotatory lactic acid, $[\alpha]_D + 3.8°$, m.p. 25–6°C, occurs in muscle tissue, while a laevorotatory form, $[\alpha]_D - 3.9°$, m.p. 25–6°C is produced by the action of **Bacillus acidi laevolactiti** on sucrose. Additionally a third form, known as racemic lactic acid, m.p. 18°C, can be isolated from sour milk. This last isomer is optically inactive. The optically active forms of lactic acid have the same physical properties apart from the **signs** of their specific optical rotations (the numerical values are equal within the limits of experimental error). They also have identical chemical properties providing that the materials with which they react are not themselves optically active and that the reactions are not catalysed by enzymes. In the presence of enzymes, marked differences in reactivity are observed, the difference often being of the 'all or none' type. For instance (+) lactic acid is oxidised to pyruvate by NAD^+ in the presence of lactate dehydrogenase (6.29) whereas (−) lactic acid is not.

What is the structural relationship between the two optically active forms of lactic acid? All the available evidence suggests that they are related as object and non-superimposable mirror image (7.8) and this relationship illustrates the fundamental basis of optical activity—**a molecule will exhibit optical activity if it is not superimposable on its mirror image**. (The reason why molecules related in this way interact differently with plane polarised light is outside the scope of this book.) A molecule need not be asymmetric, that is totally devoid of symmetry, to be non-superimposable on its mirror image. It must not, however, possess a centre or a plane of symmetry or an *n*-fold rotating axis of symmetry. The term **dissymmetric** has been used to describe such molecules but is somewhat artificial. Since a molecule and its non-superimposable mirror image are analogous to right and left hands it is now customary to say that they are **chiral*** or possess **chirality**. The most common, but not the only, structural feature which leads to chirality in a molecule is a carbon atom bearing **four different** substituents. The disposition of the four substituents about a chiral centre constitutes the **configuration** of that centre. Thus, since lactic acid contains only one chiral centre the term can also be applied to its molecule and (7.8a) can be said to describe the configuration of (−) lactic acid and (7.8b) that of (+) lactic acid. (The methods used to decide which of the optically active lactic acids has configuration (7.8a) and which (7.8b) are discussed subsequently.) Note that the two isomers differ in that the position of two substituents (the hydrogen atom and the hydroxyl group) in the chiral centre have been interchanged. In the formal operation of converting (+) lactic acid to (−) lactic acid the configuration of the former is said to be **inverted** or to undergo **inversion**. It is important to realise, however, that in practice the optically active isomers of lactic acid are not easily interconvertible because a covalent bond in the chiral centre would have to be broken and then re-formed.

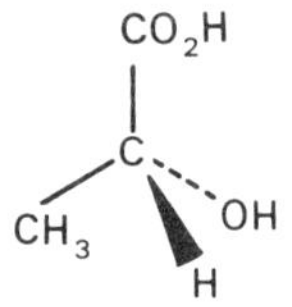

(a) (−) lactic acid

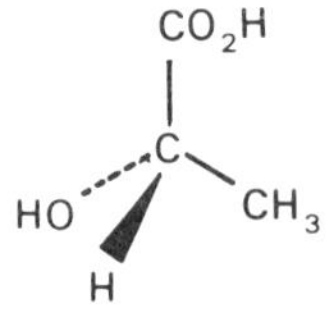

(b) (+) lactic acid

7.8 Configurations of the optically active isomers of lactic acid

The (+) and (−) forms of lactic acid are clearly stereoisomers. More specifically they are **enantiomers**, the latter being the term which is used to describe optical isomers which are related as object and non-superimposable mirror image. As has already been implied by reference to (+) and (−) lactic acid the physical properties of enantiomers are identical except that the signs of their specific rotations (which are equal in magnitude) are different. Additionally, their chemical properties are identical unless their reactions involve optically active reagents or are catalysed by enzymes.

* The word is derived from the Greek cheir, meaning hand, and denotes 'handedness'.

Fischer projection formulae

Before proceeding further it will be beneficial to explain how the configuration of a chiral centre can be depicted. 'Perspective' diagrams such as those in (7.8) can be drawn, but are not always convenient. An alternative method is to use plane diagrams such as in (7.9). These diagrams are known as Fischer projection formulae and are drawn as follows:

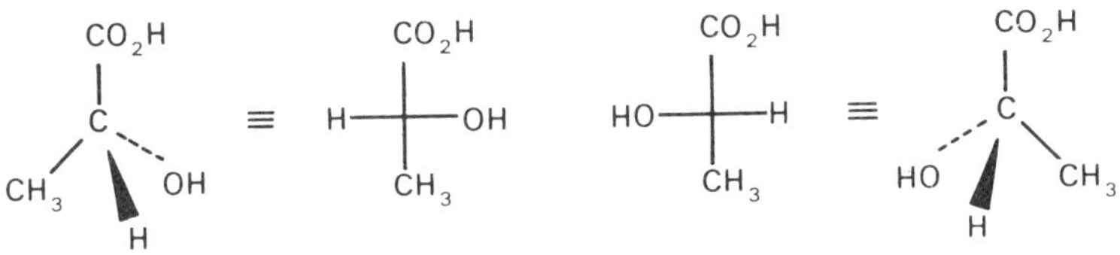

(a) (−) lactic acid

(b) (+) lactic acid

(**7.9**) Fischer projection formulae of the enantiomeric lactic acids

The chiral centre is at the intersection of the vertical and horizontal lines; the former represent bonds projecting behind the plane of the paper, the latter bonds projecting in front of the paper. The longest carbon chain is drawn in the vertical direction **with the most highly oxidised carbon at the top**. The remaining substituents at the chiral centre occupy the 'lateral' positions in the diagram. Again note that the projection formula of (−) lactic acid is converted to that of the (+) acid by interchanging the positions of only **two** substituents. In fact interchange of **any** two of the four substituents produces the same effect (7.10).

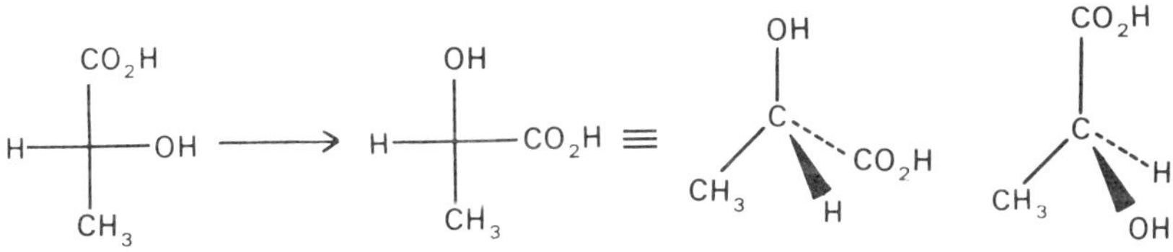

(−) lactic acid

(+) lactic acid

(**7.10**) Interconversion of projection formulae

Readers should satisfy themselves that two consecutive interchanges, each involving any two substituents, will regenerate the original configuration.

Racemic compounds

The third form of lactic acid, which is optically inactive, is known as **racemic** lactic acid. Racemates consist of equal quantities of the optically active isomers and can have one of three physical forms. They may be mechanical mixtures of equimolar quantities of the crystals of each optically active isomer. **Racemic mixtures** of this type always have lower melting points and higher solubilities than the optically active forms. Alternatively a **racemic compound** may be formed which exists in only one crystalline form, the crystals containing equal numbers of molecules of both optically active forms. Apparently equal numbers of left and right-handed objects can be packed into a crystal more efficiently than can all 'right' or all 'left-handed' objects. In keeping with this the melting point of a racemic compound is usually higher and its solubility usually lower than the corresponding values of the optically active forms. Much less commonly a racemate may consist of a solid solution of one optically active form in the other. In such a case the physical properties of the racemate are usually only slightly different from those of the enantiomers.

Other compounds which contain a single chiral centre

In addition to lactic acid many other natural products also exhibit optical activity due to the possession of a single chiral centre. Examples include other 2-hydroxyacids, 3-hydroxyacids and the 2-amino acids.

Compounds containing two, or more, chiral centres

The classic examples of optically active compounds containing two chiral centres are the 2,3,4-trihydroxybutanals, ($HOCH_2CH(OH)CH(OH)CHO$), erythrose and threose. One possible configuration of 2,3,4-trihydroxybutanal corresponds to (7.11a)

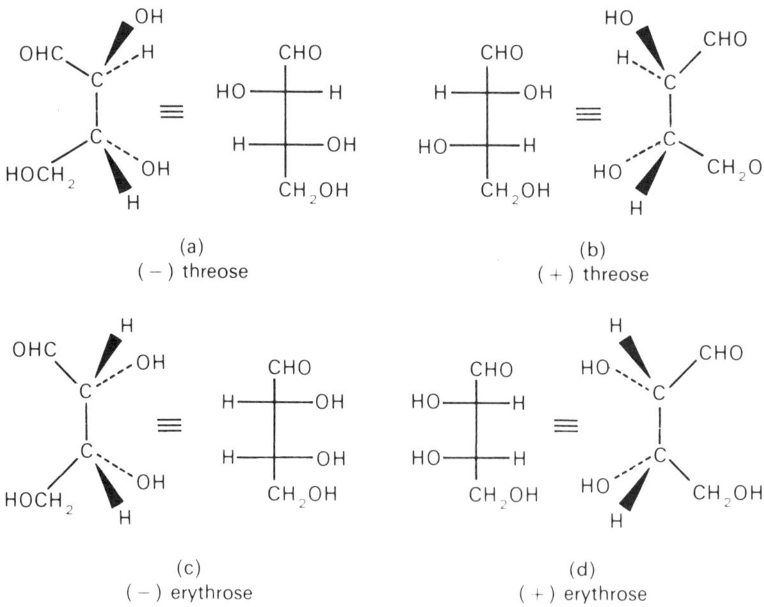

(7.11) Configurations of the optically active tetroses.

By examination of the diagrams, or better, by model building, it can be shown that (7.11a) is not superimposable on its mirror image, (7.11b). The two species are, therefore, optically active enantiomers. It has been shown that (7.11a) represents the configuration of (−) threose and (7.11b) that of (+) threose. Together these enantiomers form racemic, or (±) threose. Inversion of the configuration of C-2 in (7.11a) while retaining that of C-3 gives yet another possible configuration, (7.11c), of 2,3,4-trihydroxybutanal. It is not superimposable on its mirror image and is hence optically active. Compounds (c) and (d) are enantiomers; the former has been shown to represent (−) erythrose and the latter (+) erythrose. (The prefix erythro indicates that like substituents appear on the same side of the projection formula.) A racemic modification, (±) erythrose, of these two enantiomers is known. Neither of the erythroses has the same configuration as either of the threoses; stereoisomers of this type, which are **not** related as enantiomers, are called **diastereoisomers**. Diastereoisomers exhibit different physical properties including different specific rotations and different solubilities, and often exhibit different chemical properties.

In general a molecule which contains *n* different chiral centres is capable of existing in 2^n stereoisomeric forms. There will be 2^{n-1} pairs of enantiomers, each component of a given pair being related to the components of any other pair as a diastereoisomer. Additionally there will be a racemic form corresponding to each pair of enantiomers. This generalisation, however, requires modification if any of the isomers contain elements of symmetry which preclude chirality. The classic example of a compound which exists in less than 2^n isomeric forms due to the fact that one isomer is achiral is tartaric acid. This compound, 2,3-dihydroxybutanedioic acid, ($HO_2CCH(OH)CH(OH)CO_2H$) contains two similar chiral centres. Examination of the structure shows that two enantiomers are possible, both are known. Diagram (7.12a) is known to represent (−) tartaric acid and (7.12b) the enantiomeric (+) tartaric acid. By analogy with the discussion of the threoses and erythroses, above, it might be expected that inversion of the configuration of C-2, while retaining the configuration of C-3, in (7.12a) would formally produce an optically active diastereoisomer. The product of this formal operation (7.12 c), however, contains a plane of symmetry and is

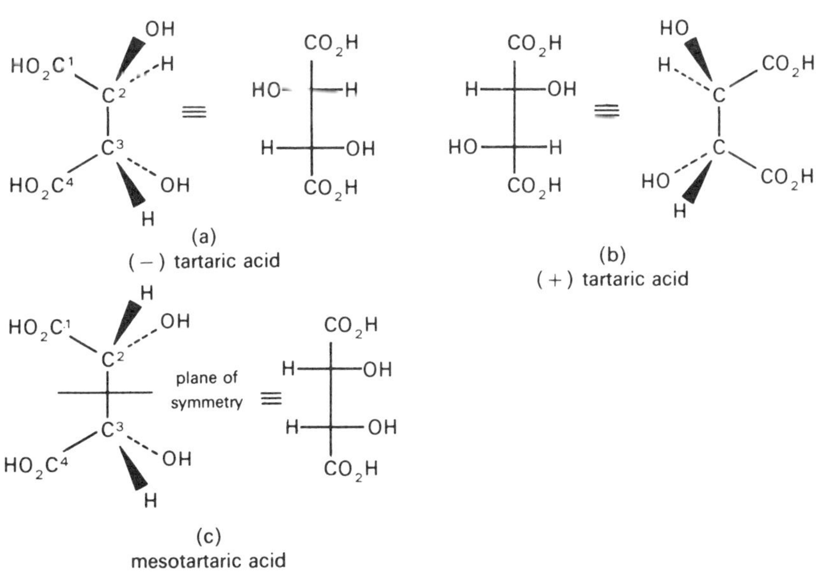

(**7.12**) Configuration of stereoisomeric tartaric acids.

optically inactive. Diagram (7.12c) represents **meso** tartaric acid which is optically inactive due to 'internal compensation', i.e. rotation produced by one chiral centre is exactly cancelled by the rotation produced by the second chiral centre which is the mirror image of the first.

The assignment and description of configuration

Relative configurations

The configurations represented by the diagrams in (7.8) to (7.12) have been assigned to specific stereoisomers. The following paragraphs outline how these assignments have been made and how configurations are described.

In the early years of organic chemistry there was no method available for determining the absolute configuration of a molecule. Configurations were therefore assigned in relation to an arbitrarily chosen standard. Since much of the early work in stereochemistry concerned carbohydrates, it is perhaps not surprising that the standard which was chosen was a simple carbohydrate, (+) glyceraldehyde. It was accepted that the two enantiomeric glyceraldehydes were related as object and non-superimposable mirror image (7.13a and b) but, as already noted, there was no method available at the time to assign the correct configuration to each enantiomer. The (+) isomer was therefore **arbitrarily** assigned configuration (7.13a) which is designated by the symbol D. The configuration of (−) glyceraldehyde is then necessarily that represented by (7.13b); it is designated by the symbol L. The relative configuration of a chiral centre in any other compound can now be deduced if it can be converted to, or derived from, one of the standards by a series of reactions which do not alter the configuration of the chiral centre. For example, none of the reactions shown in (7.14) involve rupture of a bond to the carbon at the chiral centre and all the compounds therefore have the same configuration as (+) glyceraldehyde and are D-compounds. The fact, mentioned previously, that the sign of the specific rotation of certain amino acids changes when the amino acid is protonated or de-protonated, together with the reactions shown in (7.14), indicates that there is **no correlation whatsoever between configuration and sign of rotation**.

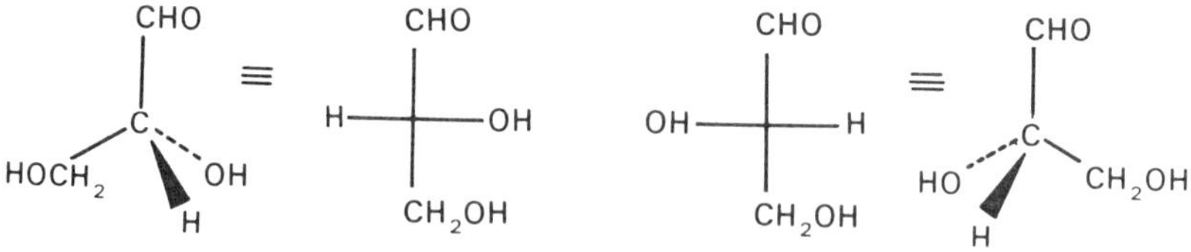

(a) D(+)glyceraldehyde (b) L(−)glyceraldehyde

(7.13) Configurations of the enantiomeric glyceraldehydes

Absolute configurations

In 1957 Bjvoet determined the absolute configuration of (+) sodium rubidium tartrate using the techniques of X-ray crystallography and found it to be the same as that which had been determined by correlation with D(+) glyceraldehyde. This means that the arbitrary choice of configuration (7.8a) for (+) glyceraldehyde had

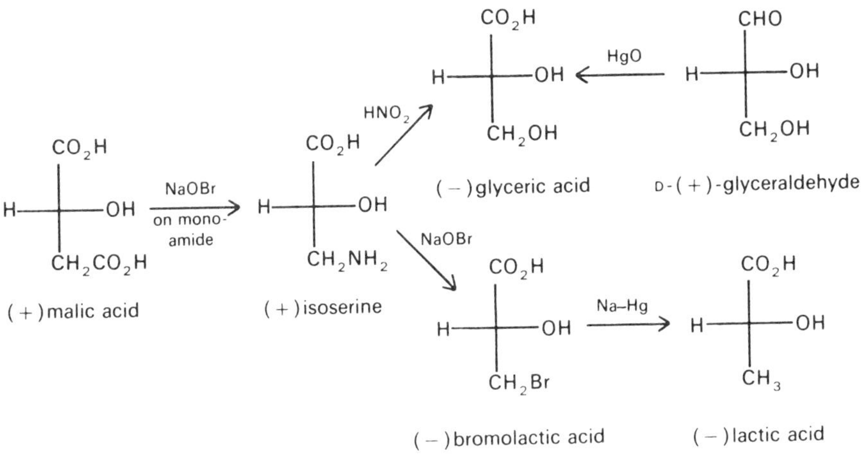

(7.14)

been correct and that all configurations which had been correctly related to it could now be said to be absolute configurations.

The Cahn-Ingold-Prelog (RS) system for describing configurations

The allocation of the D or L configuration to a compound by virtue of it having a generic stereochemical relationship to D (+) or L(−) glyceraldehyde is not entirely satisfactory. There are, fairly obviously, chemical limitations as to what types of compound can be converted to, or derived from, the glyceraldehydes by reactions which do not affect the chiral centre, or which alter it in a known manner. Additionally, ambiguities arise when two or more chiral centres are present in a molecule. For example both (+) and (−) tartaric acid can be related to D(+) glyceraldehyde (7.15); the configuration of each acid can be designated as

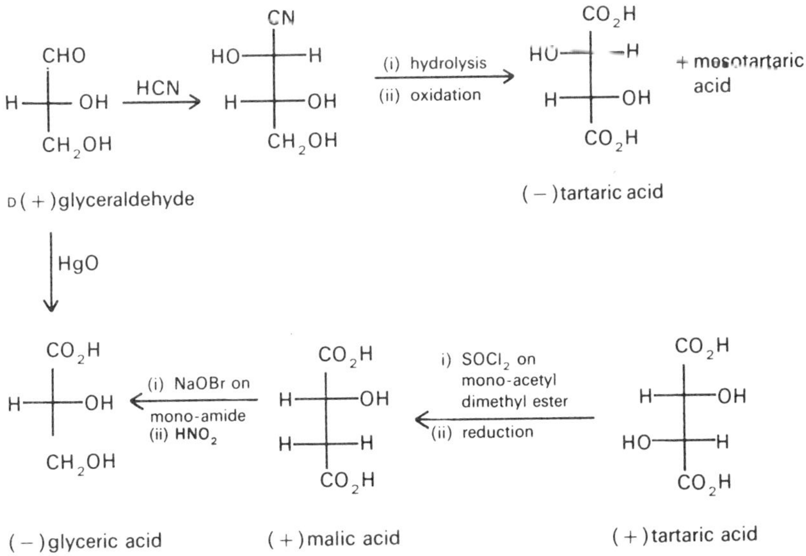

(7.15)

D or L depending on which chiral centre in the molecule is considered! A number of examples can be found in the literature where the configuration of a compound has been assigned on the basis of its projection formula, rather than on the basis of a generic relationship to (+) or (−) glyceraldehyde. An assignment of this type can only be made if the projection formula of the configuration in question has an obvious relationship to that of (+) or (−) glyceraldehyde. Again, a number of limitations are present. For instance it is not always possible to project the configuration of a compound while adhering to the appropriate rules. Thus the configurations of a compound such as 2-phenylbutanoic acid (7.16) can be projected with the most highly oxidised carbon at the top of the diagram, (7.16a), or with the longest carbon chain in the vertical direction, (7.16b), but it cannot be projected in such a way that both rules are obeyed simultaneously.

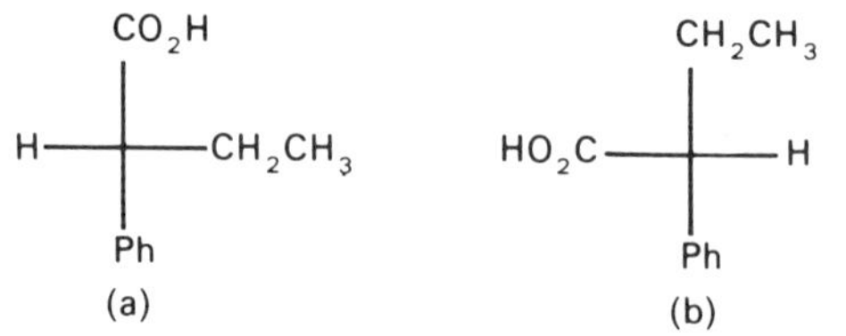

(**7.16**) Projections of 2-phenylbutanoic acid

The configuration cannot, therefore, be assigned and described unambiguously. Due to shortcomings and ambiguities, such as those described above, in making assignments of configuration on the basis of either a generic relationship or on the basis of the projection formula, Cahn, Ingold and Prelog devised an unambiguous, universally applicable, system for describing the configuration of chiral centres.

The configuration of any compound in question must, of course, be first determined experimentally. It can then be described as follows: The substituents at the chiral centre are allocated an order of priority based on decreasing atomic numbers (in the case of isotopes decreasing atomic mass); priority decreases as atomic number decreases. If two similar atoms are attached to the chiral centre, making the allocation of priority impossible on the basis of the atoms attached directly to the chiral centre, then the atomic masses of the atoms once removed, and then of those twice removed, from the chiral centre are considered and so on until a decision can be made. Atoms bound by a double bond count twice. The chiral centre is then viewed from a position **opposite** the substituent of **lowest priority**. If the remaining three substituents trace a **clockwise** order of decreasing priority then the configuration of the centre is **rectus**, signified by R (7.17a). If the

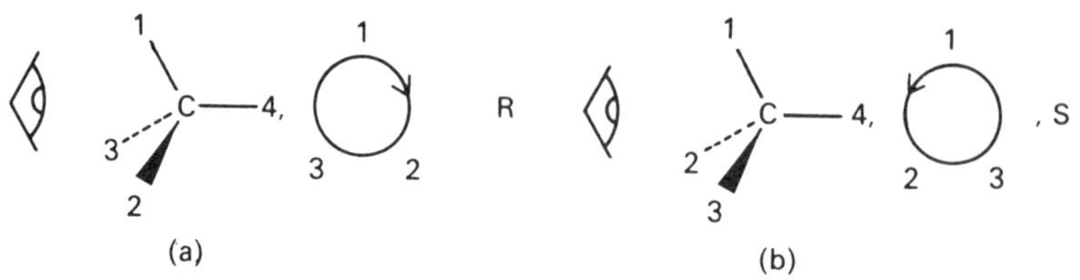

(**7.17**) Definition of R and S configurations.

order traced is **anticlockwise** then the configuration is **sinister**, signified by S (7.17b)

As an example consider (+) glyceraldehyde (7.18, Table 7.2). Structural formulae are drawn for all substituents so that the order of attachment of atoms to the chiral centre is clear; note that both doubly bound atoms in the carbonyl group are duplicated. The four sequences of atomic numbers are then determined and the order of priority is obtained. Note that the carbonyl group has a higher priority than the hydroxymethyl group by virtue of the replicated oxygen atom. When the chiral centre is viewed from a position opposite the substituent of lowest priority (the hydrogen atom in this case) the remaining substituents trace a clockwise order of decreasing priority; the configuration of (+) glyceraldehyde is therefore designated as *R*. Note that projection formulae can be used instead of perspective diagrams in order to determine the spatial sequence of substituents. Remembering that vertical lines represent bonds lying behind the plane of a projection diagram it should be clear that bringing the group with lowest priority to the bottom of the projection formula and then determining whether the remaining three substituents trace a clockwise or anticlockwise order of

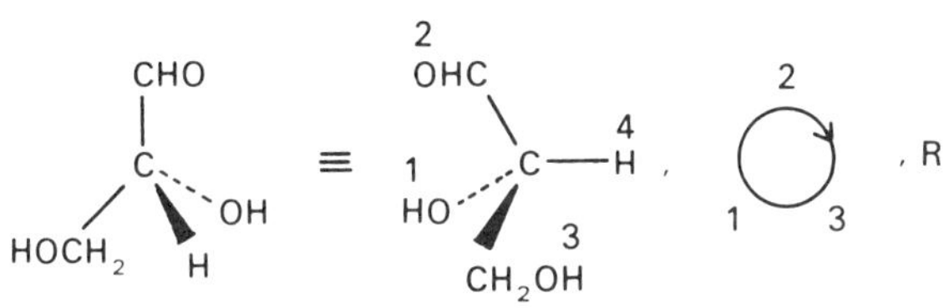

(**7.18**) Assignment of Cahn-Ingold-Prelog configuration for D(+)glyceraldehyde

Table 7.2 Assignment of priorities to substituents in the chiral centre of D(+) glyceraldehyde

Substituent		Atomic numbers of atoms in substituent, working outwards from chiral centre	Priority
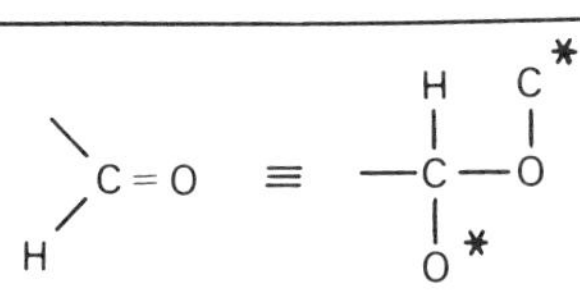		6 (8, 8, 1) 6	2
— CH_2OH,	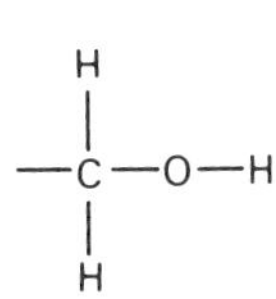	6 (8, 1, 1) 1	3
OH,	—O—H	8 (1)	1
H		1	4

* Doubly bound atoms count twice

decreasing priority is exactly equivalent to viewing the chiral centre from the side facing the substituent with lowest priority. It must be borne in mind that if it is necessary to interchange a pair of substituents in order to bring that with lowest priority to the bottom of a projection formula, then a second interchange must be performed in order to ensure that the diagram represents the same configuration (7.19).

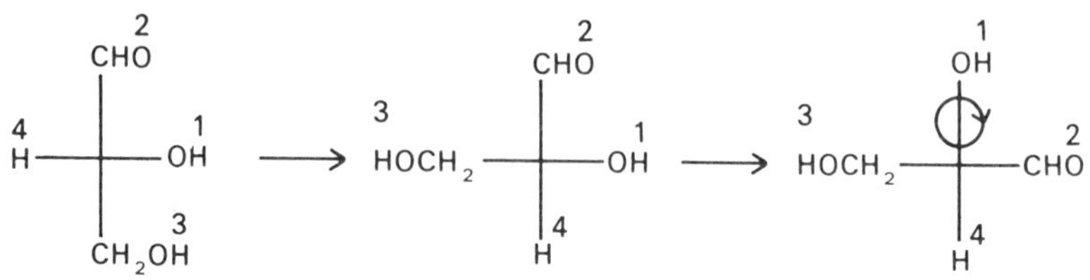

(**7.19**) Projection formulae of R(+) glyceraldehyde

When two or more chiral centres are present in a molecule the configuration of each is specified using an appropriate locant number. Thus (−) threose, (7.11a), is 2(S),3(R) (−)2,3,4-trihydroxybutanal while (−) erythrose, (7.11c), has the 2(R), 3(R) configuration. The R, S system of describing configurations, unlike the earlier D,L system, is universally applicable and is unambiguous. However, a minor disadvantage is that it obscures the generic relationship of the configurations of some compounds. For example, with the exception of glycine, all of the 2-amino acids which can be obtained by the hydrolysis of proteins have the same configuration as (−) serine (7.20 R = $HOCH_2$−).

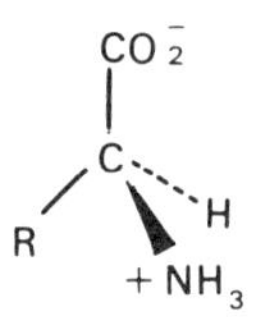

(**7.20**) Configuration of (−) serine

This configuration is known as the $L_{(s)}$ configuration* and is used as the reference configuration for 2-amino acids. The fact that all protein amino acids have the $L_{(s)}$ configuration presumably results from evolution having produced enzymes which can catalyse the incorporation of only acids with that configuration into protein. The R, S system obscures this generic relationship in that although the $L_{(s)}$ configuration of most 2-amino acids leads to their description as S-compounds, a few, for example $L_{(s)}$-cysteine, must be described as R-compounds, due solely to the value of the atomic massess of atoms in their side chains.

* The subscript refers to serine.

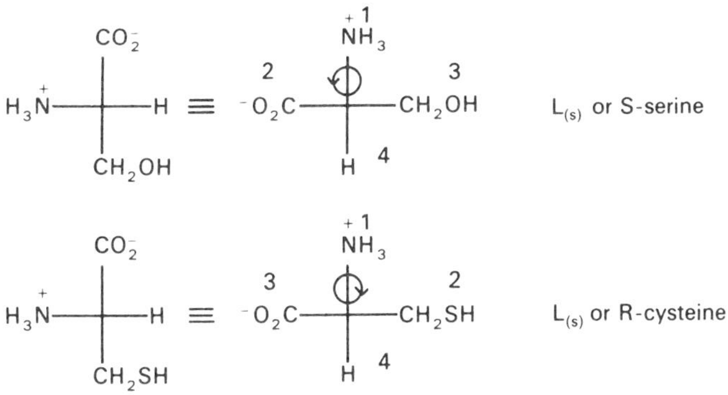

(7.21) Application of the Cahn-Ingold-Prelog rules to L(s)-serine and L(s)-cysteine.

Resolution and racemisation

Resolution

Resolution is the name given to any process by which the racemic form of a compound is separated into its optically active components. Most synthetic procedures, unless they involve an asymmetric reagent or an enzyme, produce racemic materials. Consider, for example, the addition of cyanide ions to ethanal. The carbonyl group of the aldehyde is planar and can be attacked by cyanide on each of the faces with **equal facility**. Attack on one face produces one enantiomer of 2-hydroxypropionitrile (lactonitrile) while attack on the other face produces the other enantiomer (7.22). Resolution is usually the obligatory last step in the laboratory synthesis of optically active natural products.

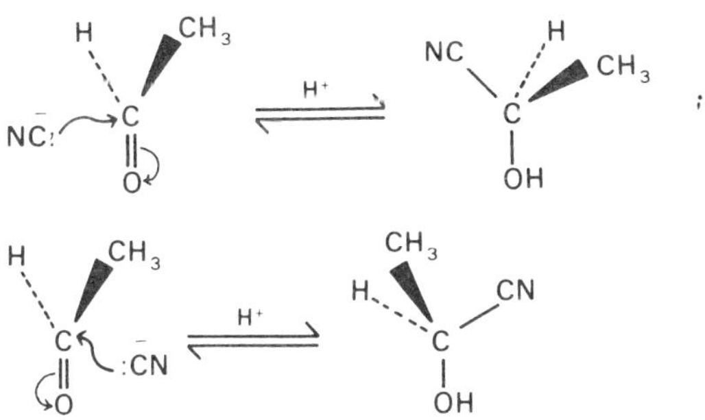

(7.22)

The classical method of resolution involves the conversion of the pair of enantiomers to be resolved into a pair of diastereoisomers which are then separated on the basis of their different physical properties. The method is particularly well suited for the resolution of enantiomeric acids and bases since they can readily be converted to, and subsequently re-generated from, diastereoisomeric salts. For example (±) tartaric acid has been resolved using the alkaloidal base (+) cinchonine (7.23) and synthetic (+) coniine, (the (+) form of the alkaloid occurs in the seeds of the spotted hemlock) using (+) tartaric acid (7.24). Some other optically active bases which have been used for the resolution of racemic acids are (−) ephedrine, (−) strychnine and (−) quinine. (−) Malic acid and (+) *N*- formylglutamic acid, among other compounds, have been used to resolve racemic bases. The choice of resolving agent is dictated by the necessity to produce the maximum difference between the physical properties of the diastereoisomeric salts.

(+) cinchonine + (±) tartaric acid ⟶ (+) cinchonine (−) tartrate + (+) cinchonine (+) tartrate

(i) separate diastereoisomers by fractional crystallisation

(ii) treat each diastereoisomer with HCl

⟶ (+) cinchonine hydrochloride + (−) tartaric acid

⟶ (+) cinchonine hydrochloride + (+) tartaric acid **(7.23)**

(±) coniine + (+) tartaric acid ⟶ (+) coniine (+) tartrate + (−) coniine (+) tartrate

(i) separate diastereoisomers by fractional crystallisation

(ii) treat each diastereoisomer with HCl

⟶ (+) tartaric acid (+) (+) coniine hydrochloride

⟶ (+) tartaric acid + (−) coniine hydrochloride **(7.24)**

Resolution by the formation of diastereoisomers is not restricted to racemic acids or bases. Racemic alcohols can be converted to acidic half esters by reaction with phthalic or succinic anhydride (7.25); the acidic half esters can then be resolved by the formation of diastereoisomeric salts and the alcohols regenerated by hydrolysis.

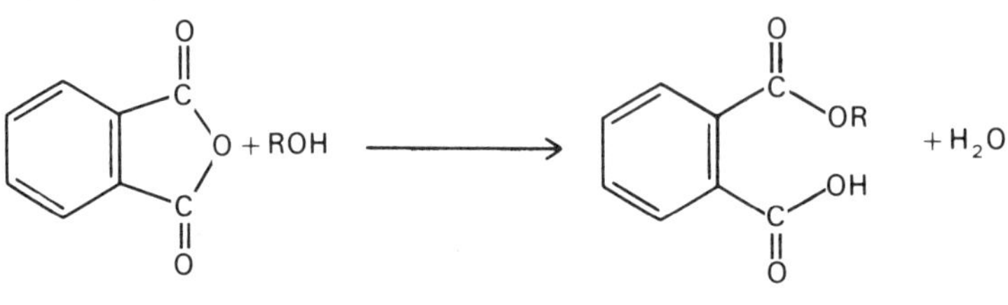

(7.25)

In fact, most racemic modifications, whatever the functional groups they contain, can be converted to an acidic or basic derivative which can then be resolved by formation of a pair of diastereoisomeric salts. The acidic or basic derivative must, of course, be such that it allows regeneration of the original components of the racemic modification in good yield, without racemisation (see next section).

Although simple in principle, resolution by the formation and separation of diastereoisomers is often technically difficult. It is frequently difficult to find a resolving agent which produces diastereoisomers with adequately different properties. Furthermore while it is usually moderately easy to isolate the least soluble diastereoisomer in pure form and in good yield it is much more difficult to isolate the more soluble diastereoisomer in pure form. To do this it is often necessary to resort to repeated fractional recrystallisation, often with changes of solvent, with consequent lowering of the yield.

In many situations only one of a pair of enantiomers is required. It is then often possible to obtain the desired compound from an enzyme catalysed reaction. Enzymes, as will be elaborated on in Sections 8.2 and 8.4 are usually highly stereospecific and will catalyse the synthesis, or degradation, of only one of a pair of enantiomers. Enzyme catalysed reactions of this type are not, strictly speaking, resolutions but nevertheless often produce the same result as required from a resolution. Two examples of asymmetric syntheses catalysed by enzymes are shown in (7.26) and (7.27).

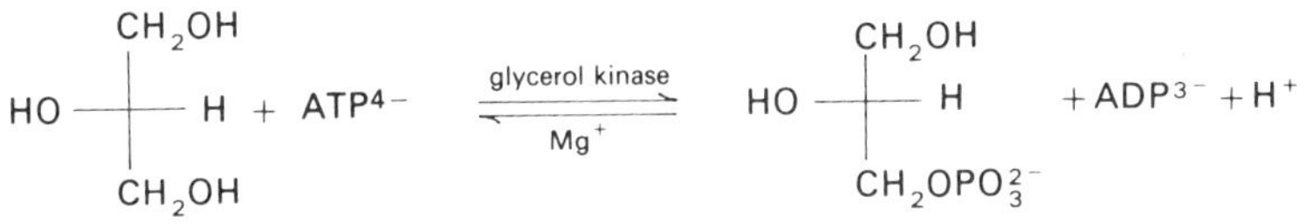

glycerol — R-3-glyceryl phosphate (7.26)

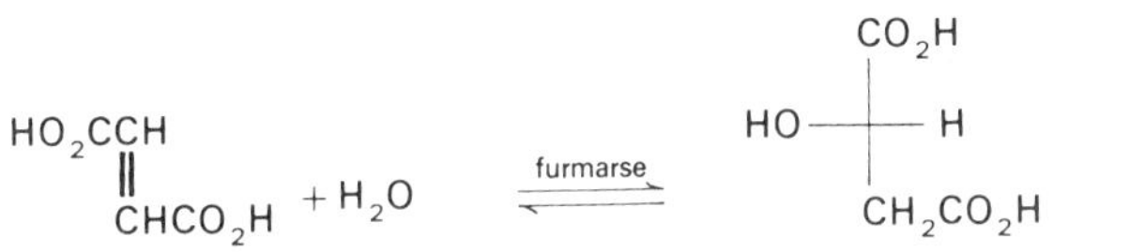

fumaric acid — L- or 2(S)- malic acid (7.27)

Asymmetric degradation has been used extensively for the commercial production of pure L-amino acids. The racemic acid is converted to the *N*-acetyl or *N*-chloroacetyl derivative which is then incubated with acylase I. This enzyme, which can be extracted from pig's kidney, catalyses the de-acylation of only the L-acyl amino acid. The mixture of L-amino-acid and D-acylamino acid which results (7.28) can be separated easily on a column of cation exchanger in the acidic form.

$$(\pm)H_3\overset{+}{N}CH(R^1)C\bar{O}_2 \xrightarrow[\text{(ii) } H_3O^+]{\text{(i) } R^2COCl/H\bar{O}} (\pm)\, R^2CONHCH(R^1)CO_2H \xrightarrow[\text{pH 8.5}]{\text{acylase I}}$$

$$\text{L-}H_3\overset{+}{N}CH(R^1)C\bar{O}_2 + \text{D-}R^2CONHCH(R^1)CO_2H$$

$$(R^2 = CH_3\text{– or } ClCH_2\text{–}) \qquad (7.28)$$

The D-acylamino acid is not retained by the column and can be recovered from the eluate; the acyl group is then removed by hydrolysis yielding the D-amino acid. The L-amino acid can be eluted from the column using a dilute solution of ammonia. As outlined, the procedure can only produce a 50% yield (on a mass basis) of L-amino acid. However, the yield can be almost doubled by racemising the D-acyl amino acid (see page 130) and recycling the racemic modification.

Determination of optical purity

The determination of the optical purity of an enantiomer is a difficult problem. If the properties of the enantiomer in question are already known then its specific optical rotation can be measured and compared with values published in the literature. However, it should be borne in mind that published values of specific optical rotations, like melting points, are merely those experimental values which are accepted by the majority of workers in the field—there are no theoretical standards. If the compound is novel, then it is essential to prepare both enantiomers. If they have numerically equal specific rotations, with opposite signs, then it is probable that both are optically pure. Probably the best method of assessing optical purity is to use an enzyme of known stereo-specificity. For example, it is known that rattlesnake venom amino acid oxidase is completely stereospecific and catalyses the oxidation of only L-amino acids (7.29).

$$H_3\overset{+}{N}CH(R)CO_2^- + H_2O + O_2 \rightleftharpoons RCOCO_2^- + \overset{+}{N}H_4 + H_2O_2 \qquad \textbf{(7.29)}$$

It is claimed that using the enzyme it is possible to detect less than 0.1 % of an L-amino acid in a supposedly pure D-amino acid; the oxygen consumed in the reaction is measured manometrically. A D-amino acid oxidase is present in pig kidney and can be used similarly to detect a D-amino acid in preparations of the L-enantiomer.

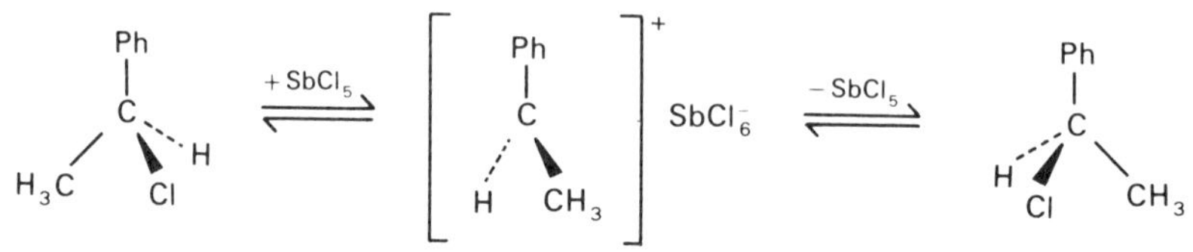

(7.30)

S-1-phenylchloroethane R-1-phenylchloroethane

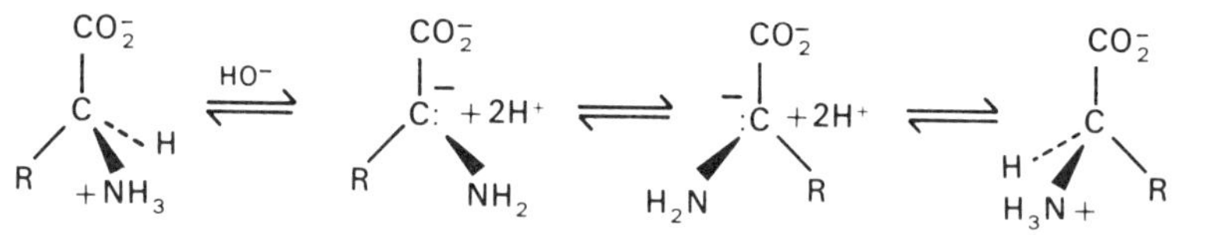

(7.31)

an L-amino acid a D-amino acid

Racemisation

Racemisation involves the conversion of an optically active compound into its racemic modification. As such, it is a process which must be avoided during synthetic procedures which utilise optically active materials. The mechanism by which racemisation occurs varies from case to case. However, all racemisations appear to involve the production of either a planar intermediate which can then undergo further reaction with equal facility on each face (7.30), or of an intermediate whose configuration is easily inverted (7.31).

7.2. The stereochemistry of alkenes and cycloalkanes

Cis-trans diastereoisomerism

In a carbon to carbon double bond the π bond prevents rotation of the carbon atoms relative to each other about the axis of the σ bond and the two carbon atoms and their substituents therefore constitute a planar system. Consequently compounds of the type xyC = Cxy can exist in two stereoisomeric forms. In one isomer like substituents are located on the **same** side of the double bond (7.32a) while in the other they are located on opposite sides of the double bond (7.32 b). Compounds of the former type are said to have *cis* configurations and those of the latter type to have *trans* configurations. In the early stereochemical literature *cis-trans* isomers were classified as geometrical isomers. However, because they are stereoisomers which are not related as enantiomers it is now accepted that it is logical to classify them as diastereoisomers.

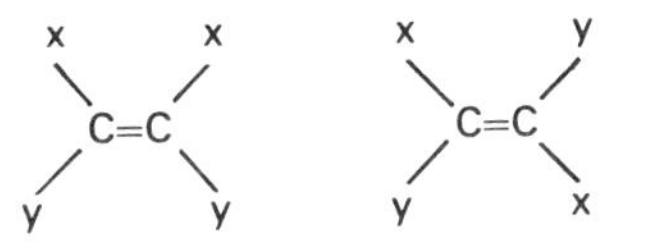

(a) *cis* isomer (b) *trans* isomer

(7.32)

Like other diastereoisomers *cis-trans* isomers have different physical and chemical properties although the differences in the latter may be small unless reactive groups are present. The classic example of *cis-trans* isomerism involves the butenedioic acids. The *cis* isomer, maleic acid, has m.p. 130–131°C, and readily forms an anhydride (7.33).

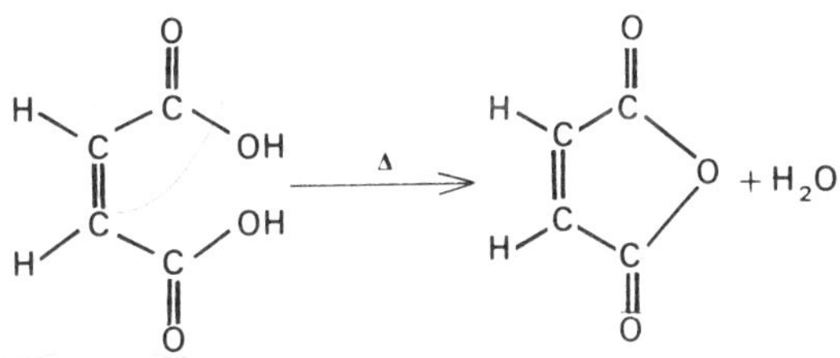

(7.33)

The *trans* isomer, fumaric acid, sublimes at 200°C and does not form an anhydride because the two carboxyl groups cannot approach each other closely enough (7.34).

In general, compounds with the *trans* configuration are more stable than the corresponding *cis* isomers because forces of repulsion between substituents are minimised in the former.

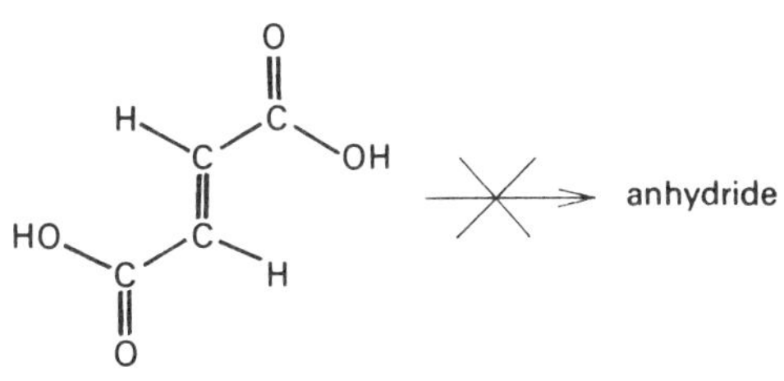

(7.34)

Despite this, both *cis* and *trans* alkenes occur naturally. It has been mentioned already that natural rubber (4.25) is all *cis* polyisoprene whereas gutta-percha (4.26) is the all *trans* compound. The double bond(s) in the majority of unsaturated fatty acids have the *cis* configuration but those present in terpenes such as β-carotene (4.23) and Vitamin A (4.24) have the *trans* configuration.

If two doubly bound carbons possess four different substituents then stereoisomerism is still possible, but the diastereoisomers (7.35) clearly cannot be described using the prefixes *cis* and *trans*. Fortunately, the Cahn-Ingold-Prelog system can be used to describe the configuration of such compounds. The two substituents on each carbon atom are assigned an order of priority based on decreasing atomic number, as described previously. If the substituent with highest priority in each group occurs on the same side of the double bond then the configuration is described as Z (zusammen = together); if they occupy positions on opposite sides of the double bond the configuration is described as E (entgegen = opposite). Two examples are shown below (7.36) and (7.37).

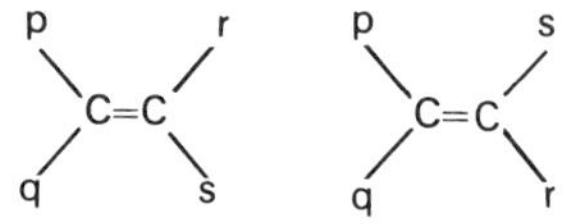

(a) (b) **(7.35)**

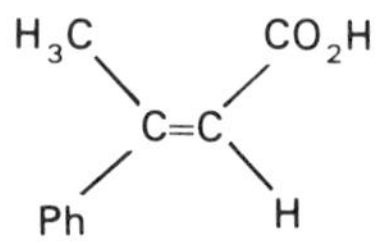

(E)-3-phenylbut-2-enoic acid

(7.36)

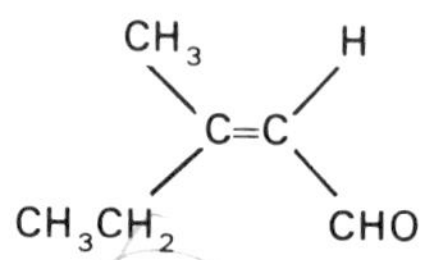

(Z)-3-methylpent-2-enal

(7.37)

Cis and *trans* isomers of alkenes exist as distinct entities because their interconversion by rotation about the carbon to carbon bond axis is prevented. The same type of isomerism is also exhibited by suitably substituted alicyclic compounds for the same reason. Thus any 1,2-disubstituted cyclopropane can, theoreticaly, exist in two diastereoisomeric forms (7.38), which are discrete by virtue of the fact that the ring structure prevents rotation about carbon to carbon bonds.

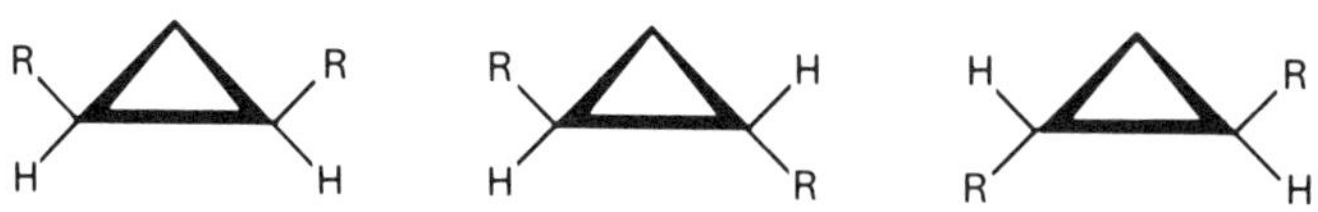

(7.38)

a *cis*-1,2-disubstituted cyclopropane (a)

(b)

enantiomeric *trans*-1,2-disubstituted cyclopropanes (c)

An additional type of isomerism is also possible with the *trans* compound in this case. Structure (7.38b) is non-superimposable on its mirror image (7.38c); these structures therefore represent a pair of optically active enantiomers. (Not all alicyclic *trans* isomers are optically active; each must be inspected to determine whether a chiral centre is present.)

Similar analyses can be made for alicyclic compounds with larger rings. For example, the structures of the isomeric disubstituted cyclobutanes which can theoretically exist are shown below (7.39).

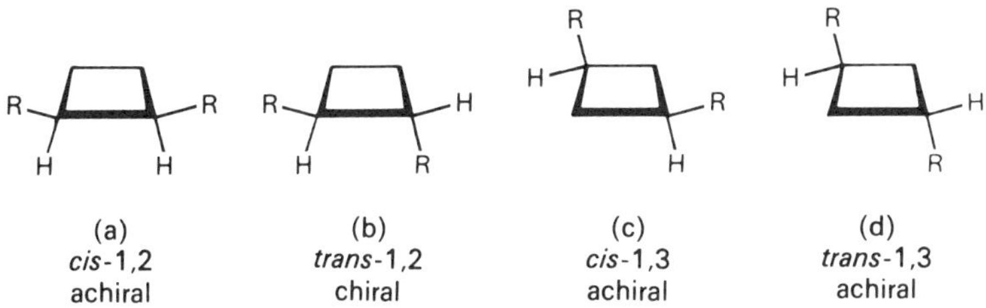

(**7.39**)

Further aspects of the stereochemistry of cyclic systems are discussed in Section 7.3.

7.3. Stereochemical phenomena which result from rotation about single bonds

All the evidence which is available suggests that, in the absence of some restraining factor, two atoms connected by a σ bond may be rotated relative to each other about the bond axis. Stereochemistry has been discussed so far without reference to any effects which result from rotation about single bonds; such effects will now be considered.

Conformational analysis

The shapes of simple molecules, for example pyramidal ammonia, can be rationalised on the basis of the existence of forces of repulsion between electron pairs. Using an extension of this basic idea, namely that substituents repel each

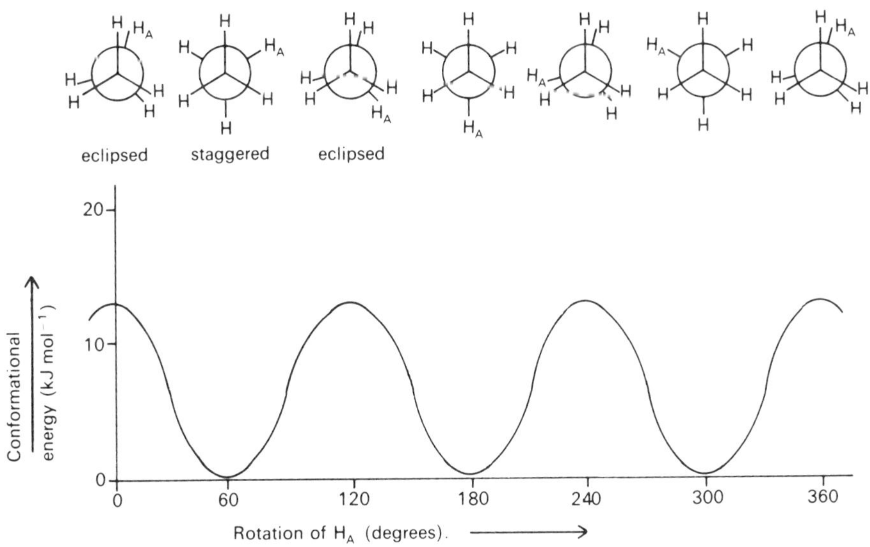

(**7.40**) Conformations of Ethane

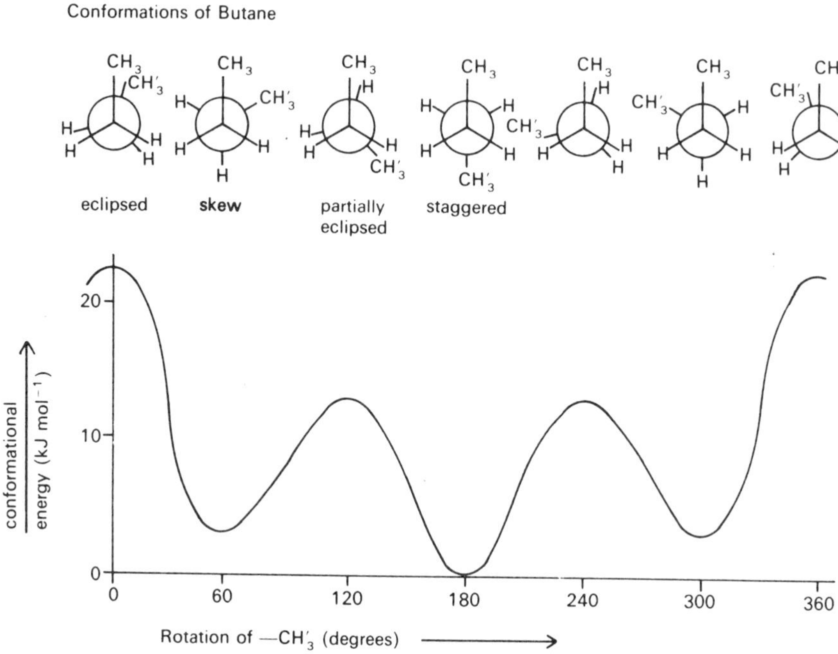

(**7.41**) Conformations of Butane

other because they are effectively 'sheathed' with electronic charge, it is possible to analyse the relative spatial distribution of the atoms in a molecule. Each spatial distribution of the atoms in a molecule resulting from rotation about single bonds is known as a **conformation** of that molecule and their analysis as **conformational analysis**.

Conformations of ethane and butane

The most stable spatial arrangement of the atoms in ethane will be that in which the hydrogen atoms are 'staggered', i.e. are as far away from each other as possible (7.42). (7.42b, known as a Newman projection, is constructed by viewing the molecule along the C–C bond and indicating the bonds to substituents on the carbon atom nearest to the viewer by full lines. The circle then represents the distal carbon atom and the short lines the bonds from it to substituents.) It has been shown that the variation in the energy of the ethane molecule as one carbon is rotated relative to the other about the carbon to carbon

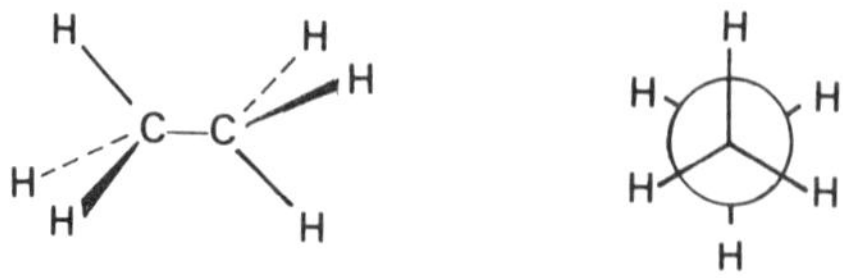

(a) (b) (**7.42**)

bond is as shown in (7.40.). Any point on the energy curve represents a possible **conformation** of ethane. At room temperature the thermal energy of the molecules is in excess of the energy barrier to free rotation and hence molecules with every possilbe conformation will be present in a sample of ethane. It follows that, in contrast to configurational isomers, conformational isomers cannot usually be isolated. The population of each conformational energy level however, will vary; that corresponding to the **staggered** conformation will be greatest and that corresponding to the **eclipsed** conformation the smallest. (7.41) shows the results of a similar analysis of the variation of energy of a butane molecule as rotation occurs about the C-2–C-3 bond. It can be seen that the fully staggered conformation is again the most stable and that the fully eclipsed conformation is the least stable. In addition, the skew and partially eclipsed forms provide, respectively, intermediate minima and maxima in the energy curve. It should now be clear that the most stable conformation of a long chain alkane will be that in which the hydrogen atoms on all adjacent carbon atoms are staggered (7.43). This has been confirmed by the results of X-ray crystallographic studies on solid, high molecular mass alkanes.

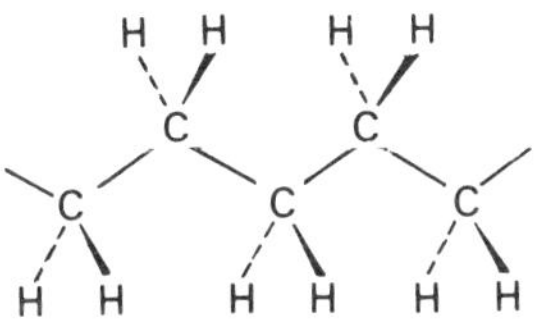

(7.43)

Conformation of cyclo-alkanes

Six membered (pyranose) and five membered (furanose) ring systems occur in carbohydrates. It is, therefore desirable to be familiar with the conformations which these and similar compounds adopt. The saturated hydrocarbons corresponding to pyran and furan are, respectively, cyclohexane and cyclopentane.

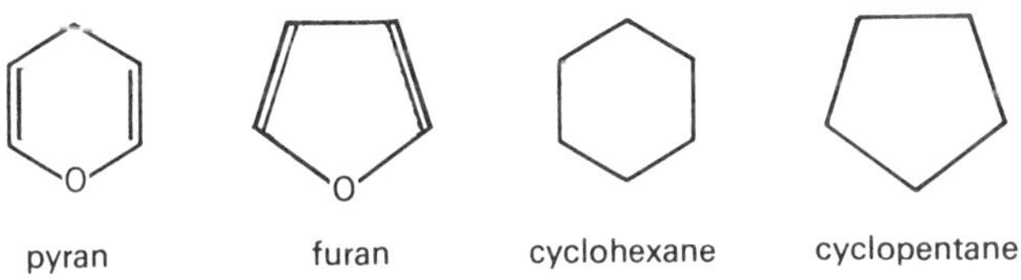

It has been shown that cycloalkanes are not planar and that their hydrogen atoms are not all eclipsed as their two dimensional formulae suggest. Cyclohexane will be most stable if the carbon to carbon bonds can retain their normal angles and if the hydrogen atoms are all staggered. This is possible if the ring is puckered as in structure (a) (7.44). This configuration of cyclohexane is known as the chair configuration.

Structure (a) can be converted to a second chair conformation, (e), since **limited** rotation is possible about the carbon to carbon bonds. In the course of this conversion cyclohexane assumes an infinite number of conformations of which (b), (c) and (d) are notable. Conformations (b) and (d) are known as half chair conformations. In them five of the six carbon atoms are co-planar and substituents on C-2 and C-3 and on C-5 and C-6 are partially eclipsed. Conformation (c) is the

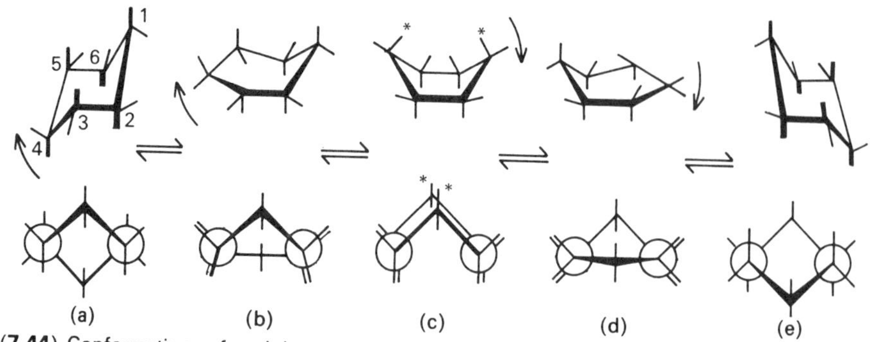

(**7.44**) Conformations of cyclohexane

boat form, in which there is a considerable amount of repulsion between the flag pole hydrogens (*in structure c) which are only 0.183 nm apart. Additionally substituents on C-2 and C-3 and on C-5 and C-6 are completely eclipsed. The force of repulsion between the flag pole hydrogens can be relieved to a limited extent by twisting C-1 and C-4 out of alignment with each other; the conformation which results is known as the 'twist boat' (c in (7.45)). The potential energy relationship between the most important conformations of cyclohexane is shown in (7.45).

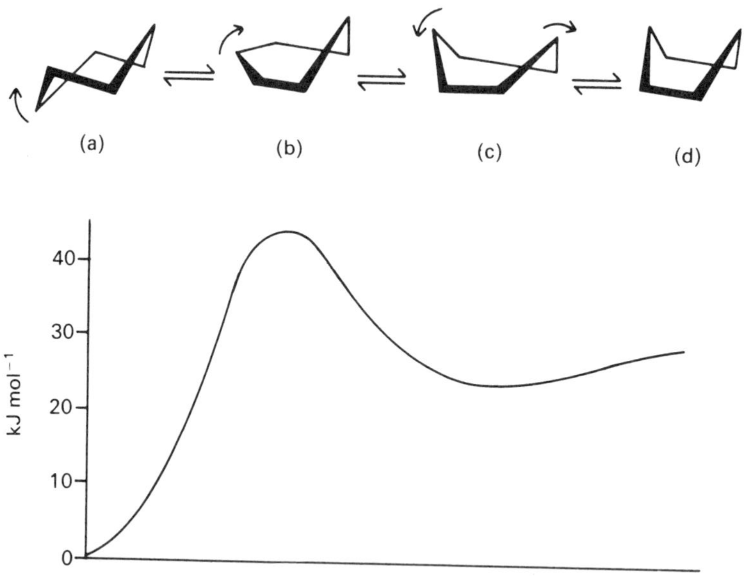

(**7.45**) Energy relationships between conformers of cyclohexane

Axial and equatorial substituents. Notice that the substituents in (7.44a), fall into two groups of six. The members of the first group occupy positions (marked by heavy lines) which approximately make a right angle with the mean plane of the ring – they are known as **axial** substituents. The members of the second group occupy positions lying approximately in the mean plane of the ring and are known as **equatorial** substituents. It appears that the forces of repulsion between two or more large groups attached to a cyclohexane ring are least when they occupy **equatorial** positions. This fact can be used to explain the relative stabilities of many isomeric substituted cycloalkanes and also variations in their reactivity. For example 2-isopropyl-5-methylcyclohexanol could theoretically exist in 8 isomeric forms. However, only two isomers occur naturally, menthol (7.46a), which has all three substituents in equatorial positions and neomenthol (7.46b) which has two equatorial and one axial substituent. An example of how differing reactivity can be explained in terms of conformation is given by the behaviour of neomenthyl chloride (7.47) and menthyl chloride (7.48) when treated with a base. It is known that the atoms or groups which are removed in an E_2 reaction must be '*anti*'* to each other and coplanar. Consequently the former compound is readily dehydrochlorinated whereas the latter reacts only slowly because it must first be converted to the unfavourable conformation in which all three substituents occupy axial positions.

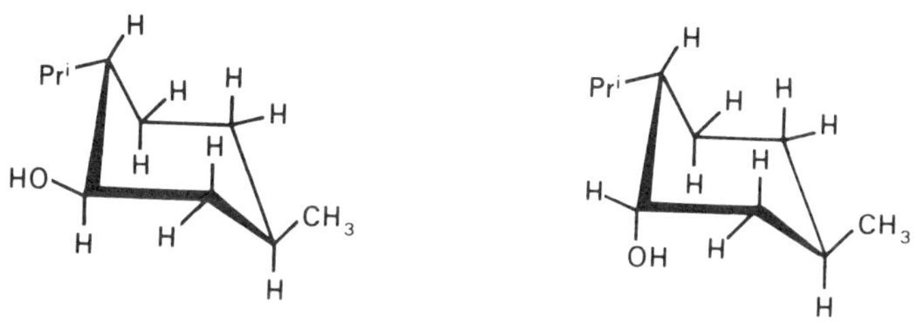

(a) menthol (b) neomenthol (7.46)

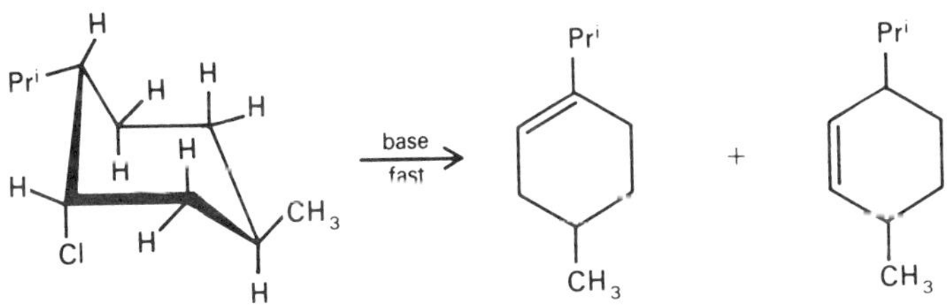

(7.47)

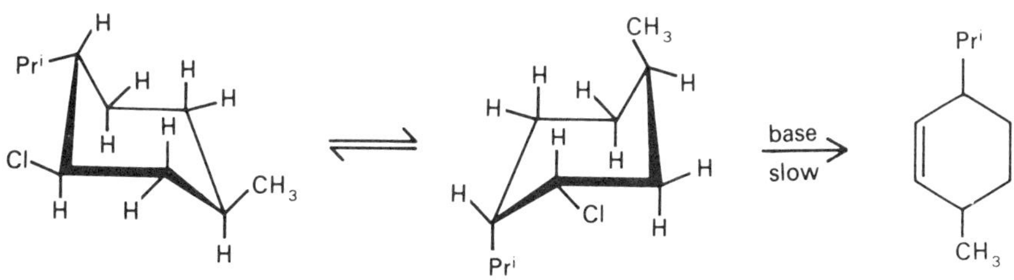

(7.48)

* The prefixes *syn* and *anti*- have meanings similar to those of *cis*- and *trans*, but are used to describe stereochemical relationships between groups during reaction, rather than in a stable molecule.

8

Enzymes

The role of enzymes as biological catalysts was introduced briefly in Chapter 3. A more detailed account of these most important constituents of living cells forms the subject of this chapter.

8.1. Classification and nomenclature of enzymes

Prior to 1961 no systematic classification of enzymes was available and their nomenclature was confused. Although many enzymes had been named in a manner which was, at least in part, indicative of their catalytic function many had been named on the basis of their initial source. For example, while the name lactate dehydrogenase describes the overall catalytic function of the enzyme in question (although it does not indicate the nature of the electron acceptor involved) the name papain indicates only that this particular enzyme (a protease) was first isolated from papaya latex. However, in 1961 the International Union of Biochemistry proposed a system of classification and nomenclature which allows any enzyme to be identified unambiguously. The class suffix is **-ase**. Each enzyme is allocated to one of six major groups, corresponding to the type of reaction which it catalyses, and is numbered accordingly. These groups are as follows:

1. Oxidoreductases

These enzymes catalyse the transfer of H, O atoms or electrons from one substrate to another, e.g. alcohol dehydrogenase:

$$RCH_2OH + NAD^+ \rightleftharpoons RCHO + NADH + H^+$$

2. Transferases

These are enzymes that catalyse the transfer of a specific group from one substrate to another, e.g. kinases, such as glucokinase, catalyse the transfer of a phosphoryl group, $-PO_3^{2-}$, from ATP to an acceptor:

$$\text{Glucose} + ATP^{4-} \rightleftharpoons \text{Glucose-6-phosphate}^{2-} + ADP^{3-} + H^+$$

3. Hydrolases

These are the hydrolytic enzymes, e.g. esterases:

$$R^1COOR^2 + H_2O \rightleftharpoons R^1COOH + R^2OH$$

4. Lyases

These are enzymes that catalyse the cleavage of C–C, C–O, C–N or C–S bonds other than by hydrolysis, e.g. decarboxylases:

$$R^1CHR^2COOH \rightleftharpoons R^1CH_2R^2 + CO_2$$

5. Isomerases

These are enzymes that catalyse an intramolecular rearrangement, e.g. racemases which catalyse reactions of the type:

L-aminoacid $\rightleftharpoons$ D-aminoacid

6. Ligases

These are enzymes which catalyse the formation of C–O, C–S, C–N or C–C bonds with the concomitant breakdown of ATP (or another nucleoside triphosphate), e.g. the aminoacid activating enzymes:

aminoacid + ATP^{4-} + tRNA $\rightleftharpoons$ aminoacyl-tRNA + pyrophosphate^{4-}

Each group is then subdivided according to the nature of the principle substrate involved. For example the oxidoreductases are divided into 13 groups.

1.1. Enzymes acting on CHOH groups as the donor
1.2. Enzymes acting on aldehyde or keto groups as the donor
1.3. Enzymes acting on CH–CH groups as the donor
1.4. Enzymes acting on CH–NH_2 groups as the donor
1.5. Enzymes acting on C–NH groups as the donor
1.6. Enzymes acting on NADH or NADPH as the donor
1.7. Enzymes acting on other nitrogenous compounds as the donor
1.8. Enzymes acting on sulphur groups as the donor
1.9. Enzymes acting on haem groups as the donor
1.10. Enzymes acting on diphenols and related substances as the donor
1.11. Enzymes acting on H_2O_2 as an acceptor
1.98. Enzymes using H_2 as a reductant
1.99. Other enzymes using O_2 as an oxidant

Further subdivisions are then made according to the nature of the other substrate. For example:

1.1.1. With NAD^+ or $NADP^+$ as acceptor
1.1.2. With a cytochrome as acceptor
1.1.3. With O_2 as acceptor
1.1.99. With other acceptors

Individual enzymes are then listed numerically in the appropriate class, e.g. E.C. 1.1.1.1. has the systematic name: alcohol: NAD oxidoreductase and the trivial name: alcohol dehydrogenase.

E.C. 1.4.3.3. has the systematic name: D-aminoacid oxidoreductase (deaminating) and the trivial name: D-aminoacid oxidase. For an unambiguous description the source of the enzyme must also be given.

8.2. Properties of enzymes

(i) All enzymes are proteins. The great majority of those which have been investigated to date are soluble, globular, proteins; others, however, are bound to membranes and are difficult to bring into solution without loss of activity. Thus enzymes possess primary, secondary, tertiary and (often) quaternary structure and are de-natured by moderate temperatures, extremes of pH, by detergents and by a number of reagents such as urea and by non-polar solvents such as ethanol and acetone.

(ii) Enzymes are extremely efficient catalysts. It has been estimated, for example, that the proteolytic enzymes of the gut accelerate the cleavage of peptide bonds by a factor of 10^5, reducing the time necessary for digestion of an average meal from 50 years to a few hours! More specifically the pseudo-first order rate constant for the hydrolysis of urea (8.1) has been estimated to be $3 \times 10^{-10}\,s^{-1}$ at 20.8°C.

$$(H_2N)_2CO + H_2O \rightleftharpoons 2NH_3 + CO_2 \qquad \textbf{(8.1)}$$

In the presence of urease the reaction is accelerated by a factor of 10^{14}, the rate constant for the rate determining step in the presence of a large excess of substrate being $3 \times 10^4\,s^{-1}$.

(iii) Most enzymes exhibit a high degree of both structural and stereochemical specificity. It appears that a few enzymes exhibit absolute structural specificity. For example, asparaginase catalyses the hydrolysis of only L-asparagine (8.2); the hydrolysis of L-glutamine, which differs from L-asparagine by only a single methylene group, is not accelerated.

$$H_3\overset{+}{N}\underset{\underset{\displaystyle CH_2CONH_2}{|}}{C}HCO_2^- + H_2O \xrightleftharpoons{\text{asparaginase}} H_3\overset{+}{N}\underset{\underset{\displaystyle CH_2CO_2^-}{|}}{C}HCO_2^- + NH_4^+ \qquad \textbf{(8.2)}$$

Similarly fumarase catalyses only the reversible interconversion of fumaric acid and L-malic acid (7.27). Most enzymes, however, will catalyse the same reaction of a small group of compounds with homologous, or closely similar, structures. In such cases specificity is reflected in the rate of reaction—that of the 'specific substrate' is usually at least an order of magnitude greater than those of others. For example, the relative rates of oxidation of a number of sugars by glucose oxidase are: β-D-glucose, 1.0, 2-deoxy-D-glucose, 0.25, D-xylose, 0.01, α-D-glucose, 0.006. Similarly the rate of chymotrypsin catalysed hydrolysis of peptide bonds is faster by a factor of about 10^2–10^3 when the carbonyl group of the peptide bond is donated by either of the aromatic amino-acids phenylalanine or tyrosine.

Another interesting aspect of structural specificity is illustrated by the enzymes catechol *N*-methyl transferase and noradrenalin *O*-methyl transferase (catechol methyltransferase). Both catalyse a reaction between the same substrates noradrenalin and *S*-adenosylmethionine. However, the former enzyme catalyses *N*-methylation of noradrenalin (5.90) whereas the latter enzyme catalyses *O*-methylation (5.93); they can be said to be **regiospecific**.

Enzymes usually exhibit absolute stereospecificity when they catalyse reactions involving chiral substrates. For example D-amino acid oxidase, which catalyses the oxidation of D-amino acids by molecular oxygen, (5.133) is completely without action on the enantiomeric compounds. Fumarase, mentioned above, not only exhibits high structural specificity but also absolute stereospecificity; D-malate cannot replace the L-isomer. Stereospecificity in enzyme catalysed reactions is discussed in more detail in Section 8.4.

(iv) Many enzymes require the presence of a non-protein co-enzyme in order to exert their catalytic effect. The nature of these materials has already been outlined on page 29, Section 3.3. The mode of action of some of the more important co-enzymes is dealt with in Chapter 9.

8.3. Mode of action of enzymes

It is important to note that enzymes affect only the rates of reactions; they accelerate the attainment of equilibrium but do not alter its position. An enzyme, being a catalyst, must accelerate a reaction by decreasing the free energy of activation, as noted in Section 3.3. In slightly more explicit terms an enzyme might raise the energy of (de-stabilise) the starting materials or decrease the energy of (stabilise) the transition state of the uncatalysed reaction, or both. Alternatively, the enzyme might provide a different, low energy pathway for the reaction, in which case it is said to provide 'alternative pathway' catalysis. Furthermore, the free energy of activation, ΔG^*, which represents the difference between the enthalpy of activation, ΔH^*, and the entropy of activation ΔS^*, can be decreased by an enzyme altering either or both the ΔH^* and ΔS^* terms.

Enzyme-substrate binding

All of the available evidence suggests that the first step in an enzyme catalysed reaction involves the substrate(s) binding to the enzyme **non-covalently** to give an enzyme-substrate complex (E·S in the generalised one substrate reaction 8.3).

$$\mathrm{E+S} \underset{k_{-1}}{\overset{k_1}{\rightleftharpoons}} \mathrm{E{\cdot}S} \underset{k_{-2}}{\overset{k_2}{\rightleftharpoons}} \mathrm{E{\cdot}P} \underset{k_{-3}}{\overset{k_3}{\rightleftharpoons}} \mathrm{E+P} \qquad \mathbf{(8.3)}$$

The binding site is usually a shallow depression, or cleft, in the surface of the enzyme and is known as the **active site**. Subsequent to substrate binding the catalytic reaction proper occurs within the E·S complex at the active site, generating product(s) and re-generating the enzyme.

Until recently it was thought that formation of an enzyme substrate complex was invariably faster than the subsequent catalytic reaction. Substrate binding is diffusion controlled; typical second order rate constants for the process lie in the range 10^5–10^9 litre mol^{-1} s^{-1}. However, an increasing body of evidence now suggests that several enzymes have evolved to a state of such high catalytic efficiency that the catalytic reaction is only partly, if at all, rate limiting. In such cases it seems that substrate binding in an exergonic reaction, or product release in an endergonic reaction, is the rate limiting step. A specific example of an enzyme of this type is triosephosphate isomerase. This enzyme catalyses the intercoversion of dihydroxyacetone phosphate (DHAP) and glyceraldehyde-3-phosphate (GAP), via an enzyme bound enediol, during glycolysis (8.4).

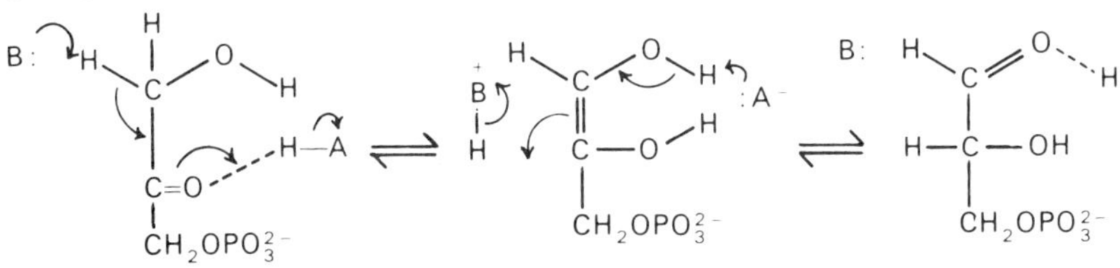

DHAP GAP **(8.4)**

The enzyme catalyses the isomerisation some 10^{10} times more efficiently than does a simple catalyst such as acetate ion and the release of GAP from the enzyme appears to be rate limiting.

Regardless of whether the step in which the substrate is converted to product limits the rate of an enzyme catalysed reaction or not, it is this step which, in contrast to the un-catalysed reaction, is accelerated and which therefore holds the key to the mode of action of the enzyme in question. In an enzyme catalysed reaction conversion of substrate to product occurs within the enzyme-substrate complex; an **intra**molecular reaction is involved, rather than an **inter**molecular reaction as in the uncatalysed case. It is logical, therefore, to look for an explanation of at least part of the catalytic effect of enzymes in terms of formation of an enzyme-substrate complex.

It is now accepted that the major effect of formation of an enzyme-substrate complex is an entropic one. Consider an un-catalysed bimolecular reaction of type (8·5).

$$A + B \rightleftharpoons A\text{----}B \rightleftharpoons A-B \qquad \textbf{(8.5)}$$

Formation of the transition complex requires a bimolecular collision to occur between the reactants, which must invariably be oriented in a highly specific way with respect to each other. This is a relatively improbable event as shown by the entropy changes listed in Table 8.1. The structure of the transition state will be very similar to that of the product and hence $\Delta S^{\ominus *}$ will undoubtedly be of the same order as $\Delta S^{\ominus}$. This means that $\Delta G^{\ominus *}$ will be large and k_r small, as is observed experimentally. If, however, the reaction is catalysed by an enzyme then the loss of translational and rotational entropy occurs in the substrate binding step(s) (8.6).

$$E + A + B \leftrightharpoons E.A.B \leftrightharpoons E + AB \qquad \textbf{(8.6)}$$

Any further decrease in entropy associated with the subsequent catalytic step will therefore be minimal, $\Delta G^{\ominus *}$ for this step will be small and k_{Cat} (k_2 in (8.3)) large. A comparison of simple intermolecular and intramolecular reactions shows that removal of the unfavourable loss of entropy from a reaction can lead to rate accelerations of the order of 10^8 (Table 8.2). The effect of enzyme-substrate binding is not, however, solely entropic. Crystallographic studies of stable enzyme-inhibitor or enzyme-pseudo substrate complexes suggest that substrates are distorted by the enzyme and bound in a configuration approaching that of the transition state for the catalytic reaction. This means that the enthalpy of activation for the subsequent catalytic step will be decreased, again increasing the rate of the reaction. There is also evidence that the converse effect operates with some enzymes, namely that substrate binding distorts the enzyme, presumably

Table 8.1 Entropy changes for a biomolecular reaction at a standard state of 1 mol l^{-1} at 20.8 K (kJ mol^{-1})

Gas phase	A	B	A-B	S
$S_{translation}$	135	135	140	−130
$S_{rotation}$	100	100	110	− 90
$S_{internal\ rotations}$	12	12	74	+ 50
S_{total}	247	247	324	−170
$-\Delta S_{vap}$	−40	−40	−60	
Solution	207	207	264	−150

Table 8.2 Relative rates of inter- and intramolecular reactions

Relative rates of anhydride formation	k_{rel}*
$CH_3COO^- + CH_3COOC_6H_4Br\,(p)$	1.0
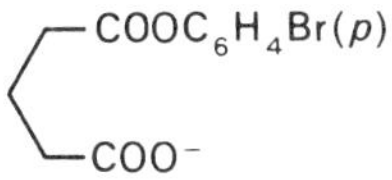	≈1 × 10^3M
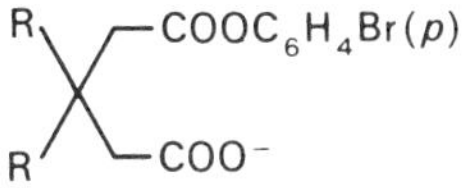	3 × 10^3M – 1.3 × 10^6M
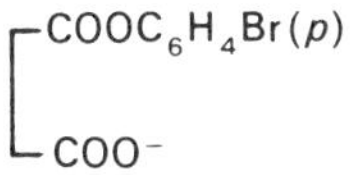	≈ 2.2 × 10^5M
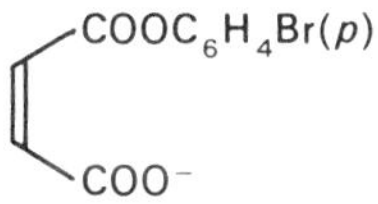	1 × 10^7M
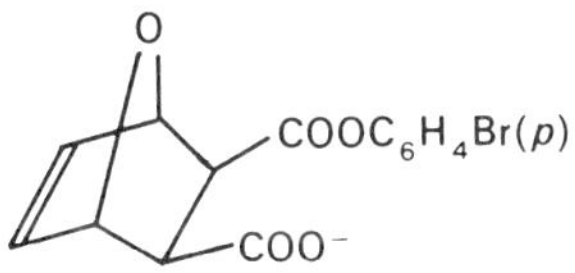	≈5 × 10^7M

Relative rates of ester formation	
$C_6H_5OH + CH_3CO_2H$	1
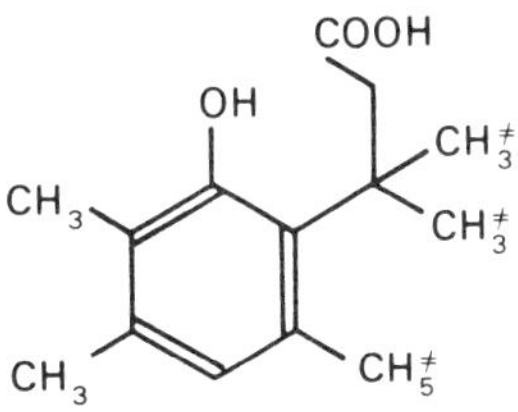	3 × 10^{15}M

* k_{rel} for a given reaction can be interpreted as the concentration to which one of the reactants in (1) must be raised in order to obtain a comparable rate.

‡ The steric hindrance provided by the methyl groups 'locks' the carboxyl group in the vicinity of the hydroxyl group.

bringing amino acid side chains into positions in which their catalytic efficiency is optimised. This effect is known as Koshland's theory of 'induced fit'.

Loss of entropy or distortion of the substrate(s) towards the configuration of the transition state during substrate binding are thermodynamically unfavourable factors. They are 'paid for' by the energy liberated by multiple noncovalent interactions between substrate(s) and enzyme. Numerous interactions, which may be co-operative, can be envisaged between an enzyme and a large substrate such as a protein or nucleic acid and in such cases it is easy to accept that the enthalpy of binding will be more than sufficient to outweigh the unfavourable entropy and de-stabilisation factors. It is less easy to accept that this is so in the case of small substrates such as urea or hydrogen peroxide. However, crystallographic studies have shown that the atoms in contact between enzyme and substrate in an enzyme-substrate complex are virtually close packed and it has been suggested that the forces of attraction between close packed atoms are much larger than previously estimated on the basis of solute-solvent or even solute-solid interactions. If this is so, comparatively few interactions may be necessary to generate the required binding energy.

The catalytic reaction

It has already been noted that enzymes probably bind their substrates in a configuration which approaches that of the transition state—i.e. that which provides for the most efficient mode of bond breakage and formation. Additionally it appears that enzymes provide most, if not all, of the features necessary for electronic catalysis of the reaction. These features are usually described as the 'mechanism of action' or, preferably, the 'reaction pathway' of the enzyme.

A study of simple organic reactions suggest that electronic catalysis by an enzyme may involve acid-base, electrophilic or nucleophilic catalysis. A reaction is said to be subject to general acid or base catalysis if it is accelerated by the presence of undissociated acids or bases, respectively. The catalysis is said to be concerted if the presence of **both** an acid and a base is more effective than either separately. The mutorotation of 2,3,4,6-tetra-*O*-methyl-D-glucose is subject to concerted general acid-base catalysis (8.7).

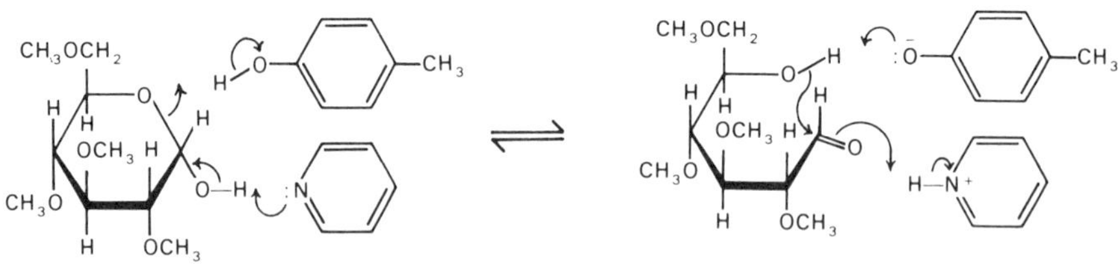

(**8.7**)

Pyridine and *p*-cresol, for example, facilitate the reaction by accepting and donating protons as shown. It is interesting to note that the rate of mutorotation is 700 times as fast in the presence of 2-hydroxypyridine (8.8) as it is in the presence of a mixture of *p*-cresol and pyridine at the same concentration. This last effect is undoubtedly largely entropic; a termolecular reaction is reduced to a bimolecular one.

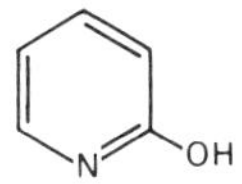

(8.8)

Electrophilic catalysis invariably involves the electrophile increasing the polarity of a reactant in such a way that reaction is facilitated. Metal ions are very effective electrophilic catalysts both in organic chemistry and in enzymology; they have the great advantage that they can function at neutral pH. For example, the hydration of pyridine-2-carboxaldehyde is catalysed by zinc ions (8.9). The zinc ions effectively remove charge from oxygen, increasing the polarity in the carbonyl group, which is thus made more susceptible to attack by the very weak nucleophile, water.

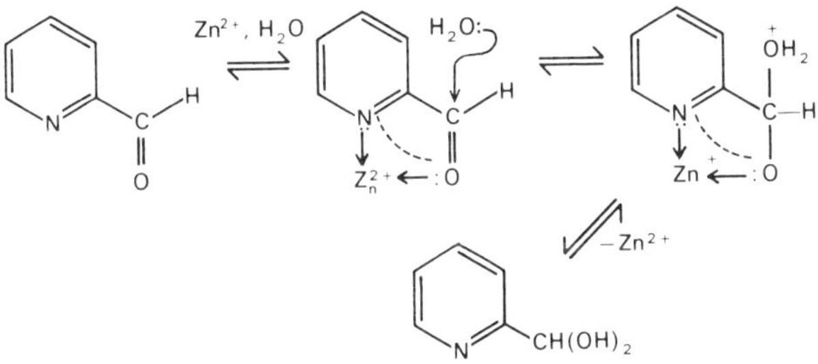

(8.9)

Catalysis by protons (specific acid catalysis) is also a type of electrophilic catalysis. Several examples of this type of catalysis, for example acetal and ester formation, have already been detailed in Chapters 5 and 6.

In nucleophilic catalysis the nucleophile usually provides an alternative pathway for the reaction which is associated with a lower energy of activation. By implication one of the reactants in a reaction which is subject to nucleophilic catalysis must itself be a nucleophile. The catalyst must be a better nucleophile than the nucleophilic reagent and must also provide a better leaving group than the reactant. These points are illustrated by the imidazole catalysed hydrolysis of *p*-nitrophenyl acetate (8.10).

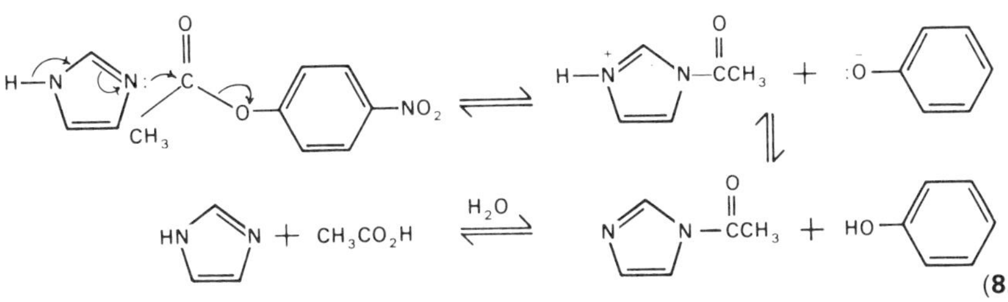

(8.10)

The involvement of some of these electronic effects in enzyme catalysis can be illustrated by reference to the postulated reaction pathways of ribonuclease A, lysozyme, carboxypeptidase A and chymotrypsin.

Ribonuclease A (RNase A, E.C. 2.7.7.16)

Ribonuclease A, isolated from bovine pancreas, catalyses the hydrolysis of ribonucleic acids. The bond which is cleaved is always such as to leave a phosphate group attached to the 3′ position of a pyrimidine ribonucleoside. RNase A is a globular protein with $M_r = 13\,700$. Both the amino acid sequence and the complete 3-dimensional structure of the enzyme are known. The molecule contains a single polypeptide chain of 124 amino acid residues in which four disulphide bridges are present. The active site of the enzyme is a shallow depression or 'waist' around the molecule. Kinetic and chemical studies suggest that two histidine residues, one protonated and one un-protonated, are essential for the activity of RNase A and two are, in fact, present in the catalytic site (His^{12} and His^{119}). The reaction pathway is thought to be as shown in (8.11). A comparison of this pathway with that of the uncatalysed reaction (8.12) shows that the two histidine residues provide concerted general acid-base catalysis of the hydrolysis. The basic histidine residue promotes the nucleophilic activity of the 2′ hydroxyl group of a pyrimidine nucleoside or of water, while the acidic histidine facilitates cleavage of P–O bonds by providing an 'electron sink' in the form of a proton. The imidazole side chain of histidine is ideal for providing enzymic acid-base catalysis because the pK_a of the ring nitrogens (6.0) is such that the probabilities that they may be protonated or un-protonated at physiological pH are approximately equal.

Lysozyme (E.C. 3.2.1.17)

Lysozyme was discovered by Fleming in 1922. It is an enzyme which occurs ubiquitously, e.g. in egg white, skin and in tears. It catalyses the hydrolysis of an alternating *N*-acetylglucosamine-*N*-acetylmuramic acid co-polymer which is present in bacterial cell walls causing the cells to undergo lysis.

Lysozyme was the second protein, and the first enzyme, to have its complete three-dimensional structure determined by X-ray diffraction analysis. It is a globular protein, $M_r = 14\,600$, containing a single polypeptide chain of 129 amino acid residues with four disulphide bridges present. As in RNase A, the active site of lysozyme is a shallow groove in the vicinity of the equator of the molecule. The reaction pathway which the enzyme is postulated to use is based on the results of kinetic, chemical, X-ray diffraction and model building studies and is as follows.

Six monosaccharide residues (A to F, 8.13) of the polymer are bound at the active site; at least thirteen hydrogen bonds are involved in the binding. There is some evidence that ring D, which is an *N*-acetylmuramic acid residue, is bound in the half-chair configuration. As will be seen, the half-chair configuration approximates to that of the transition state for the reaction. It appears, therefore, that part of the catalytic effect of lysozyme arises from ΔG^* being decreased by raising the energy of the substrate towards that of the transition state. The enzyme catalyses the cleavage of the α-(1 → 4) glycosidic bond associated with ring D. The carboxylate group in the side chain of Glu^{35}, which is in a non-polar

* The evidence is not conclusive and has recently been questioned.

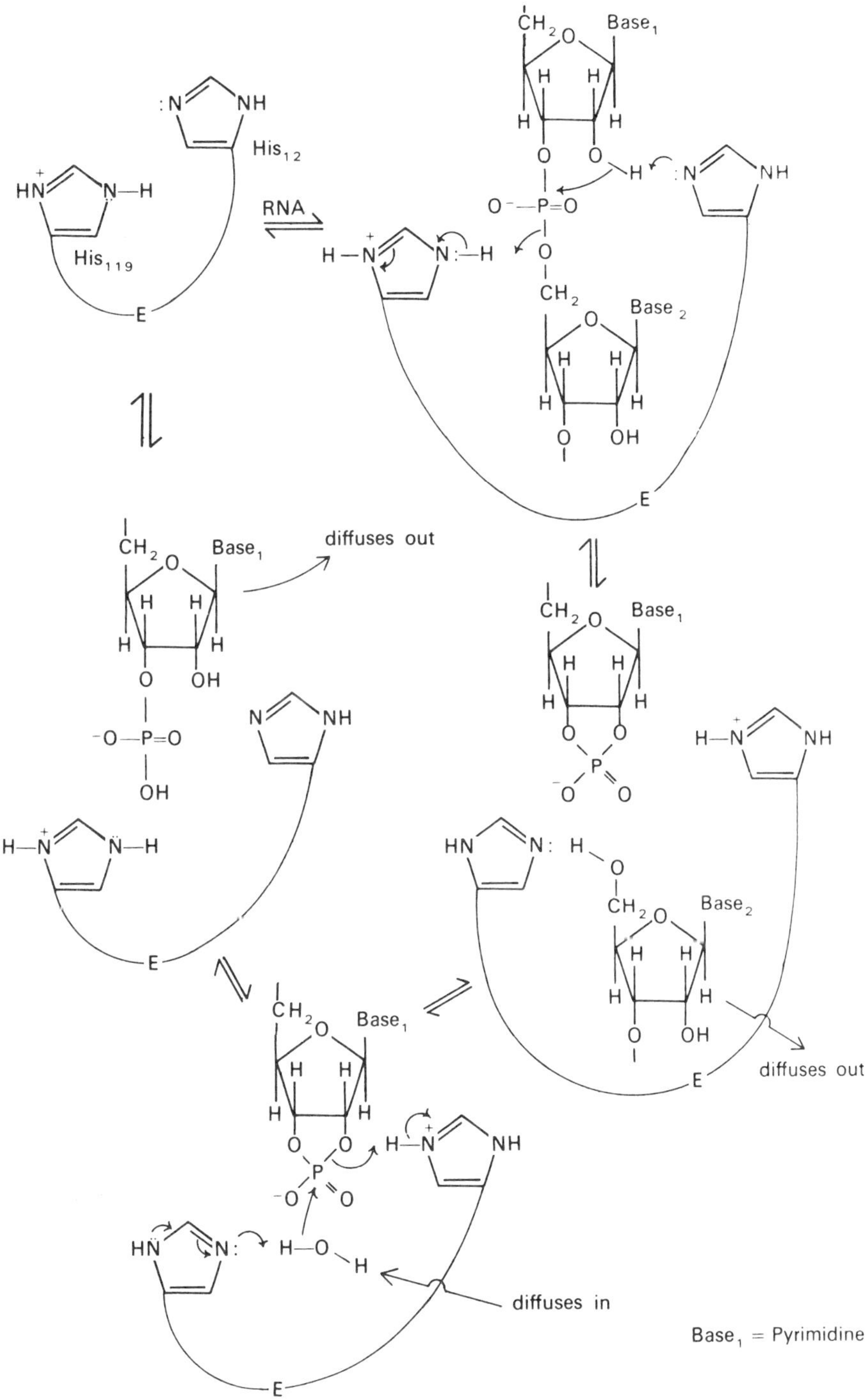

(**8.11**) Postulated reaction pathway for the hydrolysis of RNA catalysed by RNase A

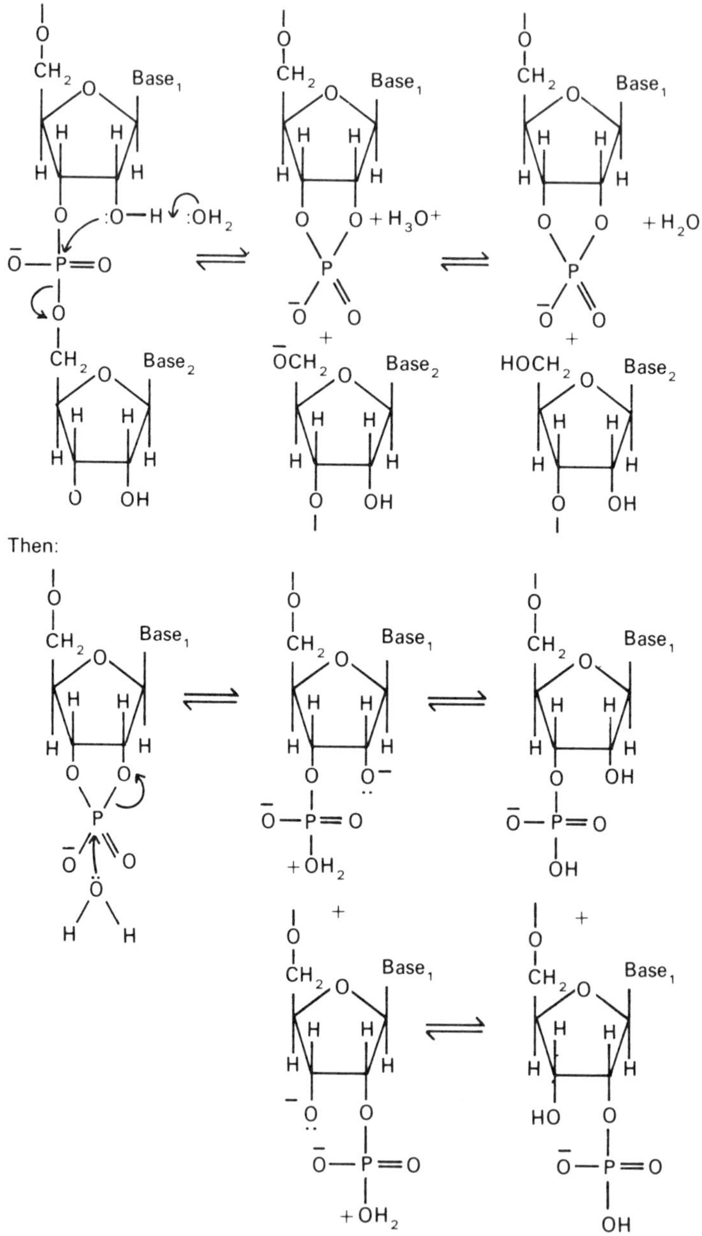

(**8.12**) Mechanism of uncatalysed hydrolysis of a polyribonucleotide

environment and therefore normally not ionised, provides acidic catalysis for ionisation of the substrate at the 1-position of ring D. The carbonium ion which results could be stabilised by (i) charge delocalisation involving the p electrons of the ring oxygen. This is only possible when the oxygen atom and C-1 are co-

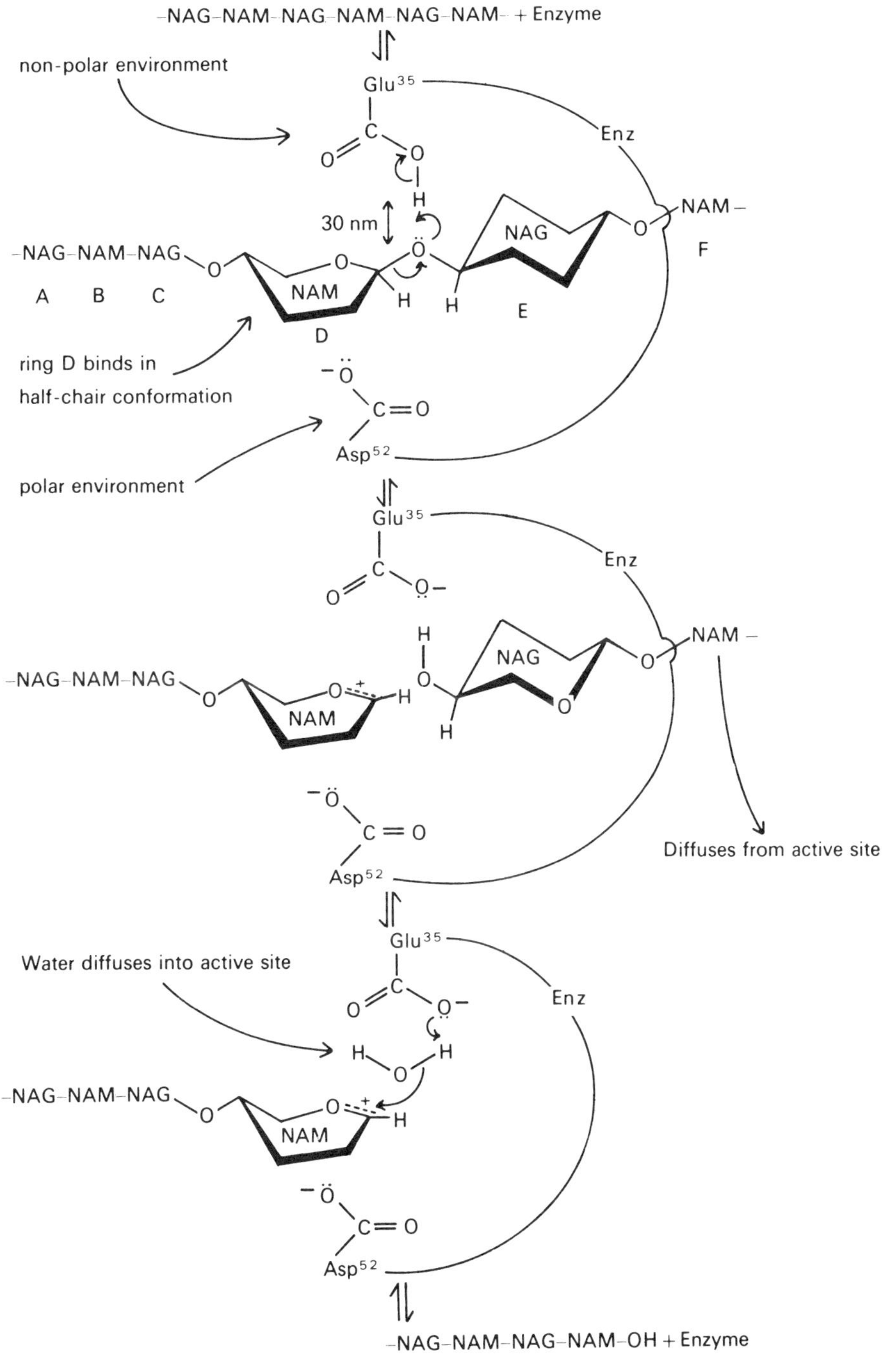

(**8.13**) Possible reaction pathway of lysozyme. NAG = *N*-acetylglucosamine, NAM = *N*-acetylmuramic acid; substituents on residues D and E have been omitted for clarity.

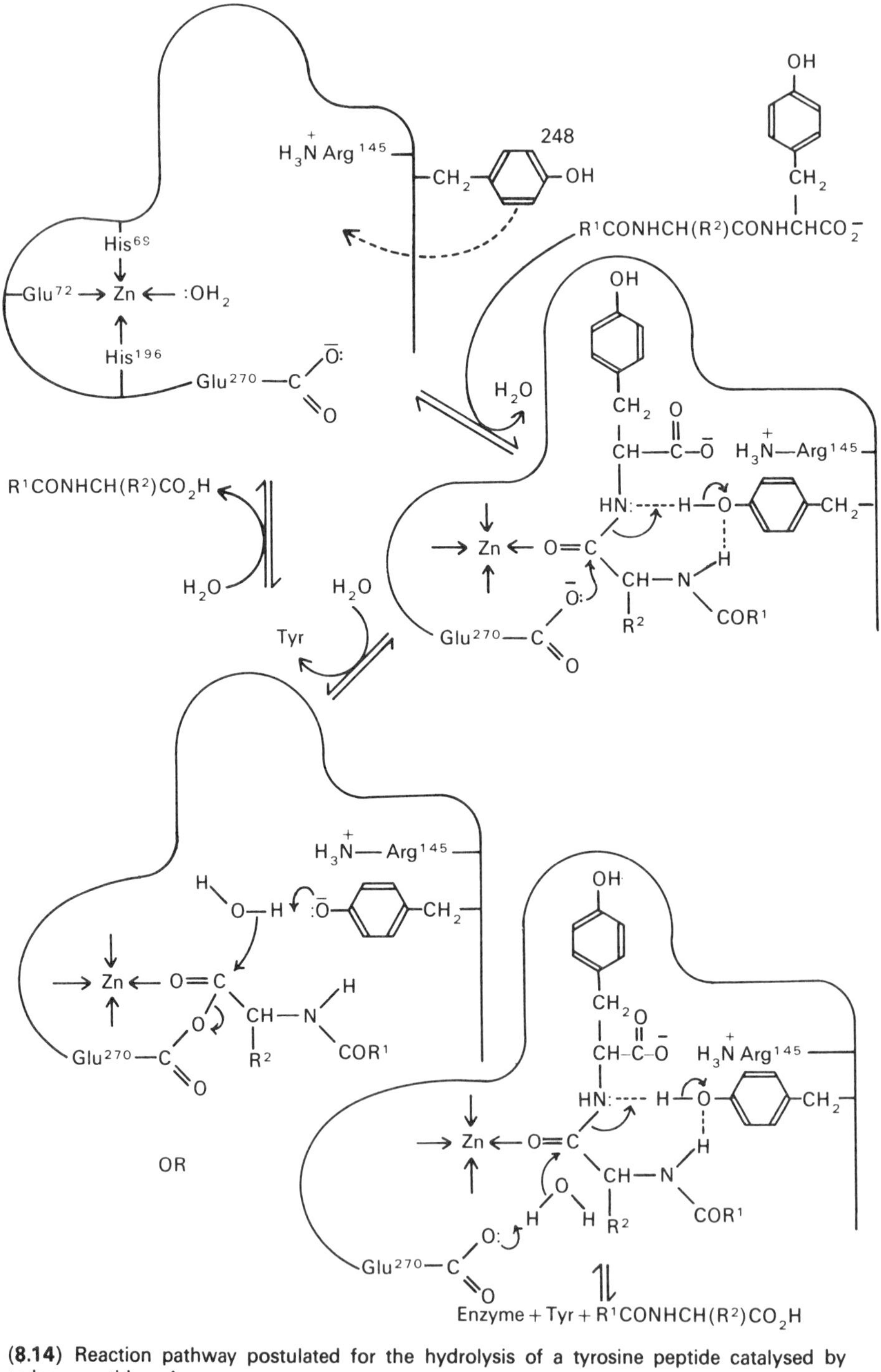

(8.14) Reaction pathway postulated for the hydrolysis of a tyrosine peptide catalysed by carboxypeptidase A

planar, i.e. when the ring has the half-chair configuration, (ii) by interaction with the ionised side chain of Asp^{52}. Water then replaces residues E and F in the active centre, and its nucleophilic activity is increased by the now ionised Glu^{35} side chain ensuring rapid formation of the second product of the hydrolysis. The pathway of the enzyme catalysed reaction should be compared with that of the uncatalysed reaction (5.80).

Carboxypeptidase A (E.C. 3.4.2.1)

Carboxypeptidase A is produced from its inactive pancreatic precursor, procarboxypeptidase (a zymogen), by the action of trypsin. As its name suggests carboxypeptidase A is an exopeptidase, catalysing the hydrolytic removal of amino acids from the carboxyl terminus of a polypeptide chain; the rate of cleavage is highest when the C-terminal amino acid has an aromatic or branched aliphatic side chain.

The enzyme has $M_r = 34\,600$. It has been sequenced and its three-dimensional structure has been determined by X-ray crystallography. The molecules contain one polypeptide chain of 370 amino acid residues and one disulphide bridge. The active site of the enzyme (8.14) has been shown, by a combination of techniques, to contain a hydrophobic pocket which accommodates the hydrophobic side chain of the substrate's C-terminal residue. A zinc ion, which binds and presumably therefore **polarises** (electrophilic catalysis) the carbonyl group of the peptide bond which is cleaved, is also present as is a protonated arginine (Arg^{145}) side chain and an ionised glutamic acid (Glu^{270}) side chain. It is suggested that when the substrate binds to the enzyme the side chain of Tyr^{248} is caused to move the relatively large distance of 1.2 nm bringing its hydroxyl group into juxtaposition with the labile peptide bond. Its function in the active centre is possibly to form a hydrogen bond with the leaving group portion of the substrate. At present there is insufficient evidence to distinguish between direct attack on the substrate by Glu^{270}, forming a mixed anhydride which is then decomposed by water, and indirect attack via a water molecule.

α-Chymotrypsin (E.C. 3.4.4.5)

α-Chymotrypsin, a serine hydrolase*, is probably the most thoroughly investigated of all enzymes. It is produced in the duodenum by the action of trypsin (aided by autolysis) on an inactive, pancreatic precursor, chymotrypsinogen. The enzyme is an endopeptidase which catalyses the hydrolysis of, particularly, peptide bonds in which the carbonyl group is associated with an amino acid with an aromatic side chain; it also exhibits esterase activity. It has been established beyond doubt that during the catalytic reaction the acyl group of the substrate is transferred to the hydroxyl group of a serine residue in the active site of the enzyme; the acyl enzyme (b, 8.15) is then hydrolysed, catalytically, by water.

It appears, therefore, that chymotrypsin provides nucleophilic catalysis for the reaction. However, alcoholic hydroxyl groups are only weakly nucleophilic and

* The serine hydrolases are a group of proteases and esterases having several common features. They all contain a reactive serine in their catalytic sites and all are irreversibly inhibited by organophosphorus compounds such as di-isopropyl phosphorofluoridate. Those serine hydrolases whose structures are known have catalytic sites identical to that of chymotrypsin, although their overall primary and tertiary structures may differ considerably. Examples are chymotrypsin, trypsin, thrombin, elastase, subtilisin, α-lytic protease and acetylcholinesterase.

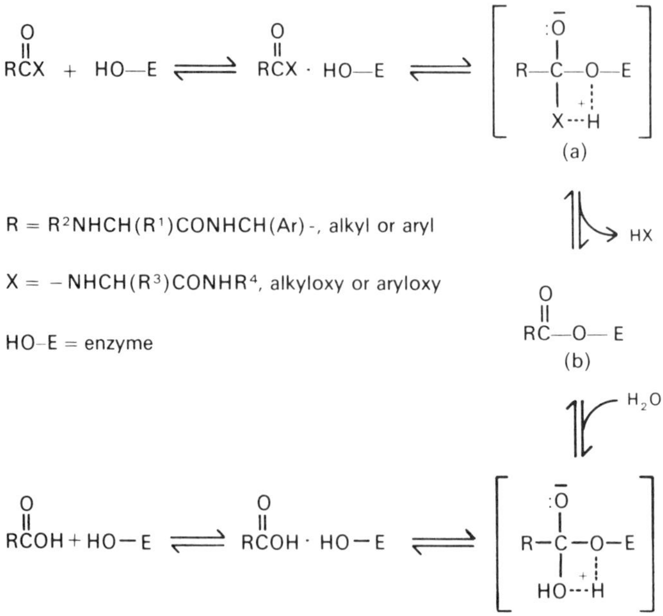

(8.15)

cannot attack carbonyl groups directly (Sections 5.1 and 6.2). The structure of the enzyme must, therefore, be such that the nucleophilic character of the serine hydroxyl group is enahnced and/or the enzyme must increase the susceptibility of the substrate's carbonyl group to nucleophilic attack.

The complete 3-dimensional structure of the enzyme is known from crystallographic studies. The molecules (M_r = 25 000), which are roughly ovoid, are composed of a single polypeptide chain containing five disulphide bridges. The active site contains three amino acids: Ser^{195}, His^{57} and Asp^{102}, which are essential for activity. Additionally a 'hydrophobic pocket' is present which accepts the aromatic group of the substrate. It has been suggested that the side chains of Ser^{195}, His^{57} and Asp^{102} (the latter is ionised at physiological pH and is located in a hydrophobic region of the enzyme) are hydrogen bonded in such a way that the nucleophilic character of the serine hydroxyl group is increased (8.16); the bonded residues are said to constitute a 'charge relay system'.

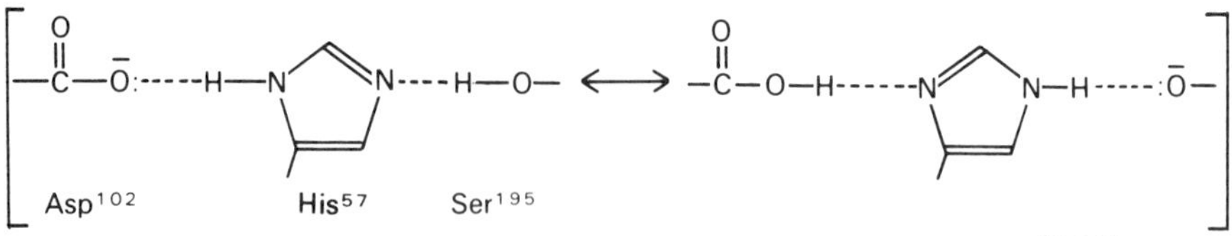

(8.16)

The reaction pathway for the chymotrypsin catalysed hydrolysis of amides and esters might then be as shown in (8.17).

However, the suggestion that the charge relay increases the nucleophilic activity of Ser^{195} has been questioned on several grounds. For instance, refinement of the X-ray diffraction data has shown that the hydrogen bond between His^{57} and Ser^{195} is very weak, or even non-existent. It now appears

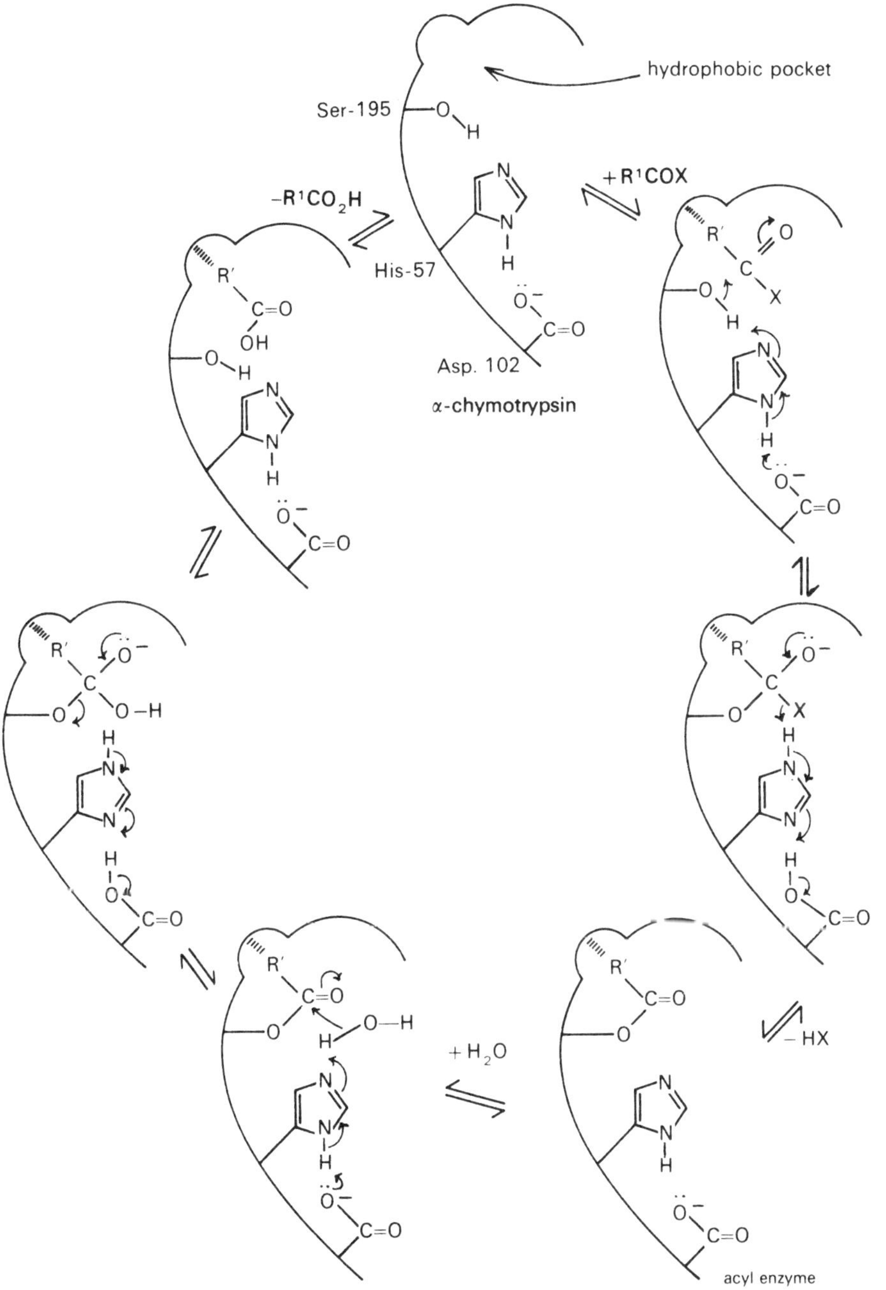

(8.17) Postulated reaction pathway for the chymotrypsin catalysed hydrolysis of amides and esters. X = OR^2 or $-NHR^3$ R^1 = $-NHCH(Ar)-$, R^2 = alkyl or aryl, R^3 = H or a peptide chain

that the enzyme may perform the alternative function of activating the substrate towards nucleophilic attack. Investigations of the binding of both natural and synthetic inhibitors of chymotrypsin show that they bind so that the carbonyl (or other) group which corresponds to the one which is attacked in the substrate has a tetrahedral disposition in relation to the γ-oxygen of serine^{195}. That is, the substrate binds in a conformation approximating to that of the transition state (tetrahedral oxyanions are accepted as being intermediates in uncatalysed acyl transfer reactions). The oxygen of the carbonyl group in the labile peptide bond appears to form two hydrogen bonds with N–H groups of the enzyme in a region known as the 'oxyanion hole'. This type of hydrogen bonding will increase the polarity of the carbonyl group and, together with the geometrical distortion produced by binding, may make it susceptible to rapid attack by the γ-oxygen of the serine residue. The role of the hydrogen bonded Asp^{104}, His^{57} and Ser^{195} residues may then be to facilitate the transfer of a proton between the γ-oxygen of the serine, the leaving group and water, rather than to increase the activity of the serine hydroxyl group.

Summary

It has already been stated that the 'entropic effect' of enzymes may cause rate accelerations of the order of 10^8. It is estimated that electronic catalysis might contribute a further factor of the order of 10^2. It appears, therefore, that these two effects coupled, possibly, with 'strain' effects can adequately explain the magnitude of rate accelerations produced by enzymes. However, although the principles of enzyme action can be understood in these terms it has been possible to postulate reaction pathways for only a very few enzymes. Even where pathways have been suggested much uncertainty still exists about their details. For example, it must be borne in mind that the results of crystallographic studies, from which much has been inferred regarding mechanistic pathways, give only a static picture of an enzyme or derivative at a point which is often not on the catalytic pathway. Unfortunately evidence relating to dynamic aspects of enzyme catalysed reactions, such as changes in the conformations of both enzyme and substrate, is difficult both to obtain and interpret.

8.4. Some stereochemical aspects of enzyme catalysed reaction

It has already been noted that enzymes usually exhibit virtually absolute stereospecificity when they catalyse the reactions of chiral substrates. Stereospecificity of this type can be explained by assuming that the substrate is bound at the active site by **three** of the substituents in the chiral centre. Consider, for example, the general case of a chiral substrate of type (8.18a) and suppose that the active site of the enzyme contains binding sites for groups q, r and s disposed on the enzyme surface as shown in (8.18b). Clearly (8.18a) can bind effectively to the

(a) (b) (c) (d) **(8.18)**

sites, but its enantiomer (8.18c) cannot. Only two groups of (8.18c) can bind to the enzyme and in view of the fact that the reacting groups in an enzyme catalysed reaction undoubtedly must be oriented with great precision it is reasonable to suppose that the reaction of (8.18c) will not be accelerated.

Prochirality in enzyme catalysed reactions

The ability of many enzymes to distinguish between two chemically equivalent groups can also be explained on the basis that the substrate is bound to the active site by three substituents. Consider the general case of a molecule of type (8.18d) containing a carbon atom bearing only three different groups and suppose that the binding sites in the active centre of the enzyme are again disposed as shown in (8.18b). If the equivalent groups are arbitrarily numbered as shown, it can be seen that the only efficient mode of binding of the substrate to the enzyme involves group r^2 which will therefore undergo reaction rather than group r^1 which projects away from the enzyme surface. A molecule such as (8.18d) which although not possessing a chiral centre, can give rise to one in an enzyme catalysed reaction or by reaction with a chiral reagent, is said to be **prochiral**. For example glycerol is a prochiral molecule and glycerol kinase, which catalyses reaction (7.26) is an enzyme which is able to distinguish between chemically equivalent groups, in this case the hydroxymethyl groups.

The chemically equivalent groups in a prochiral centre are often hydrogen atoms as in ethanol (8.19a) and this molecule can be used to illustrate how such groups are designated. Consider the effect of substituting deuterium (^2H) for one of the hydrogen atoms (8.19b). Hydrogen and deuterium are chemically equivalent, but are stereochemically distinct because deuterium has higher priority in the Cahn-Ingold-Prelog system for designating configuration than hydrogen has (p. 123 section 7.1). The configuration of the chiral centre in (8.19b) is therefore S while that of the chiral centre in (8.19c) is R; the chemically equivalent hydrogen atoms in (8.19a) can therefore be designated as pro-S and pro-R (8.19d).

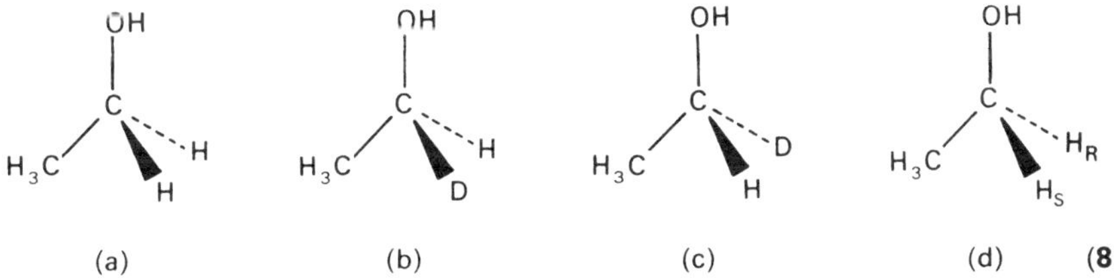

(8.19)

The reaction catalysed by alcohol dehydrogenase (ADH) provides another good example of one in which an enzyme differentiates between two chemically equivalent hydrogen atoms, in this case the C-1 hydrogens of ethanol. It is found that when deuterioacetaldehyde is reduced by NADH under the catalytic influence of the enzyme only one enantiomer, S-1-deuterioethanol, of the deuterated alcohol is formed (8.20). As the equation indicates the enzyme is also stereospecific with respect to the co-enzyme. The 4-position of the nicotinamide ring constitutes a prochiral centre from which the pro-R hydrogen is removed.* This is

* The face from which the pro-R hydrogen atom is removed is known as the *re*- or A face; the other face of the ring is the *si*- or B face.

confirmed by the fact that the co-enzyme, stereo-specifically labelled with deuterium in the 4-R position, transfers the label to the pro-R position of ethanol (8.21).

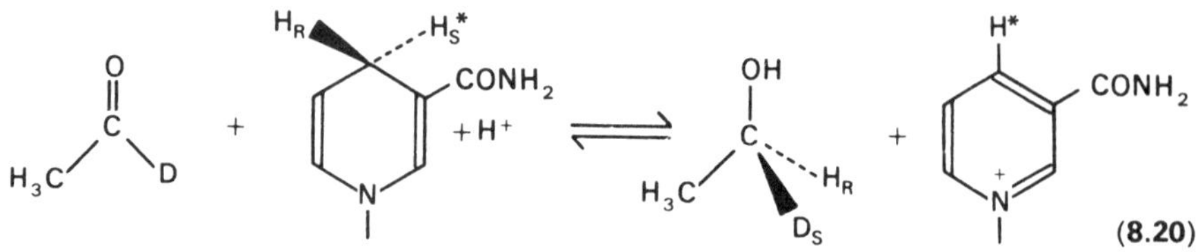

(8.20)

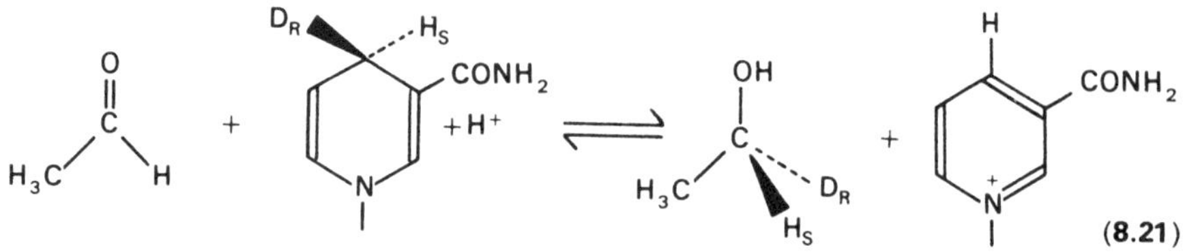

(8.21)

A final example of enzyme stereospecificity is given by the reaction in the citric acid cycle catalysed by aconitase (8.22). This enzyme distinguishes between chemically equivalent carboxymethyl groups, and also between the chemically equivalent hydrogen atoms within the reactive methylene group. Furthermore, re-addition of water to the double bond of the intermediate, *cis*-aconitate, is also stereospecific. These, and other examples of stereospecificity in enzyme catalysed reactions, are a reflection of the fact that during the reaction the substrates are immobilised on the enzyme surface at the active site. The stereochemistry of removal of groups from, or addition of groups to, substrates is then governed by the structure of the active site which, all the available evidence suggests, is precisely defined.

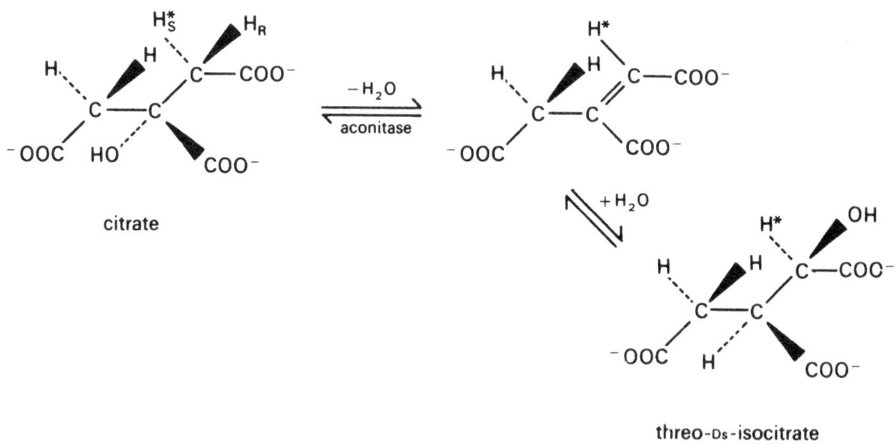

(8.22)

8.5. Enzyme inhibitors

Many compounds are known which decrease the rate of a particular enzyme catalysed reaction. Such compounds are known as enzyme inhibitors and some are of considerable importance as commercial products and as research tools. Enzyme inhibitors can be divided into two types, viz. those which act in an irreversible manner and those which act in a reversible manner.

Irreversible inhibition of enzymes

(i) Active-site directed irreversible inhibitors.
A number of compounds are known which, when they react with certain enzymes, form a covalent bond to the active site and prevent the enzyme from functioning; such compounds are known as active site directed inhibitors. Because a covalent bond is formed the reaction is stoichiometric (8.23), the rate of inhibition is relatively slow (8.24) and the process is not reversed by removing excess inhibitor by, for example, dialysis or chromatography.

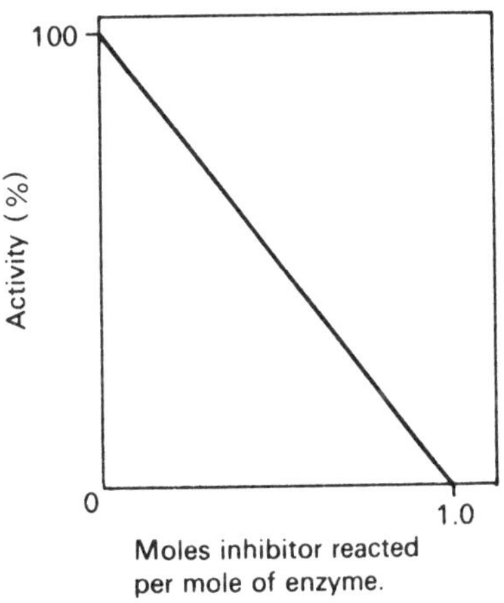

(**8.23**) Graph showing stoichiometry of reaction of di-ethyl *p*-nitrophenyl phosphate with chymotrypsin

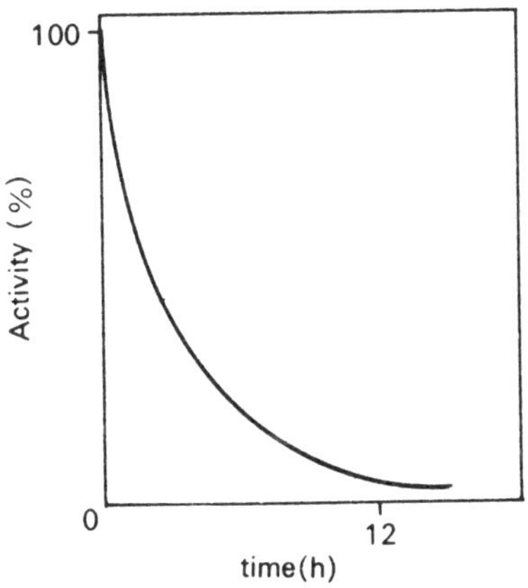

(**8.24**) Time course of inhibition of chymotrypsin by di-ethyl *p*-nitrophenyl phosphate

This type of inhibition is exemplified by the effect of the organophosphorus compounds and carbamates on the activity of acetylcholinesterase. The substrate of this enzyme, acetylcholine, is a transmitter substance in the nervous system of both insects and mammals. If the passage of nerve impulses across synaptic clefts and neuromuscular junctions is to be controlled the transmitter must be destroyed after it has reacted with its receptor site on the distal side of the cleft or junction. The role of acetylcholinesterase is to catalyse the destruction of acetylcholine by hydrolysis; the integrity of the enzyme is therefore essential to the proper functioning of the nervous system. The enzyme, a serine hydrolase, uses a reaction pathway which is similar to that used by chymotrypsin (8.17) for the hydrolysis of esters and peptides. The substrate acetylates a single serine residue in the active centre of the enzyme and the acetyl-enzyme which results then undergoes catalysed hydrolysis (8.25).

$$CH_3COOCH_2CH_2\overset{+}{N}(CH_3)_3 + HOCH_2\text{—enz} \rightleftharpoons CH_3COOCH_2\text{—enz} + HOCH_2CH_2\overset{+}{N}(CH_3)_3;$$

$$CH_3COOCH_2\text{—enz} + H_2O \rightleftharpoons CH_3CO_2H + HOCH_2\text{—enz} \qquad \textbf{(8.25)}$$

It appears that organophosphorus compounds of type (a) in (8.26) are sufficiently similar to the natural substrate to form an enzyme-substrate complex and to then phosphorylate the active serine residue (8.26).

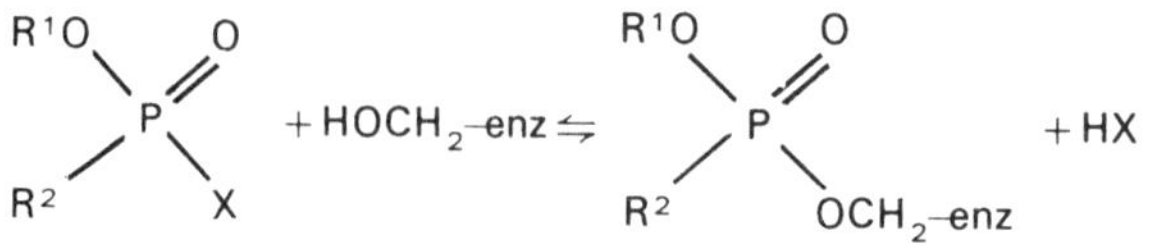

(8.26)

(a)

R^1 = alkyl or aryl
R^2 = alkyloxy, aryloxy, alky, aryl or substituted amino

X = a good leaving group, e.g. —F, —CN, —O— 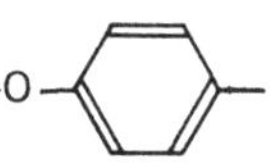 —NO_2, —$\overset{O}{\overset{\|}{O P}}(OR)_2$

However, whereas the acetylated enzyme undergoes rapid catalysed hydrolysis the phosphoryl derivatives of the enzyme are stable and are hydrolysed at a negligible rate. Consequently as phosphorylation proceeds the activity of the enzyme decreases and reaches zero when all of the enzyme has reacted. The stability of phosphorylated derivatives of the enzyme has not, as yet, been satisfactorily explained. It is possilbe that the relatively large substitued phosphoryl group distorts the active centre of the enzyme sufficiently to prevent catalysis occurring.

Organophosphorus inhibitors of acetylcholinesterase were originally developed as chemical warfare agents ('nerve gases') and the early compounds were highly toxic to both mammals and insects. The L.D.$_{50}$ values, in man, for di-isopropyl phosphorofluoridate (8.26a, R′ = $(CH_3)_2CH$, R^2 = $(CH_3)_2CHO$, X = F) and Sarin (8.26a, R′ = CH_3CH_2, R^2 = CH_3, X = F) have been estimated to be in the range 0.01 to 0.1 mg kg^{-1}. Subsequently, however, compounds which showed selective toxicity towards insects were developed, some of which have been widely used as insecticides. Among those still used, particularly in horticulture, are parathion (8.27, R = CH_3CH_2, X = S, Y = OC_6H_4 *p*-NO_2) and malathion (8.27), R = CH_3, X = S, Y = $-\underset{CH_2CO_2CH_2CH_3}{\underset{|}{SCHCO_2CH_2CH_3}}$).

$$(RO)_2\overset{X}{\overset{\|}{P}}—Y \qquad \textbf{(8.27)}$$

These compounds are themselves not highly toxic to either mammals or insects. The latter, however, contain enzymes in their gut, nerve cord and fat bodies which catalyse the oxidative de-sulphurisation of parathion and malathion, producing para-oxon (8.27, R = CH_3CH_2, X = O, Y = $OC_6H_4NO_2$) and malaoxon (8.27, R = CH_3, X = O, Y = $\left.—\underset{CH_2CO_2CH_2CH_3}{\underset{|}{SCHCO_2CH_2CH_3}}\right)$ respectively, both of which are much more effective acetylcholinesterase inhibitors. In contrast the major metabolic reactions of parathion and malathion in mammals involve de-esterification (hydrolysis) catalysed by non-specific phosphatases and carboxyesterases present in plasma and liver; the products of hydrolysis are non-toxic and are excreted.

A number of carbamates, for example Carbaryl (8.28), are also used as insecticides. These compounds are pseudo-substrates for acetylcholinesterase.

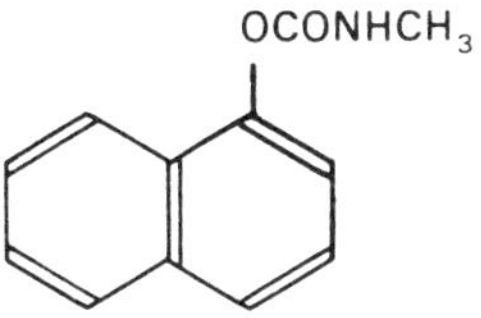

(**8.28**)

A carbamoylated enzyme is formed (8.29) which then undergoes enzyme catalysed hydrolysis at a slow, but finite rate.

$$R^1NHCOOR^2 + HOCH_2\text{–enz} \rightleftharpoons R^1NHCOOCH_2\text{–enz} + HOR^2 \quad (\mathbf{8.29})$$

Penicillins, as exemplified by penicillin G, (8.30) are also thought to be irreversible enzyme inhibitors.

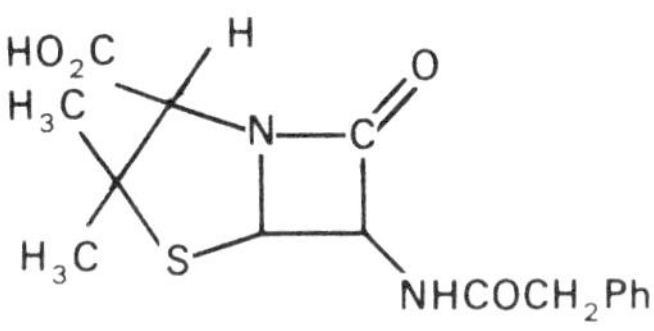

(**8.30**)

The main morphological effect of penicillin G on pencillin sensitive bacterial cells is to cause them to undergo lysis. The internal pressure inside bacterial cells is high (2.03 mN m^{-2} for gram positive cells, 0.5 mN m^{-2} for gram negative cells) and is opposed by a strong cell wall. The wall is composed of heteropolysaccharide chains cross-linked by short polypeptide chains; the last step in its biosynthesis, in **S. aureus** cells, can be represented by (8.31).

Molecular models show that penicillin G is remarkably similar in shape and dimensions to the terminal D-alanyl-D-alanine residues which are involved in the cross-linking reactions. Furthermore, it is known that penicillins are good acylating agents due, at least in part, to the fact that steric strain is relieved when the *β*-lactam ring is opened (8.32).

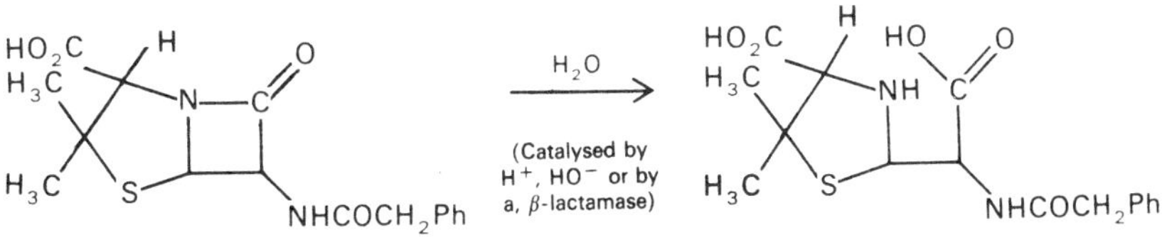

(**8.32**)

It therefore seems reasonable to accept the evidence which suggests that penicillin G binds to the active centre of glycopeptide transpeptidase and inhibits it by acylating an essential group in its active centre. Clearly inhibition of this type

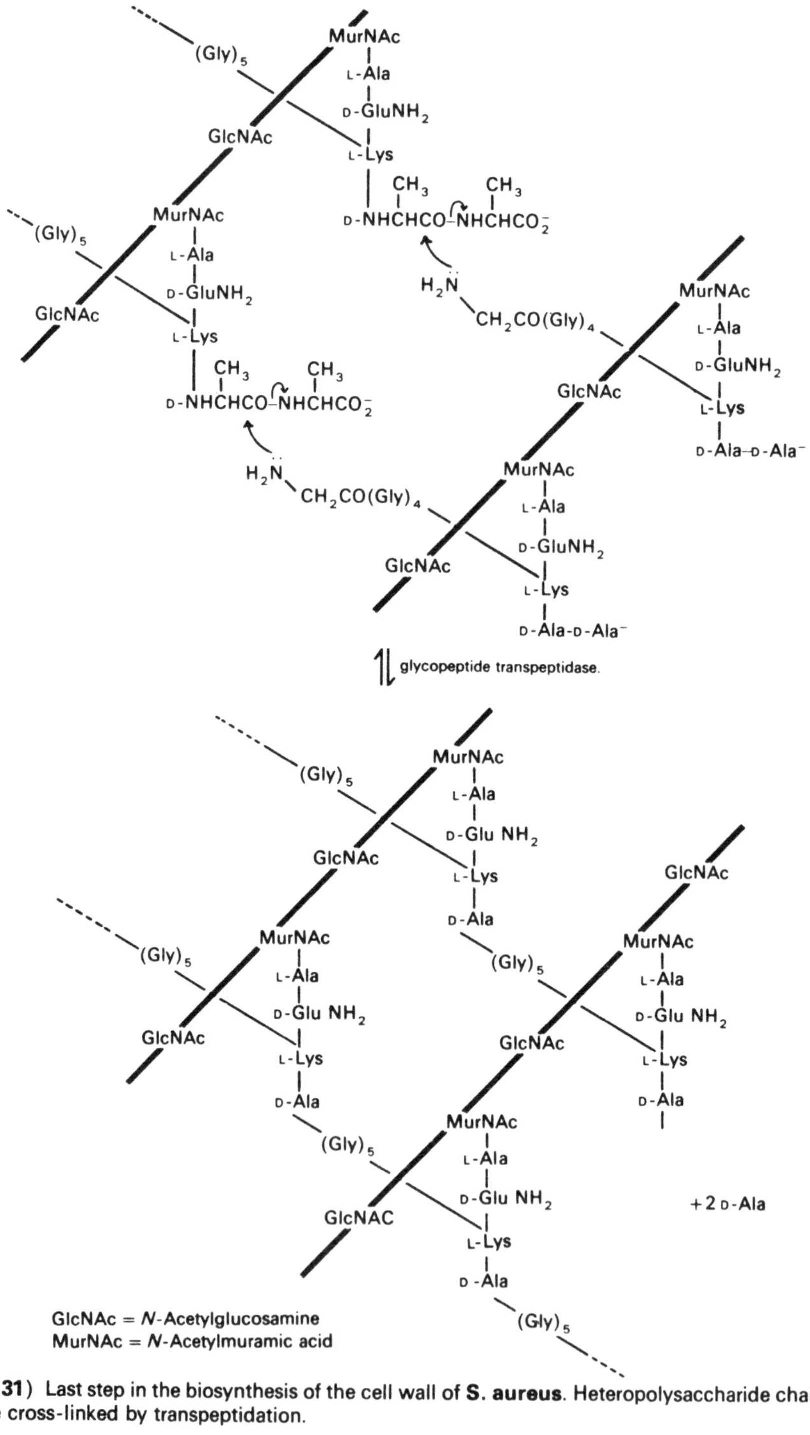

(8.31) Last step in the biosynthesis of the cell wall of **S. aureus**. Heteropolysaccharide chains are cross-linked by transpeptidation.

would lead to a defective cell wall which would be unable to withstand the high pressure inside the cell which would therefore rupture. The situation is, in fact, more complex than outlined above. Different penicillins produce different morphological effects on cells and there is much evidence to suggest that bacteria contain several penicillin sensitive enzymes, the inhibition of each contributing to cell lysis.

Finally it is worth noting that the food industry spends considerable sums of money on research into ways of inactivating certain enzymes present in stored foodstuffs. For example polyphenol oxidase is the enzyme which is principally responsible for catalysing the 'browning' of stored fruit and vegetables. It is clearly desirable to inactivate this enzyme without impairing the quality of the foodstuff. At present this is done by adding small quantities of sulphur dioxide but the exact mechanism of action of this inhibitor is not known.

The organophosphorus compounds, carbamates and penicillin are all known as 'active site directed' inhibitors. It is not essential however, that the inhibitor forms a covalent bond with the active centre of the enzyme for the latter to be irreversibly inactivated. It is conceivable that reaction at a site some distance from the active site could cause an irreversible conformational change in the enzyme molecule which would destroy activity.

(ii) 'Suicide' or 'enzyme-activated irreversible inhibitors'.

Active-site directed irreversible inhibitors are usually group specific. For example, as noted previously, all of the serine hydrolases are inhibited by di-isopropyl phosphorofluoridate and several penicillin sensitive enzymes are inhibited by penicillins and similar β-lactam antibiotics.

In attempts to obtain greater inhibitor specificity, which is obviously highly desirable if inhibitors are to be clinically useful, much effort has been expended to produce 'suicide' or 'enzyme-activated irreversible inhibitors'. This type of inhibitor is chemically inert but after being accepted by the enzyme as a substrate, is converted via the catalytic reaction, to a reactive species which then inhibits the enzyme. Because pyridoxal phosphate (PLP, Section 9.3) dependant enzymes are involved in the metabolism of a variety of physiologically important compounds, much attention has been focused on the development of specific inhibitors of this type. A good example of an enzyme-activated irreversible inhibitor of a PLP dependant enzyme is D-3-fluoro-alanine (8.33) which inhibits alanine racemase.

$$FCH_2CH(\overset{+}{N}H_3)CO_2^- \qquad \textbf{(8.33)}$$

This compound shows interesting antibacterial activity without host toxicity by blocking the biosynthesis of D-alanine, an essential constituent of bacterial cell walls. The evidence which is available suggests that the mechanism of inhibition is as shown in (8.34).

Here the enzyme abstracts a proton from C-2 of the amino-acid residue in the PLP-amino-acid Schiff base, as it does when the normal substrate is present (Section 9.3). The electron pair of the C-2 to H bond then displaces fluoride, giving rise to an enamine which appears to be attacked, in a Michael type addition reaction, by a nucleophilic group of the enzyme; the alkylated enzyme which results is inactive. Although fluorine is difficult to displace in S_N2 reactions, it is a good leaving group in eliminations and its small size avoids problems due to steric hindrance. For these reasons, many enzyme-activated irreversible inhibitors of

PLP dependant enzymes incorporate fluorine as a group which is eliminated during activation of the inhibitor.

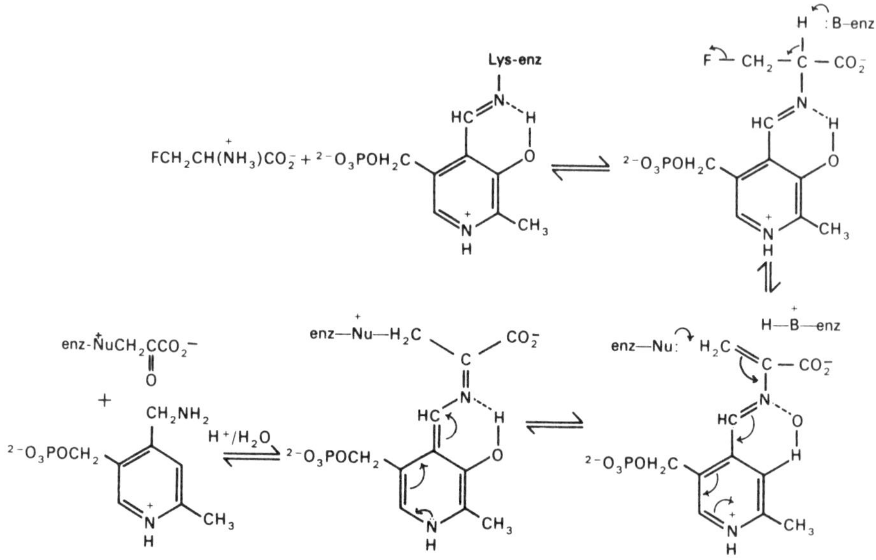

(8.34)

Reversible inhibition of enzymes

A large number of compounds are known which inhibit certain enzymes by processes which are reversed if the inhibitor is removed by, for example, dialysis or chromatography. Logically, these compounds are known as reversible enzyme inhibitors. Their common characteristic is that they bind to the enzyme, not necessarily at the active site, **non-covalently**. Inhibition is therefore established rapidly and there is no simple relationship between the amount of inhibitor bound and the degree of inhibition produced (cf. irreversible inhibitors). Reversible inhibitors can be divided into two types.

(i) **Competitive inhibitors.**

Competitive inhibitors are compounds which compete with substrate in binding to the active site but when bound do not undergo a catalytic reaction or, if they do, react more slowly than substrate. Logically the structure of a competitive inhibitor must be similar enough to that of the appropriate substrate to allow it to be bound to the active site, but sufficiently different to ensure that no catalytic reaction occurs. The classic example of this type of inhibitor is malonic acid which competes with succinic acid for the catalytic site of succinate dehydrogenase (8.35). Malonate is bound to the active site of the enzyme but, since its structure does not allow formation of a double bond, no catalytic reaction occurs.

$$\text{E-FAD} + \bar{\text{O}}\text{CCH}_2\text{CH}_2\text{CO}_2^- \rightleftharpoons \text{E-FA}\bar{\text{D}}.\text{O}_2\text{CCH}_2\text{CH}_2\text{C}\bar{\text{O}}_2 \rightleftharpoons \text{E-FADH}_2 + {}^-\text{O}_2\text{CCH{=}CHCO}_2^-$$

succinate — fumarate

$\downarrow\uparrow$ ${}^-\text{O}_2\text{CCH}_2\text{CO}_2^-$ (malonate)

$$\text{E-FAD.O}_2\text{CCH}_2\text{CO}_2^-$$

(8.35)

A number of competitive inhibitors of enzymes are important in the medical field. The sulphonamide drugs (8.36a) are thought to act as competitive inhibitors, with respect to 4-aminobenzoic acid (8.36b), of dihydropteroate synthetase. In bacteria, but not in mammals, this enzyme catalyses the formation of 7,8-dihydropteroic acid (8.37), an intermediate in the biosynthesis of folic acid.

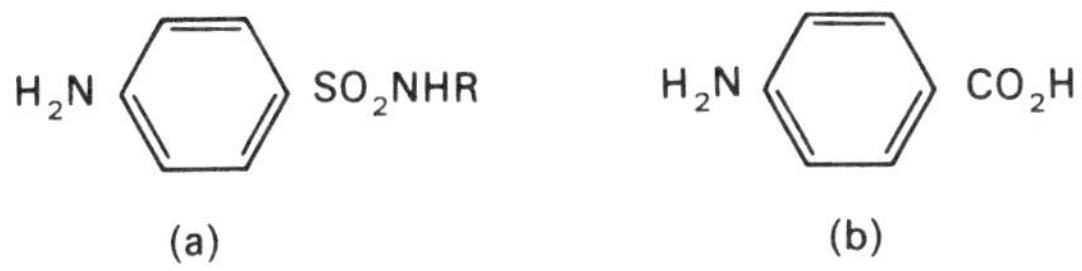

(8.36)

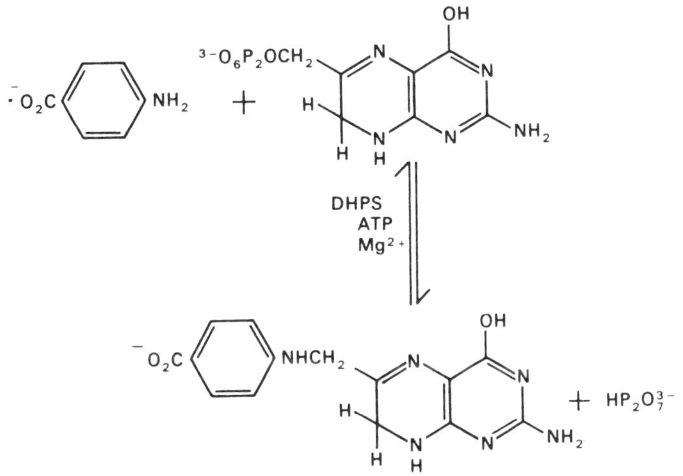

(8.37)

DHPS = dihydropteroate synthetase

The latter is a co-enzyme which is involved with reactions in which a group containing one carbon atom is transferred between substrates (section 9.4). Mammals must ingest preformed folic acid because they cannot synthesise it, whereas bacteria must synthesise their supply. The sulphonamides, therefore, show selective toxicity to bacteria by disrupting their supply of folic acid. Additionally there is some evidence that bacteria can incorporate sulphonamides into defective folic acids; these drugs may not, therefore, act **solely** by competitive enzyme inhibition.

Another example of the use of competitive inhibitors of enzymes in medicine is given by Pargyline ($PhCH_2N(CH_3)CH_2C{\equiv}CH$, *N*-methyl-*N*-propynylbenzylamine). This compound, and similar acetylenic amines, is used as an anti-hypertensive agent in psychiatric medicine. It acts by reversibly inhibiting mono-amine oxidase, an enzyme which is involved in catalysing the degradation of the neuro-transmitter substance, noradrenalin.

Acetylcholinesterase is inhibited reversibly by a large number of quaternary ammonium compounds. Of these Edrophonium chloride (8.38a) is used clinically as a short acting muscle relaxant and as an antidote to Curare poisons and both physostigmine (8.38b) and neostigmine (8.38c) are used in the treatment of glaucoma and myaethenia gravis. The latter two substances exhibit characteristics of both irreversible and reversible inhibitors. Their initial action is as reversible inhibitors because they are both rapidly bound to the active site of the enzyme, in competition with acetylcholine, by their positively charged nitrogens. Their final action, however, is as irreversible inhibitors, because once the compounds are bound they slowly carbamoylate the active serine residue of the enzyme. The mode of action of 1R:2R fluorocitrate on aconitase is similar. Fluorocitrate (8.39 1R:2R-1-fluro-2-hydroxy-1,2,3-propane tricarboxylic acid) is the highly toxic metabolite of fluoro-acetic acid.

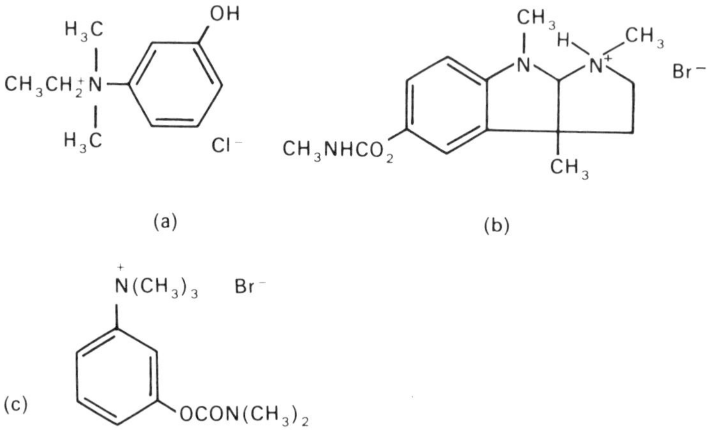

(8.38)

$CHFCO_2H$
|
$C(OH)CO_2H$
|
CH_2CO_2H

(8.39)

Initially it inhibits aconitase reversibly. The initial, rapid reversible inhibition is then followed by slow, irreversible inhibition as a reactive group within the active site of the enzyme is alkylated.

(ii) Non-competitive and un-competitive inhibitors

Non-competitive and un-competitive enzyme inhibitors are compounds which bind reversibly (non-covalently) to a site **other than** the substrate binding site. Non-competitive inhibitors can bind both to the free enzyme and to the enzyme-substrate complex, whereas un-competitive inhibitors can bind **only** to the latter. In contrast to irreversible inhibitors and competitive, reversible inhibitors commercial applications of non-competitive and un-competitive inhibitors do not appear to have been found; their utility appears to be restricted to the study of enzyme kinetics and they will not be discussed further here.

(iii) Transition state analogues as enzyme inhibitors

It has already been stated (Section 8.3) that a considerable body of evidence suggests that an enzyme binds its substrate (s) in a configuration approaching that of the transition state of the reaction which it catalyses. More specifically, application of the transition state theory of reaction rates to reactions catalysed by enzymes predicts that the dissociation constant (K_T, 8.40) which describes binding of the transition state by an enzyme, will lie in the range 10^{-12} to 10^{-16}.

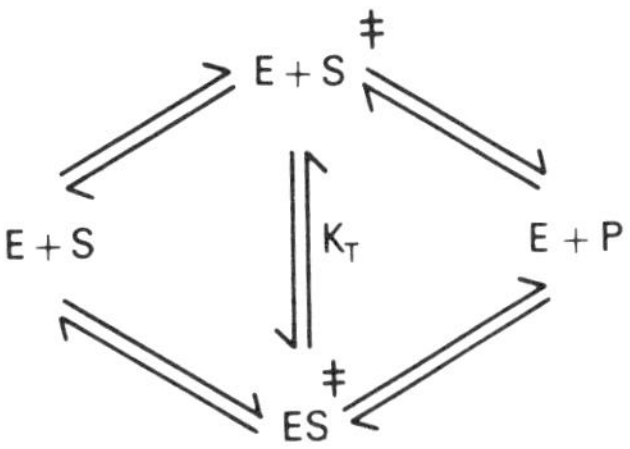

(8.40)

Thus the affinity of an enzyme for the transition state of the corresponding uncatalysed reaction is expected to be enormous. On this basis it would further be expected that unreactive analogues of the transition state would also bind to an enzyme very tightly at the catalytic site and would therefore be very effective and very specific inhibitors of the enzyme. Unfortunately, precise analogues of transition states cannot be constructed because the precise structures of the latter are unknown—they are species in which some covalent bonds originally present in the starting materials of the reaction are partially broken and other bonds which are present in the products are partially formed. It is logical to assume, however, that any metastable intermediate produced in a reaction will have a structure and energy very similar to those of the transition states by which it is formed and removed. Analogues of metastable reaction intermediates would therefore be expected to approximate to true transition state analogues. They also should bind extremely tightly to the appropriate enzyme and act as very effective inhibitors of it. These predictions have been verified in practice. Many transition state inhibitors are now known; some examples are given below.

As will be shown in (9.74) one of the meta-stable intermediates which is thought to be formed during the decarboxylation of 2-oxoacids is an enanamine. Specifically in the case illustrated it is the anion of hydroxyethylthiamine pyrophosphate (HETPP). In this intermediate the thiazole ring is formally uncharged and the bond from C-2 to the exocyclic group has some double bond character. On the basis of these observations it is not surprising that the thiazolone analogue of the HETPP anion (8.41) is a very effective inhibitor of the pyruvate dehydrogenase complex of **E. Coli**. The thiazolone binds so tightly, presumably to the de-carboxylase component of the enzyme complex, that it has been possible only to estimate an upper limit for its dissociation constant (5×10^{-10}).

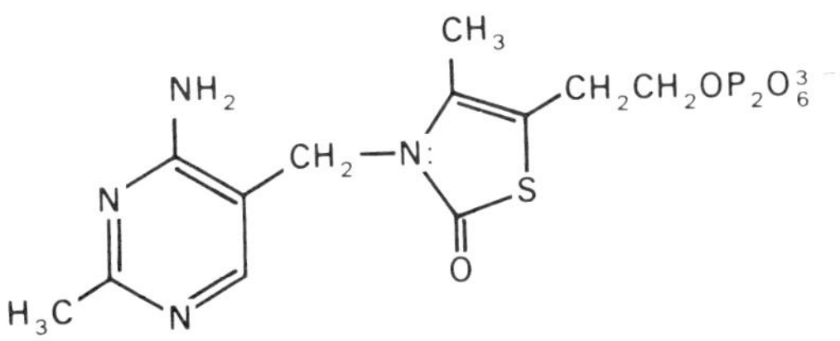

(8.41)

Acylation reactions are thought to involve a tetrahedral oxygen anion as a metastable intermediate (6.84) and a number of enzymes which catalyse acyl transfer reactions are inhibited by analogues of a tetrahedral anion.

Chymotrypsin, for example, is strongly inhibited by phenylethaneboronic acid (8.42).

$PhCH_2CH_2B(OH)_2$ **(8.42)**

It is known that boronic acids combine with hydroxide ions as shown by (8.43) and that the anions which are formed have tetrahedral configurations.

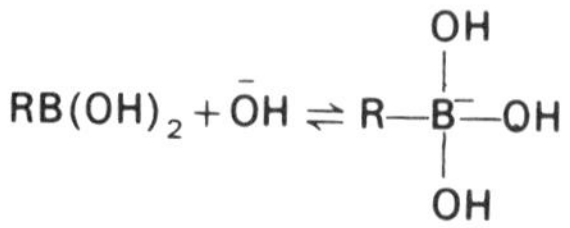

(8.43)

It seems likely, therefore, that phenylboronic acid inhibits chymotrypsin by forming a tetrahedral, anionic, adduct with the hydroxyl group of the serine residue in the catalytic site ((8.44) cf. (8.15), (8.17)). The phenyl group presumably binds in the hydrophobic pocket known to be present in the active site of the enzyme.

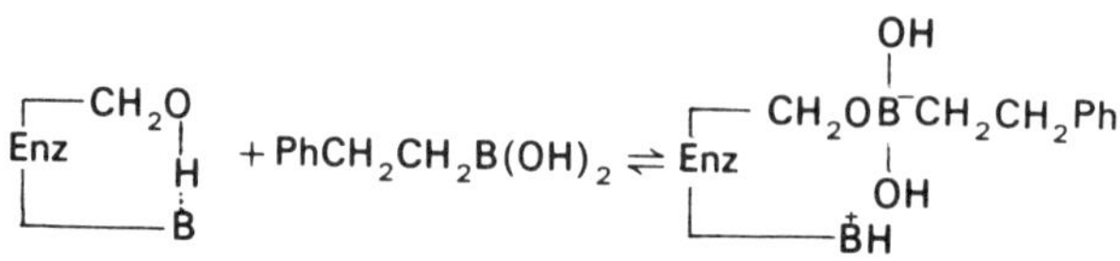

(8.44)

It is interesting to note that a group of protease inhibitors, known as the leupeptins, which are produced by at least seventeen species of **Streptomyces** appear to function as transition state analogue inhibitors. The leupeptins (8.45) are tripeptides containing a C-terminal arginal residue.

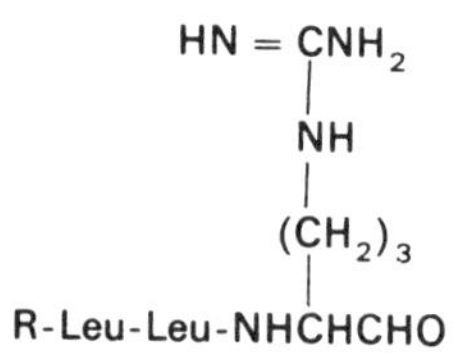

R = CH_3CO- or CH_3CH_2CO-, Leu may be replaced by Ile or Val. **(8.45)**

They inhibit a number of proteases apparently by forming a tetrahedral hemi-acetal or hemi-thioacetal with a seryl or cysteinyl side chain in the active centres of the enzymes,

$$EnzCH_2XH + RCHO \rightleftharpoons EnzCH_2XCH(R)OH, \quad X = O \text{ or } S$$

The potential of transition state analogue inhibitors of enzymes as drugs is currently being investigated in several laboratories.

9

Co-enzymes

In Section 3.3, it was noted that co-enzymes are small non-protein moieties, some of which act as obligatory second substrates of their apo-enzyme while others act as activators of the substrate of their apo-enzyme. They can be regarded as nature's 'bench reagents'. If a chemist wishes to oxidise an alcoholic hydroxyl group he/she reaches for a bottle of acidic dichromate solution; the living cell, however, usually employs either nicotinamide adenine dinucleotide or flavin adenine dinucleotide. Co-enzymes therefore provide a logical 'bridge' between carbon chemistry in the laboratory and in the living cell. Some of the more important co-enzymes are the subject of this section.

9.1. Adenosine triphosphate (ATP)

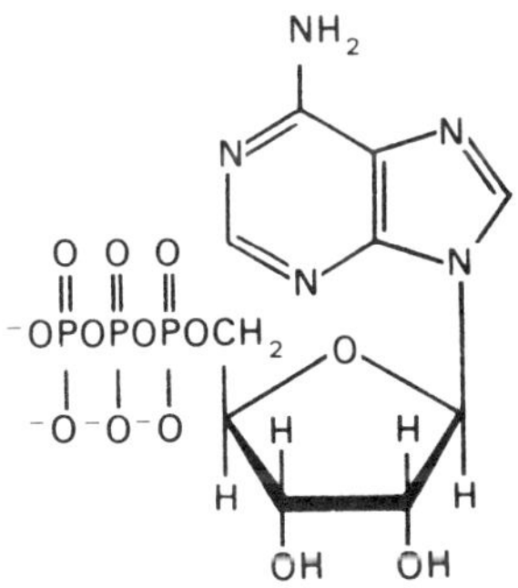

(9.1)

Adenosine triphosphate (9.1) acts as co-enzyme to a very large number of enzymes. Indeed, the nucleoside triphosphate is of such importance in living cells that their metabolism is often described in terms of its production, coupled to oxidative reactions, and its consumption in, principally, biosynthetic processes. The main roles of ATP as a co-enzyme are as follows:

i) As a mediator of condensation reactions

The major, and many minor, constituents of living cells—proteins, polysaccharides, nucleic acids, acylglycerols, etc.,—are, formally, products of condensation reactions. The equilibrium constants of condensation reactions which occur in aqueous solution are usually much less than unity, due at least in part to the fact that the water produced in such a reaction must be liberated into a medium in which its chemical potential is already high (its mole fraction is approximately unity). Reactions with small equilibrium constants do participate in a number of metabolic pathways in living cells, but only in situations where a significant flux of material through the reaction is ensured by virtue of the fact that one of its products is rapidly removed by a subsequent process. For example the reaction (6.17) in which fructose-1,6-diphosphate is cleaved to yield one molecule each of glyceraldehyde-3-phosphate and dihydroxyacetone phosphate has $K = 1 \times 10^{-5}$. Despite this the reaction occurs as a component of the glycolytic pathway because the two three carbon products can be equilibrated and one of them, glyceraldehyde-3-phosphate, rapidly removed by conversion to 1,3-diphosphoglyceric acid. In contrast, reactions with small equilibrium constants could not be utilised by living cells in any situation where the products of the reaction must be **accumulated** in significant quantities. Proteins, polysaccharides, nucleic acids, acylglycerols and certain other condensation products must be accumulated; it is not surprising, therefore, to find that these materials are synthesised indirectly in cells by way of processes which have, overall, large equilibrium constants.

More specifically, these indirect processes employ coupled reactions involving ATP. Two reactions are said to be coupled if the second utilises a product of the first as a starting material; the common product-starting material is referred to as the 'common intermediate' of the two reactions. The involvement of ATP in such reactions can be illustrated by reference to one of the ways (previously outlined in Section 6.2) in which acetylcoenzyme A is synthesised in bacterial cells. First, acetate is phosphorylated by ATP under the catalytic influence of acetate kinase (9.2); acetylphosphate then acetylates co-enzyme A under the catalytic influence of phosphotransacetylase (9.3).

$$CH_3CO_2^- + ATP^{4-} \xrightleftharpoons{\text{acetate kinase, } Mg^{2+}} CH_3COOPO_3^{2-} + ADP^{3-} \quad \textbf{(9.2)}$$

$$CH_3COOPO_3^{2-} + HSCoA \xrightleftharpoons{\text{phosphotransacetylase, } Mg^{2+}} CH_3CoSCoA + HOPO_3^{2-} \quad \textbf{(9.3)}$$

$$\textbf{Sum: } CH_3CO_2^- + ATP^{4-} + HSCoA \rightleftharpoons CH_3COSCoA + ADP^{3-} + HPO_4^{2-} \quad \textbf{(9.4)}$$

The equilibrium constants for the reactions are 8×10^{-3} and 74 respectively, and hence that for the overall reaction (9.4) is $8 \times 10^{-3} \times 74 = 0.59$. This figure is still less than unity but should be compared with that for direct esterification, 6.6×10^{-5}. Other examples of coupled reactions involving ATP are given by the biosynthesis of nucleotides (9.5, 9.6 and 9.7), the activation of amino-acids as esters of a specific tRNA prior to their incorporation into polypeptides (9.8, 9.9, and 9.10), and the biosynthesis of polysaccharides such as glycogen (9.11). In higher animals this last biosynthetic process utilises both ATP and UTP. However, since nucleoside triphosphates can be equilibrated under the catalytic action of nucleoside diphosphate kinase (NDK, myokinase) the addition of each glucose unit to a growing glycogen molecule effectively requires two molecules of ATP.

Why is the equilibrium constant for reaction (9.4) so much larger than that for the direct condensation reaction? In an attempt to answer this question it should be noticed that any set of coupled reactions, above, removes the **elements** of water (not a molecule of water as such) from the reactants of the condensation and incorporates them into phosphate, pyrophosphate or adenylate. For example,

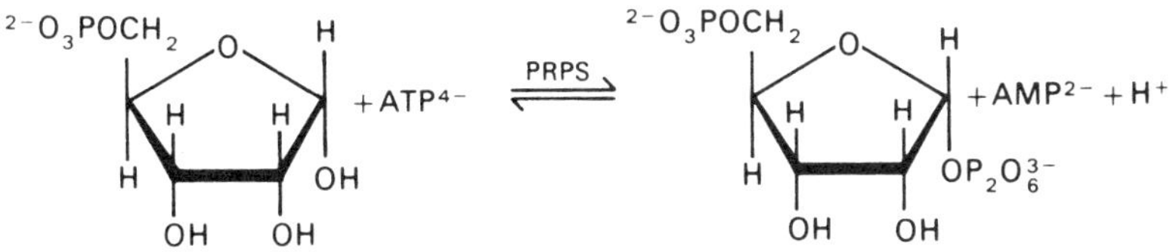

(9.5)

α-D-ribose-5-phosphate

1-pyrophosphorylribose-5-phosphate

PRPS = 1-pyrophosphorylribose-5-phosphate synthetase

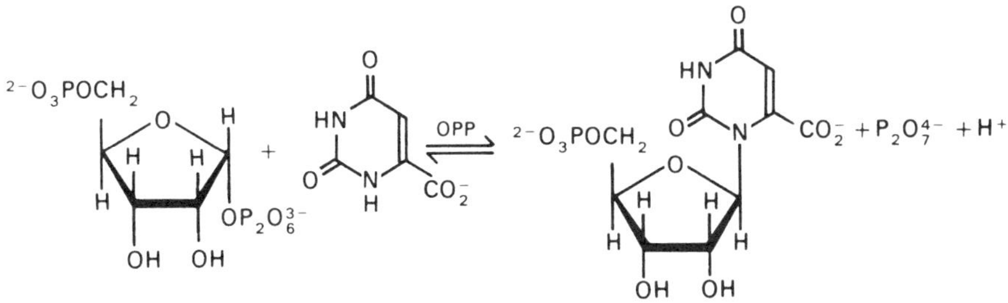

OPP = orotidine-5-phosphate pyrophosphorylase (9.6)

Sum: α-D-ribose-5-phosphate^{2-} + ATP^{4-} + orotate^{-} $\rightleftharpoons$ orotidine-5′-phosphate^{3-} + AMP^{2-} + $P_2O_7^{4-}$ + 2H^{+} (9.7)

$$H_3\overset{+}{N}CH(R)CO_2^- + ATP^{4-} \underset{}{\overset{Mg^{2+},\ AATS}{\rightleftharpoons}} H_2NCH(R)COO\overset{\overset{O}{\|}}{\underset{\underset{O^-}{|}}{P}}OAd + P_2O_7^{4-} + H^+ \qquad (9.8)$$

an amino-acyladenylate

$$H_2NCH(R)COO\overset{\overset{O}{\|}}{\underset{\underset{O^-}{|}}{P}}OAd + HO\text{-}tRNA \overset{Mg^{2+},\ AATS}{\rightleftharpoons} H_2NCH(R)COO\text{-}tRNA + AMP^{2-} + H^+ \qquad (9.9)$$

AATS = an amino-acyl-tRNA synthetase. Ad = adenosinyl.

Sum: $H_3\overset{+}{N}CH(R)CO_2^-$ + ATP^{4-} + HO-tRNA $\rightleftharpoons$ $H_2NCH(R)COO$-tRNA + AMP^{2-} + $P_2O_7^{4-}$ + 2H^{+} (9.10)

making allowance for ionisation, in (9.2) a hydroxyl group is removed from acetic acid and liberated as part of ADP, while in (9.3) a proton is removed from co-enzyme A and liberated as part of phosphoric acid. One might therefore suspect, intuitively, that formation of the condensation product would be facilitated because water is being removed and liberated in the form of species which will have mole fractions much less than unity. Notice, also, that the net reaction corresponding to any set of coupled reactions is not only the sum of the component reactions but also the sum of the direct condensation reaction and the

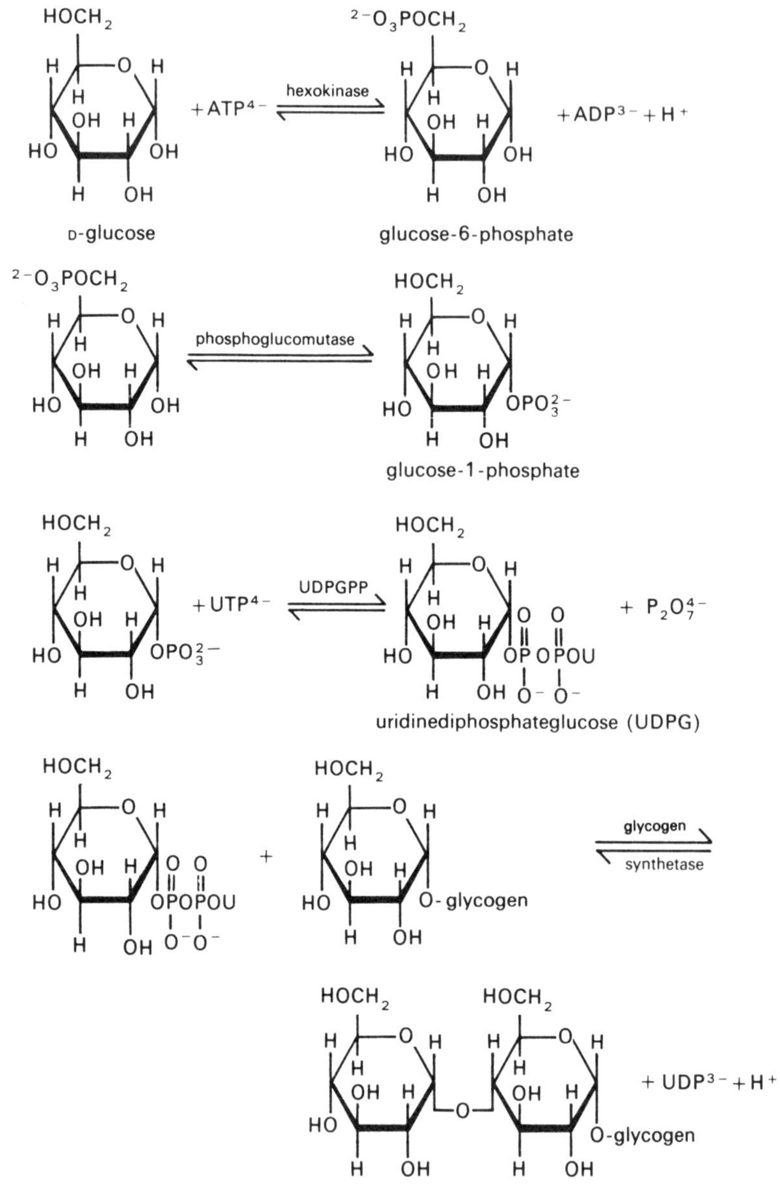

UDPGPP = uridinediphosphoglucosepyrophosphorylase

$$UDP^{3-} + ATP^{4-} \underset{}{\overset{NDK}{\rightleftharpoons}} UTP^{4-} + ADP^{3-}$$

NDK = nucleosidediphosphate kinase

Sum: glucosyl-OH + 2ATP^{4-} + H$_2$O + HO-glycogen $\rightleftharpoons$ glucosyl-O-glycogen + 2ADP^{3-} + 2HPO$_4^{2-}$ + 2H$^+$ **(9.11)**

hydrolysis of ATP. This means that the equilibrium constant for the net reaction is not only equal to the product of the constants of the component reactions, but also to the product of the constants for the direct condensation reaction and the hydrolysis of ATP as well. Now the equilibrium constant for the hydrolysis of ATP has been estimated as 1×10^6. This means that if ATP is used as a reactant in, and is cleaved by, a set of coupled reactions leading to a condensation product, then the equilibrium constant for the net reaction will be 1×10^6 times that of the direct condensation. It is immaterial that ATP is not hydrolysed directly by water during the coupled reaction; it is a fundamental thermodynamic property of chemical reactions that the equilibrium constant reflects only the equilibrium concentrations of reactants and products and is completely independent of the pathway which the reaction follows.

The size of the equilibrium constant for the (direct or indirect) hydrolysis of ATP indicates that a solution of ATP is thermodynamically very much less stable than a solution of its hydrolysis products. It has been suggested that the large difference in stability reflects the fact that strong forces of electrical repulsion between negatively charged oxygen atoms are reduced when ATP is cleaved and also the fact that negative charge in ADP and phosphate can be de-localised more effectively than it can in ATP.

The inclusion of ATP as a reactant in a condensation greatly enhances the thermodynamic potential for reaction. This increased potential, however, would be of no value if it could not be realised by way of kinetically viable (that is low energy) pathways. Since a large number of condensation reactions involving ATP are observed in living cells it is apparent that the central role of the nucleotide in metabolism is due as much to its kinetic versatility as to its thermodynamic properties.

ATP does not react with water at an appreciable rate at physiological pH and normal temperatures; it is kinetically stable towards attack by water and, by inference, other weak nucleophiles. In contrast, in the presence of divalent metal ions, usually magnesium but occasionally manganese, and an appropriate enzyme, ATP is kinetically unstable towards many nucleophiles. Note, however, that direct enzyme catalysed hydrolysis of ATP is observed in cells only under special circumstances. This is logical, because if ATP is to function as a mediator of condensation reactions and in other roles its concentration in cells must be maintained at an adequate level—it must **not** be destroyed by direct hydrolysis. ATP undergoes three types of enzyme catalysed nucleophilic displacement reaction in which

(i) the nucleophile attacks the γ-phosphorus atom of ATP displacing ADP; a phosphoryl group is transferred from ATP to the nucleophile. These reactions are catalysed by **kinases** and are exemplified by (9.2).

(ii) the nucleophile attacks the β-phosphorus atom of ATP displacing AMP; a pyrophosphoryl group is transferred to the nucleophile. The synthesis of 1-pyrophosphorylribose-5-phosphate (9.5) is an example of this type of reaction, which is much less common than type (i) or (iii).

(iii) the nucleophile attacks the α-phosphorus atom of ATP displacing pyrophosphate; an AMP group is transferred to the nucleophile. Nucleotidyl transferase and ligases such as amino-acyl-tRNA synthetases (9.8) catalyse this type of reaction.

In coupled reactions leading to condensation products the reactions just described are followed by a further reaction in which the phosphorylated, pyrophosphorylated or adenylated product of the first reaction is itself attacked by a nucleophile; a phosphate, pyrophosphate or adenylate group is displaced as the condensation product is formed. (9.3, 9.6 and 9.9, respectively). Now phosphate, pyrophosphate, adenylate and ADP groups can all be displaced by a nucleophile as stable, weakly basic anions; they are all intrinsically good leaving groups. Current research suggests that ATP binds divalent cations in different modes which, together with enzyme catalysis, further enhances the properties of the aforementioned groups as leaving groups. Metal ion binding neutralises negative charge and effectively decreases the basicity of the leaving groups; binding of the oxygens of ATP to cationic centres in an enzyme would be expected to have the same effect. It is also probable that metal ion binding determines and facilitates, at least in part, attack by the nucleophile at a specific phosphorus atom. Again, enzyme-substrate binding undoubtedly contributes to this feature of the reaction. More specifically it appears that when displacement occurs at the α-phosphorus of ATP (catalysed by some kinases, tRNA synthetases, DNA polymerase, etc.) the leaving group (pyrophosphate) is liganded to the metal ion by the oxygens of the γ-phosphorus. When AMP is the leaving group, as in the reaction catalysed by 1-pyrophosphorylribose-5-phosphate synthetase (9.5) the group which is attacked is liganded via the β- and γ-oxygens. In contrast, many kinases, e.g. pyruvate kinase, hexokinase and arginine kinase, catalyse reactions in which both leaving group (ADP) and the phosphoryl group which is attacked are liganded to magnesium; the β and γ or alternatively the α, β and γ oxygens are involved.

Finally it is worth noting that condensation products are often prepared in the laboratory in a manner exactly analogous to that observed in living cells. For example esters and amides are often prepared indirectly using two coupled reactions involving a common intermediate such as an acid chloride (9.12, 9.13 and 9.14).

$$\underset{\text{benzoic acid}}{PhCO_2H} + PCl_5 \rightleftharpoons \underset{\text{benzoyl chloride}}{PhCOCl} + POCl_3 + HCl \qquad \textbf{(9.12)}$$

$$PhCOCl + HSCH_2CH_3 \rightleftharpoons \underset{S\text{-ethyl thiobenzoate}}{PhCOSCH_2CH_3} + HCl \qquad \textbf{(9.13)}$$

$$\textbf{Sum: } PhCO_2H + PCl_5 + HSCH_2CH_3 \rightleftharpoons PhCOSCH_2CH_3 + POCl_3 + 2HCl \qquad \textbf{(9.14)}$$

While quantitative data is not available in all cases, there is no doubt that the inclusion of a compound such as PCl_5 as a reactant improves the thermodynamic potential for reaction. PCl_5 and the common intermediates derived from it are also very reactive in a kinetic sense.

ii) As a mediator of the conversion of chemical energy to electrical, osmotic or mechanical energy

ATP is involved, *in vivo*, in a number of processes in which chemical energy is converted to electrical, osmotic or mechanical energy. For example an ATP dependent ion pump maintains the osmotic gradients of sodium and potassium ions across the membrane of red blood cells ($[K^+]_{rbc} = 150$ mM, $[K^+]_{plasma} = 5.4$ mM; $[Na^+]_{rbc} = 12$–20 mM, $[Na^+]_{plasma} = 144$ mM) and ATP is claimed to be the immediate mediator of muscle contraction. Coupled reactions of the types described in (i), above, are probably involved in these processes. However, the nature of the reactions is unknown, precluding further discussion here.

iii) As a mediator of the elimination of water

A number of reactions which involve the elimination of a molecule of water are known to occur in living cells; examples have already been quoted in Sections 4.2 and 5.1 (see also, 7.27). The reactions cited are used for both degradative and biosynthetic processes. In contrast, a number of dehydrations associated only with biosynthetic pathways are mediated by ATP. For example one step in the biosynthesis of inosinic acid involves cyclisation of α-*N*-formylglycinimide ribonucleotide by elimination of water (9.15)

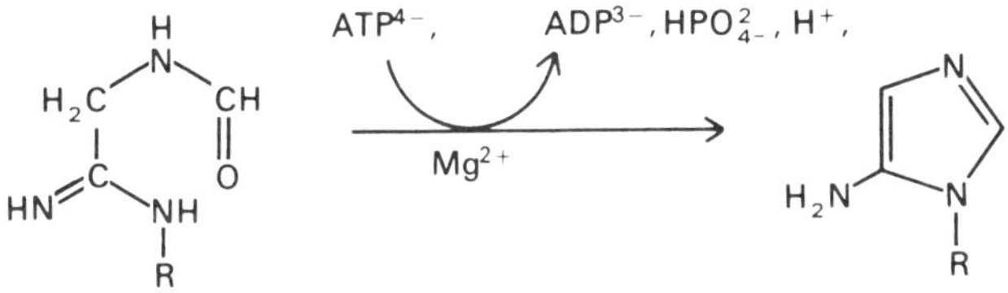

(9.15)

R = 1′-ribosyl-5′-phosphate

This reaction probably involves nucleophilic attack by the acetal nitrogen on the carbonyl group. The hydroxy derivative which is formed is then phosphorylated by ATP and phosphoric acid is eliminated to generate the 2, 3 double bond. Finally pseudo-aromatic character is generated in the product by tautomerisation of the exocyclic imino group (9.16)

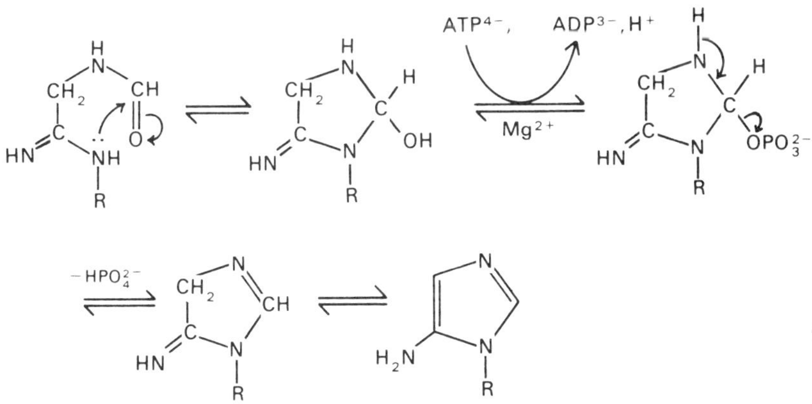

(9.16)

R = 1′-ribosyl-5′-phosphate

Carbonyl functions in naturally occurring heterocyclic compounds such as uridine triphosphate (UTP) are often converted to amino functions by analogous

reactions (9.17)

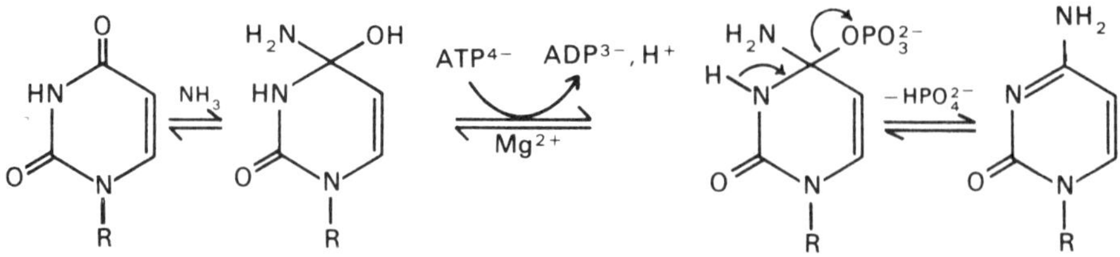

R = 1′-ribosyl-5′-triphosphate **(9.17)**

Reference to Section 5.1 indicates that in the type of reaction just cited ATP is facilitating the elimination of water by converting the hydroxyl group to a better leaving group. The reactions are exactly analogous to those used in the laboratory to eliminate water from an alcohol via the acetate or toulene-*p*-sulphonate ester. A variation on this theme is seen in the biosynthesis of terpenes. 3R-Mevalonic acid, derived from acetate, is both phosphorylated and pyrophosphorylated, the latter process involving two consecutive phosphorylations. A double bond is then introduced by elimination of phosphate **and** carbon dioxide. The unsaturated material, isopentenylpyrophosphate is isomerised, producing dimethylallylpyrophosphate. The former compound then acts as a nucleophile towards the latter in an S_N2 reaction in which pyrophosphate acts as leaving group. (9.18)

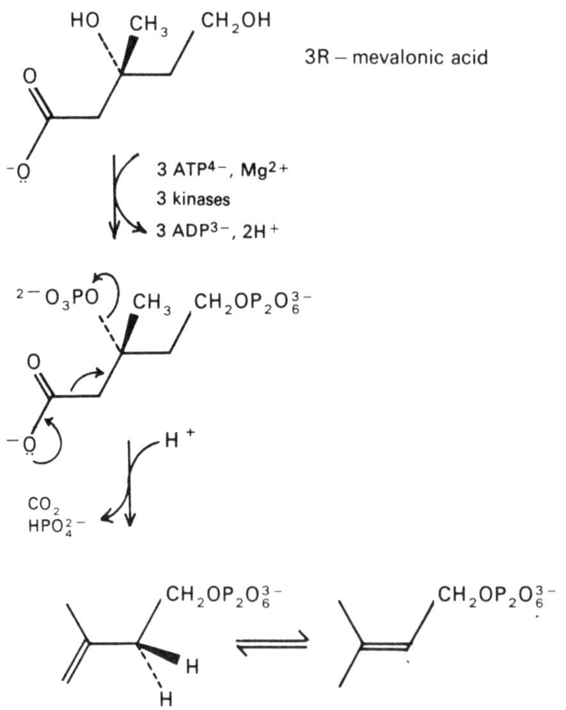

isopentylpyrophosphate dimethylalkylpyrophosphate **(9.18 contd)**

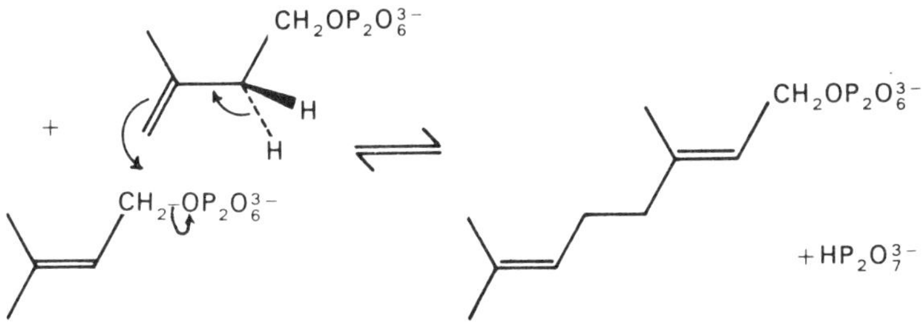

trans-geranylpyrophosphate **(9.18)**

This reaction sequence gives a very good illustration of the similarities, outlined in Section 5.1, between nucleophilic substitution and β-elimination. Thus two hydroxyl groups in mevalonic acid are activated, in a very similar manner, one towards substitution and the other towards elimination; enzyme function and specificity clearly determines the course of reaction of each modified hydroxyl group.

9.2. Redox co-enzymes

Virtually all enzymes which catalyse redox reactions utilise one of two types of co-enzyme—either a pyridine nucleotide or a flavin.

The pyridine nucleotides

Two pyridine nucleotides, nicotinamide adenine dinucleotide (NAD^+9.19a) and its 2′-phosphate ester ($NADP^+$, 9.19b) together with their reduced forms (NADH and NADPH) constitute the mobile pool of redox reagents in living cells. The nucleotides act as diffusible co-enzymes to a large number of dehydrogenases which catalyse a range of reactions illustrated by equations (9.20) to (9.24). Two generalisations can be made:

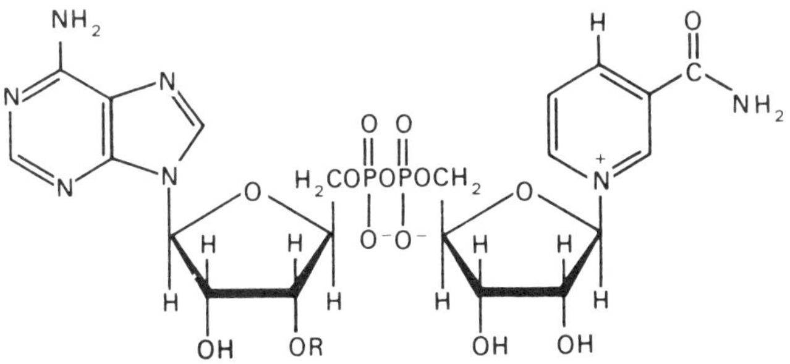

(**9.19**, a, R = H; b, $R = \overset{O}{\overset{\|}{P}}(OH)_2$) **(9.19)**

$$CH_3CH_2OH + \overset{+}{NAD} \overset{ADH}{\rightleftharpoons} CH_3CHO + NADH + \overset{+}{H} \qquad \textbf{(9.20)}$$

ethanol ethanal

$$\underset{\text{glutamate}}{H_3\overset{+}{N}CH(CH_2CH_2CO_2^-)CO_2^-} + NA\overset{+}{D} + H_2O \overset{GDH}{\rightleftharpoons} \underset{\text{2-oxoglutarate}}{O{=}C(CH_2CH_2CO_2^-)CO_2^-} + NADH + H^+ \quad (9.21)$$

$$\underset{\text{glyceraldehyde-3-phosphate}}{{}^{2-}O_3POCH_2CH(OH)CHO} + NAD^+ + HPO_4^{2-} \overset{GPDH}{\rightleftharpoons} \underset{\text{1,3-diphosphoglyceric acid}}{{}^{2-}O_3POCH_2CH(OH)CO_2PO_3^{2-}} + NADH + H^+ \quad (9.22)$$

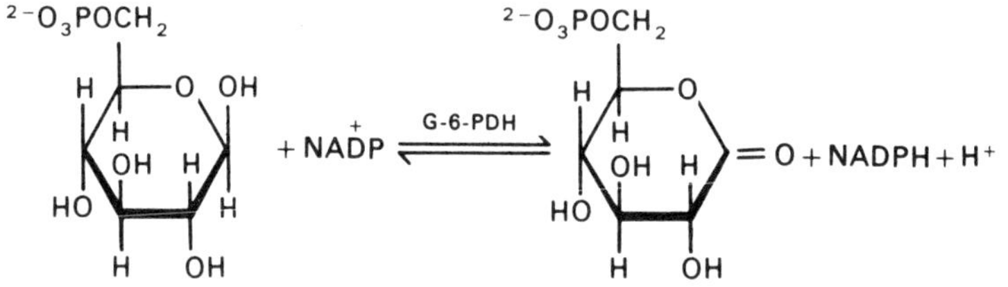

D-glucose glucose-6-phosphate (9.23)

$$\underset{\text{crotonylACP}}{CH_3CH{=}CH\overset{O}{\overset{\|}{C}}SACP} + NADPH + H^+ \overset{CR}{\rightleftharpoons} \underset{\text{butyrylACP}}{CH_3CH_2CH_2\overset{O}{\overset{\|}{C}}SACP} + NADP^+ \quad (9.24)$$

ADH = alcohol dehydrogenase; GPDH = glyceraldehyde-3-phosphate dehydrogenase; GDH = glutamate dehydrogenase; G-6-PDH = glucose-6-phosphate dehydrogenase; CR = crotonylACP reductase.

i) the reactions in which the nucleotides participate usually involve interconversion of $>$CHOH and $>$C=O, $>$CH–NH_2 and $>$C=NH, (the latter are usually hydrolysed to $>$C=O groups) or $_{H}{>}$C=O and $_{HO}{>}$C=O groups; additionally–CH_2–CH_2– and –CH=CH–groups can be interconverted provided they are activated by an adjacent carbonyl group.

ii) reactions in which NAD^+/NADH participate are usually components of degradative pathways leading to the synthesis of ATP. In contrast, reactions involving $NADP^+$/NADPH are usually components of biosynthetic pathways.

Mode of action of the pyridine nucleotides as redox reagents.

All the evidence which is available indicates that, in terms of their mechanism of action, the functional part of both $NAD^+$and NADP is the quaternised pyridine ring.

Investigations of the reduction of relevant simple 1-substituted pyridines, and of their reactions with nucleophiles, together with studies of enzyme catalysed reactions involving $NAD(P)^+$ suggest that the co-enzymes undergo 2 electron reduction at the 4 position of the pyridine ring (9.25). It is probable, but not proven beyond doubt, that the reduction is mediated by direct transfer of a hydride ion from the substrate, SH, to the co-enzyme. The positively charged nitrogen in $NAD(P)^+$ is presumably able to delocalise the negative charge introduced by the hydride ion and thus stabilise the transition state involved in the reaction.

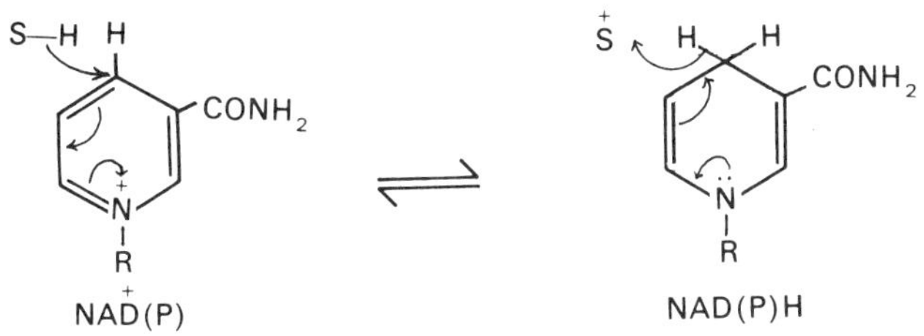

(9.25)

In keeping with the formulation of the redox reaction it has been demonstrated that a number of simple 1-substituted-1,4-dihydropyridines analogous to NAD(P)H are competent donors of hydride ion to electrophilic centres, although the conditions under which these model reactions occur are very different from those employed for NADH dependant reactions catalysed by enzymes.

The mechanism of action and stereochemistry of reaction (9.20), catalysed by alcohol dehydrogenase (ADH), has been investigated in considerable detail. Liver ADH contains zinc which is essential for catalytic activity. On the basis of X-ray diffraction studies it appears that the zinc binds both substrate and product. It presumably also polarises the latter, facilitating transfer of the hydride ion to the carbonyl group in the reverse reaction (9.26). The stereochemistry of this redox reaction has provided a striking example of enzyme stereospecificity as discussed in Section (8.4).

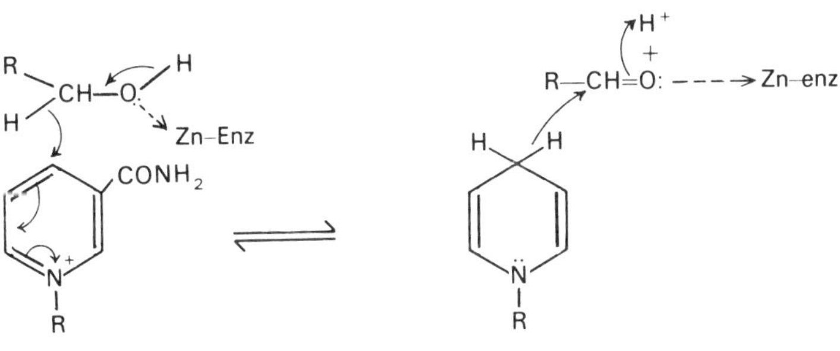

(9.26)

The steps in the interconversion of glutamate and 2-oxoglutarate by reaction (9.21) have been outlined previously in (5.132). The mechanism of the first step involving interconversion of a $>$CH–NH_2 and a $>$C=NH group is undoubtedly analogous to that involved in the oxidation of alcohols by $NAD(P)^+$. The steps involved in (9.22) have been given in (6.38) and need no further elaboration here. The most likely mechanism of the reduction of 2,3-unsaturated esters of ACP (9.24), or of co-enzyme A, by NAD(P)H involves protonation of the substrate followed by hydride transfer from the co-enzyme.

The Flavin Co-enzymes

The flavin co-enzymes, flavin mononucleotide (FMN, 9.27a) and flavin adenine dinucleotide (FAD, 9.27b) are both derivatives of Vitamin B_2 (riboflavin).

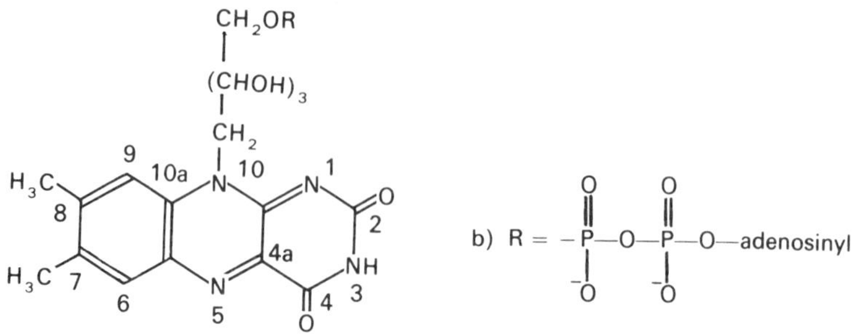

(9.27)

The mononucleotide is riboflavin-5′-phosphate and the dinucleotide is the adenylyl derivative of the mononucleotide. FMN and FAD appear to function in the same way, acting as co-enzymes for a wide variety of enzymes which catalyse redox reactions. The flavin nucleotides, however, differ from the other main type of cellular redox reagent, the pyridine nucleotides, in three important respects. First, flavins are always bound tightly, although not invariably covalently, to their apo-enzymes. This means that after reaction a reduced flavin must be re-oxidised before the holo-enzyme can participate in further catalysis. In contrast NADH diffuses away, after reaction, from a catalytically active enzyme which must only bind NAD^+ in order to catalyse further oxidation. A consequence of this is that flavins do not appear explicitly in the stoichiometry of the redox reactions in which they are involved (9.28)

$SH_2 + \text{Enz–Fl} \rightleftharpoons S + \text{Enz–Fl}H_2$ (reductive half reaction)

$\text{Enz–Fl}H_2 + X \rightleftharpoons \text{Enz–Fl} + XH_2$ (oxidative half reaction)

Sum: $SH_2 + X \rightleftharpoons S + XH_2$ **(9.28)**

Secondly, flavins are more versatile redox reagents than the pyridine nucleotides. The latter appear to be concerned only with **concerted** two electron transfers, whereas the former can participate in both concerted two electron transfers or in two sequential one electron transfers. Both one and two electron reduction of flavins involve the iso-alloxazine system; the product of one electron reduction is the stable flavin semi-quinone (9.29).

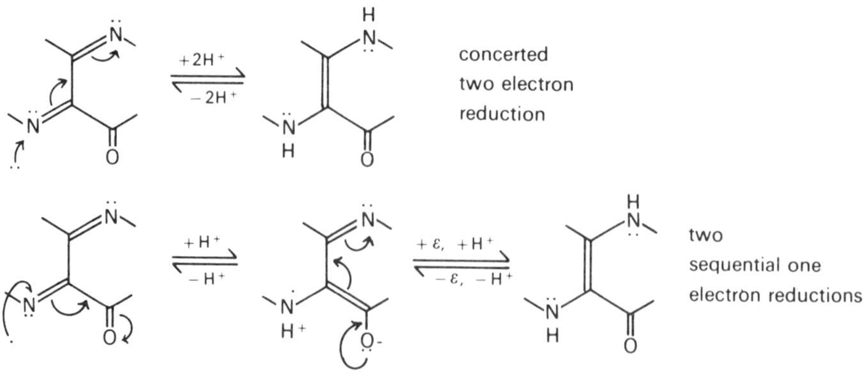

flavin semiquinone **(9.29)**

Thirdly, whereas reduced pyridine co-enzymes cannot be re-oxidised directly by molecular oxygen, dihydroflavins undergo rapid autoxidation and the intermediate(s) produced in the process are probably involved in the oxygen insertions catalysed by certain flavoenzymes.

Roles and modes of action of the flavin co-enzymes

As the chemistry of flavin mediated redox reactions is extremely complex, and because no obvious analogies with simple laboratory reactions can be made, only the major roles of the co-enzymes will be discussed.

Flavoenzymes have four major roles in living cells:

(*i*) *The oxidation of NAD(P)H.* Mitochondrial NADH dehydrogenase is an enzyme complex containing an FMN dependant dehydrogenase and several non-haem iron proteins. The complex catalyses the oxidation of NADH at the expense of FMN. The reduced flavin is oxidised by co-enzyme Q (ubiquinone), the electrons probably being transferred one at a time via non-haem iron proteins (9.30)

$$\mathrm{NADH} + \mathrm{H}^+ + \mathrm{Enz{-}FMN} \rightleftharpoons \mathrm{NAD}^+ + \mathrm{Enz{-}FMNH_2};$$

$$\mathrm{Enz{-}FMNH_2} + \mathrm{CoQ} \underset{\text{iron proteins}}{\overset{\text{non-haem}}{\rightleftharpoons}} \mathrm{Enz{-}FMN} + \mathrm{CoQH_2} \qquad \textbf{(9.30)}$$

Soluble enzymes which catalyse the transfer of hydrogen between NAD^+ and NADPH have been isolated from *Pseudomonas fluorescens* and from certain species of *Azotobacter*. These enzymes, unlike the membrane bound FMN dependant transhydrogenase of mitochondria, contain tightly bound FAD. The co-enzyme appears to undergo sequential oxidation and reduction during the reaction (9.31).

$$\mathrm{NAD}^+ + \mathrm{Enz{-}FADH_2} \rightleftharpoons \mathrm{NADH} + \mathrm{H}^+ + \mathrm{Enz{-}FAD};$$

$$\mathrm{Enz\ FAD} + \mathrm{NADPH} + \mathrm{H}^+ \rightleftharpoons \mathrm{Enz{-}FADH_2} + \mathrm{NAD}^+\mathrm{P} \qquad \textbf{(9.31)}$$

In addition to the two enzymes above, a number of other flavoproteins are reduced by NAD(P)H during the course of the reaction which they catalyse. For example, glutathione reductase appears to be a pyridine nucleotide dependent dehydrogenase (9.32) as do bacterial orotate dehydrogenase (9.33) and several mono-oxygenases such as microsomal amine *N*-oxidase (9.34). These enzymes, however, are flavoproteins and the role of NAD(P)H is to reduce the flavin to the dihydroform which then reacts with substrate or with oxygen. (See later paragraphs of this section).

$$\mathrm{G{-}S{-}S{-}G} + \mathrm{NADPH} + \mathrm{H}^+ \underset{\text{reductase}}{\overset{\text{glutathione}}{\rightleftharpoons}} \mathrm{2GSH} + \mathrm{NADP}^+ \qquad \textbf{(9.32)}$$

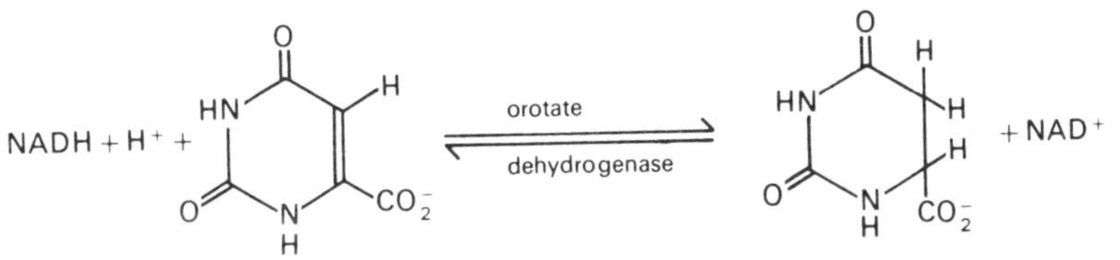

(9.33)

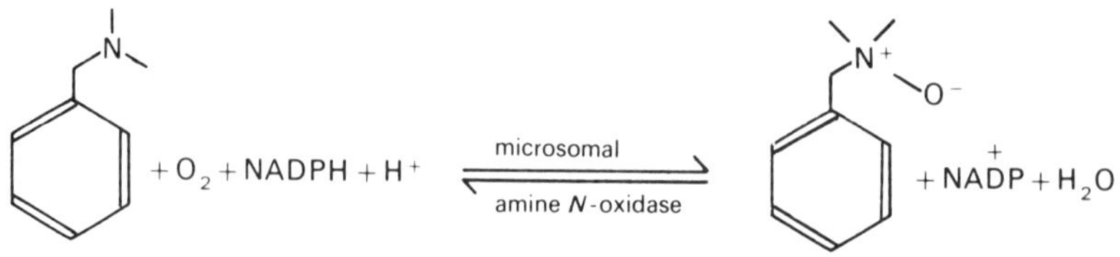

(9.34)

The evidence which is available suggests that the reduction of flavins by NAD(P)H involves direct, stereospecific, transfer of a hydride ion from C-4 of the nicotinamide to N-5 of the flavin (9.35)

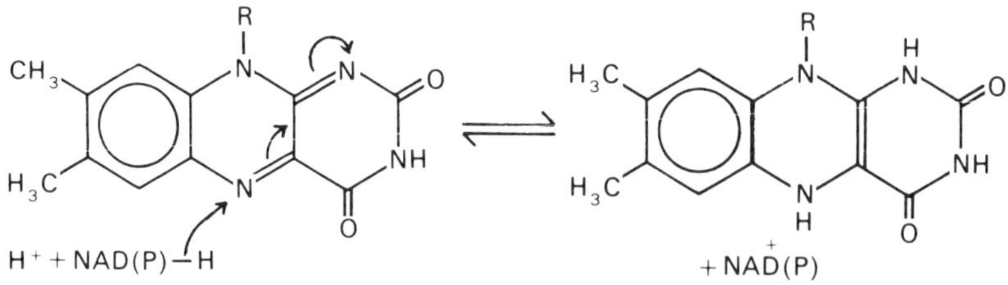

$R = -CH_2(CHOH)_3CH_2OPO_3^{2-}$ or $-CH_2(CHOH)_3CH_2OP(=O)(O^-)-O-P(=O)(O^-)-O-$Adenosinyl **(9.35)**

(ii) *Interconversion of* $-CH_2-CH_2-$ *and* $-CH{=}CH-$ *groups adjacent to a carbonyl group.* Although a few enzymes which catalyse this type of conversion utilise pyridine nucleotide co-enzymes, most are flavo-enzymes. Thus bacterial orotate dehydrogenase, succinate dehydrogenase and the acylcoenzyme A dehydrogenases which catalyse reactions (9.33), (9.36) and (9.37), respectively, are all flavo-enzymes.

$$^{-}O_2CCH_2CH_2CO_2^- + \text{Enz–FAD} \rightleftharpoons {}^{-}O_2CCH{=}CHCO_2^- + \text{Enz–FADH}_2 \quad \textbf{(9.36)}$$

$$RCH{=}CHCOSCoA + \text{Enz–FADH}_2 \rightleftharpoons RCH_2CH_2COSCoA + \text{Enz–FAD} \quad \textbf{(9.37)}$$

It appears that these dehydrogenations proceed by a stereospecific proton abstraction-hydride transfer mechanism (9.38).

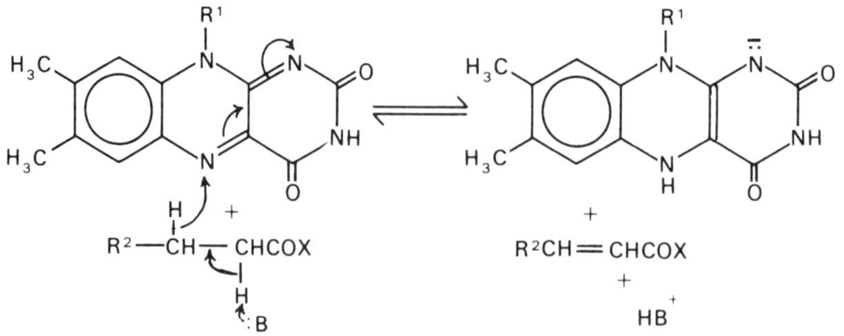

$R^1 = -CH_2CH(OH)_3CH_2OPO_3^{2-}$ OR $-CH_2(CHOH)_3CH_2OP(=O)(O^-)OP(=O)(O^-)O$ Adenosinyl **(9.38)**

The β-hydrogen atoms in the reduced substrates are weakly acidic due to the $-M$, $-I$ polarisation effects of the adjacent carbonyl group; the departing hydride ion is transferred to N_5 of the flavin as described previously for transfer from NAD(P)H. The reduced flavo-enzymes which are produced in these dehydrogenations are re-oxidised by non-haem iron proteins.

(*iii*) *Oxidation of sugars, amines, amino-acids and hydroxyacids.* Several types of compound, including simple sugars, amino-acids, hydroxyacids and some amines can be oxidised in living cells, apparently by direct interaction with molecular oxygen (9.39 and 9.40)

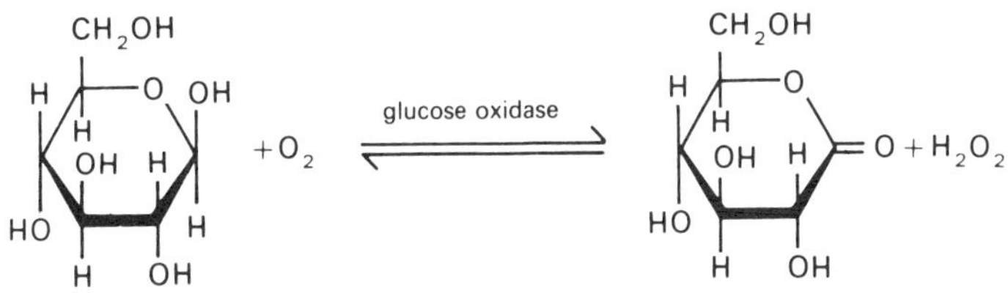

(**9.39**)

β-D-glucose D-gluconolactone

$$CH_3CH(OH)CO_2^- + O_2 \xrightleftharpoons{\text{lactate oxidase}} CH_3COCO_2^- + H_2O_2 \qquad (\mathbf{9.40})$$

lactate pyruvate

The enzymes which catalyse these reactions, however, are flavo-proteins and the net reaction is, in fact, the sum of two half reactions. The first half reaction (9.41a) involves oxidation of the substrate by the flavo-enzyme and the second (9.41b) oxidation of the reduced flavin by molecular oxygen.

$$\text{Enz–Fl} + SH_2 \rightleftharpoons \text{Enz–FlH}_2 + S \qquad (\mathbf{a})$$

$$\text{Enz–FlH}_2 + O_2 \rightleftharpoons \text{Enz–Fl} + H_2O_2 \qquad (\mathbf{b})$$

$$\text{Sum: } SH_2 + O_2 \rightleftharpoons S + H_2O_2 \qquad (\mathbf{9.41})$$

It has been suggested, largely on the basis of studies of non-physiological substrates, that the first step in such oxidations involves the production of a carbanionic transition state or intermediate; this is in keeping with the fact that the physiological substrates possess very weakly acidic hydrogen atoms attached to the carbon which is oxidised. There is no agreement, however, as to how the carbanion is oxidised. Oxidation may involve two electron or one electron transfer to the flavin, alternatively a covalent adduct may be formed from carbanion and flavin which then 'collapses' by an oxidative reaction. As can be seen from equation (9.41) it is the reduced flavin containing oxidase, rather than the substrate, which is oxidised by molecular oxygen. It has been suggested that in this oxidative reaction one electron is transferred from ground state oxygen ($^3O_2$) to the flavin, thus producing the superoxide radical-anion and the flavin semiquinone radical. These radicals may then combine to give the flavin 4a-hydroperoxide which decomposes to give oxidised flavin and hydrogen peroxide (9.42).

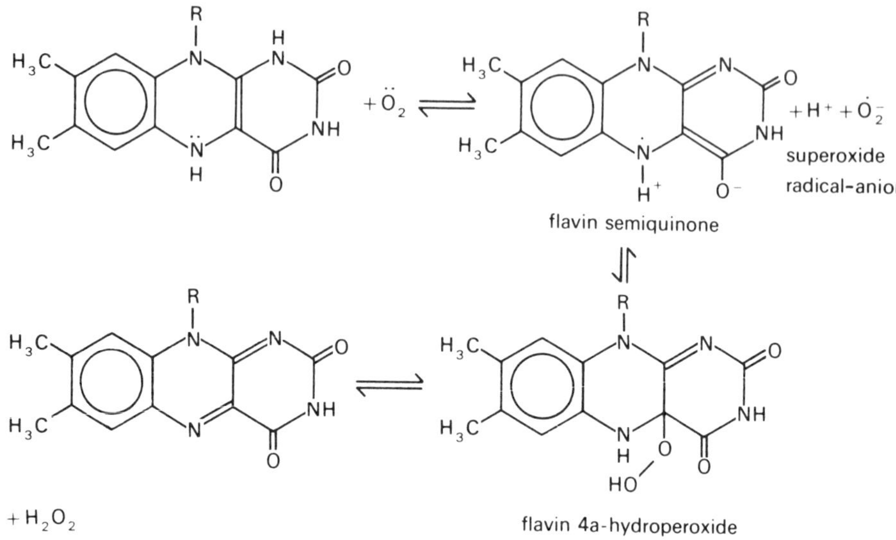

$R = -CH_2(CHOH)_3CH_2OPO_3^{2-}$ OR $-CH_2(CHOH)_3CH_2OP(=O)(O^-)OP(=O)(O^-)O$–Adenosinyl **(9.42)**

(*iv*) *Insertion of oxygen.* In living cells a number of oxygen insertion reactions, exemplified by (9.43), are catalysed by flavoenzymes known as mixed function oxidases or, better, as mono-oxygenases.

$$SH + O_2 + NADPH + \overset{+}{H} \rightleftharpoons SOH + H_2O + NA\overset{+}{D}P \qquad \textbf{(9.43)}$$

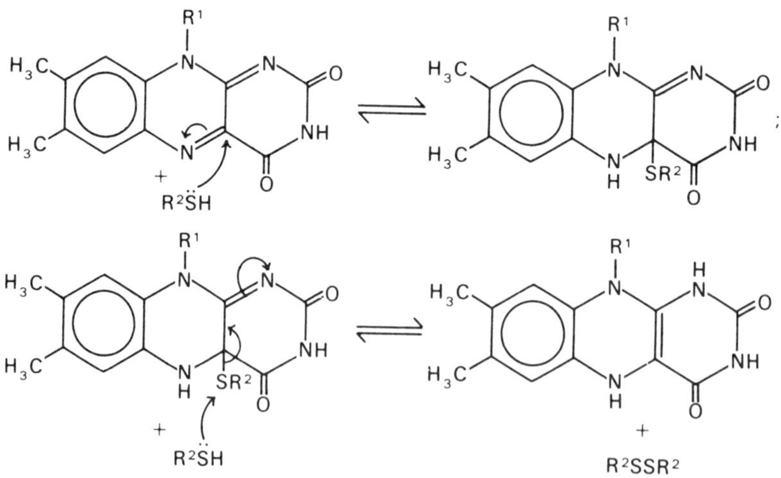

$R' = -CH_2(CHOH)_3CH_2OPO_3^{2-}$ OR $-CH_2(CHOH)_3CH_2OP(=O)(O^-)OP(=O)(O^-)O$–Adenosinyl **(9.44)**

As mentioned previously (p. 179) the role of NAD(P)H in this type of reaction is to reduce the flavo-enzyme rather than the substrate or oxygen. The reduced flavin is thought to react with molecular oxygen to give the 4a-hydroperoxide as described in (9.42); this intermediate then oxidises the substrate. Model studies have shown that flavin peroxides can, indeed, oxidise thiols and amines to give the same products as those produced in enzyme catalysed reactions; the mechanism of this type of oxidation is still under debate.

The role of C-4a as an electrophilic centre is supported by the fact that thiol-disulphide interconversions catalysed by flavo-enzymes also appear to proceed via a C-4a adduct (9.44).

9.3. Pyridoxal phosphate, thiamine pyrophosphate and biotin

Both pyridoxal phosphate and thiamine pyrophosphate are co-enzymes which modify the reactivity of the substrates of their apo-enzymes. Biotin is probably a co-enzyme of the same type, although it is not clear exactly how it exerts its effect.

Pyridoxal phosphate (PLP)

Virtually all of the enzymes which catalyse the reactions of 2-amino-acids in living cells require pyridoxal phosphate (9.45) as a co-factor.

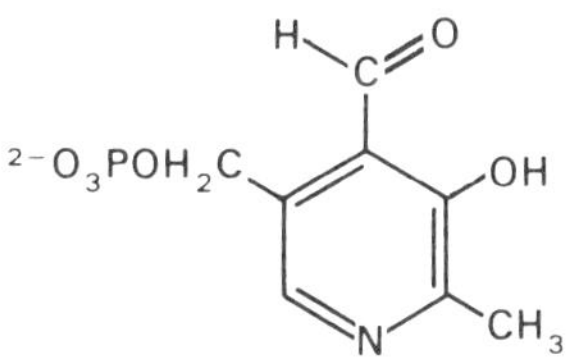

(9.45)

The aldehyde, which is the phosphorylated oxidation product of pyridoxol (Vit B_6) is one of the most versatile co-enzymes known. The apo-enzymes with which it combines catalyse reactions of 2-amino-acids involving transamination (9.46), racemisation (9.47) de-carboxylation (9.48), $\alpha\beta$ or $\beta\gamma$ elimination from, or β or γ

$$\underset{\text{glutamate}}{\overset{+}{H_3N}CH(CH_2CH_2CO_2^-)CO_2^-} + \underset{\text{oxaloacetate}}{O{=}C(CH_2CO_2^-)CO_2^-} \underset{}{\overset{\text{GOT}}{\rightleftharpoons}} \underset{\text{2-oxoglutarate}}{O{=}C(CH_2CH_2CO_2^-)CO_2^-} + \underset{\text{aspartate}}{\overset{+}{H_3N}CH(CH_2CO_2^-)CO_2^-} \quad (9.46)$$

$$\text{L-alanine} \underset{\text{alanine racemase}}{\overset{\text{bacterial}}{\rightleftharpoons}} \text{D-alanine} \quad (9.47)$$

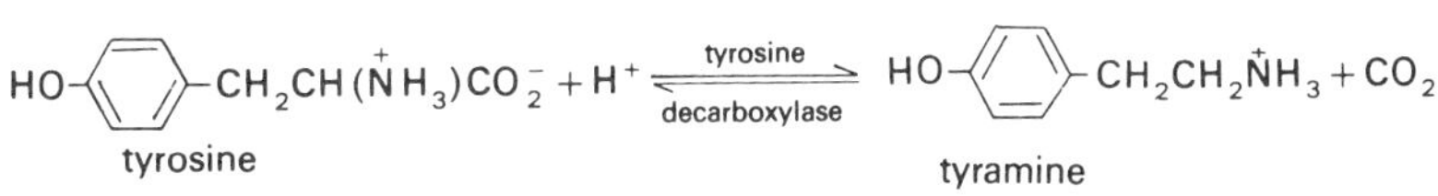

(9.48)

$$\underset{\text{serine}}{HOCH_2CH(\overset{+}{N}H_3)CO_2^-} \underset{\text{dehydrase}}{\overset{\text{serine}}{\rightleftharpoons}} H_2O + [CH_2{=}C(\overset{+}{N}H_3)CO_2^-] \rightarrow \underset{\text{pyruvate}}{CH_3COCO_2^-} + \overset{+}{N}H_4 \quad (9.49)$$

substitution in, side chains (9.49 to 9.52, respectively) and side chain cleavage (9.53).

$$\underset{\text{methionine}}{CH_3SCH_2CH_2CH(\overset{+}{N}H_3)CO_2^-} + HO_2 \xrightleftharpoons{\text{methioninase}} CH_3SH + \underset{\text{2-oxobutanoate}}{CH_3CH_2COCO_2^-} + \overset{+}{N}H_4 \quad (9.50)$$

$$\underset{O\text{-acetyl serine}}{CH_3COOCH_2CH(\overset{+}{N}H_3)CO_2^-} + H_2S \xrightleftharpoons[\text{cysteine synthetase}]{\text{bacterial}} \underset{\text{cysteine}}{HSCH_2CH(\overset{+}{N}H_3)CO_2^-} + CH_3CO_2^- + H^+ \quad (9.51)$$

$$\underset{O\text{-succinylhomoserine}}{{}^-O_2CCH_2CH_2COOCH_2CH_2CH(\overset{+}{N}H_3)CO_2^-} + HSCH_2CH(\overset{+}{N}H_3)CO_2^- \xrightleftharpoons[\text{synthetase}]{\text{cystathionine}}$$

$${}^-O_2CCH_2CH_2CO_2^- + \overset{+}{H} + \underset{\text{cystathionine}}{{}^-O_2CCH(\overset{+}{N}H_3)CH_2SCH_2CH_2CH(\overset{+}{N}H_3)CO_2^-} \quad (9.52)$$

$$\underset{\text{serine}}{HOCH_2CH(\overset{+}{N}H_3)CO_2^-} \xrightleftharpoons[\text{transhydroxymethylase}]{\text{serine}} HCHO + \underset{\text{glycine}}{H_2C(\overset{+}{N}H_3)CO_2^-} \quad (9.53)$$

It is worth noting that there is one type of reaction of 2-amino-acids which is not catalysed by pyridoxal dependant enzymes. This type of reaction involves the elimination of ammonia and is catalysed by enzymes known as ammonia lyases (9.54).

$$\underset{\text{phenylalanine}}{PhCH_2CH(\overset{+}{N}H_3)CO_2^-} \xrightleftharpoons[\text{ammonia lyase}]{\text{phenylalanine}} \underset{trans\text{-cinnamate}}{PhCH{=}CHCO_2^-} + \overset{+}{N}H_4 \quad (9.54)$$

The fact that pyridoxal phosphate could not act as co-enzyme in this type of reaction should be evident from the subsequent discussion of its mode of action.

Mode of action of pyridoxal phosphate

In all of the PLP dependant enzymes which have been examined to date the co-enzyme is bound covalently at the active site by an imino function involving the 6-amino group of a lysine residue (9.55).

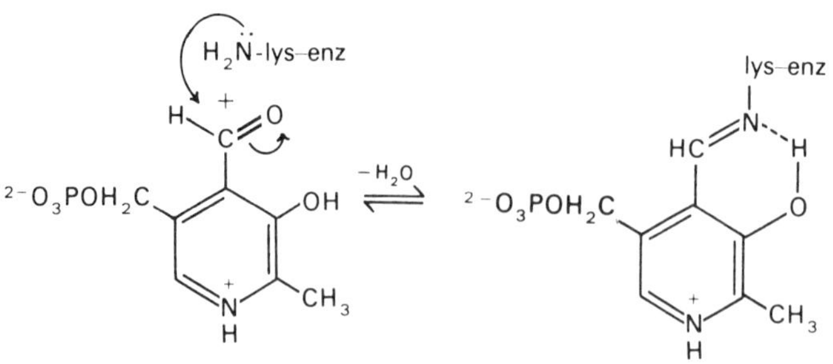

(9.55)

The first step in all of the reactions catalysed is thought to be a **transimination**, producing an enzyme bound PLP-amino acid Schiff base. (9.56).

It appears that when the Schiff base is formed the apo-enzyme, depending on its specificity, orientates the C-2–H, C-2–CO_2^- or the C-2–C-3 bond so that it lies in a plane perpendicular to that containing both the imino group and the pyridinium ring. The apo-enzyme then induces the cleavage of the bond so orientated. The structure of the amino-acid-co-enzyme Schiff base is such that the charge in the carbanionic product of bond cleavage can be extensively de-localised (9.57).

$$H_3\overset{+}{N}CH(R)CO_2^- \rightleftharpoons H_2NCH(R)CO_2^- + H^+;$$

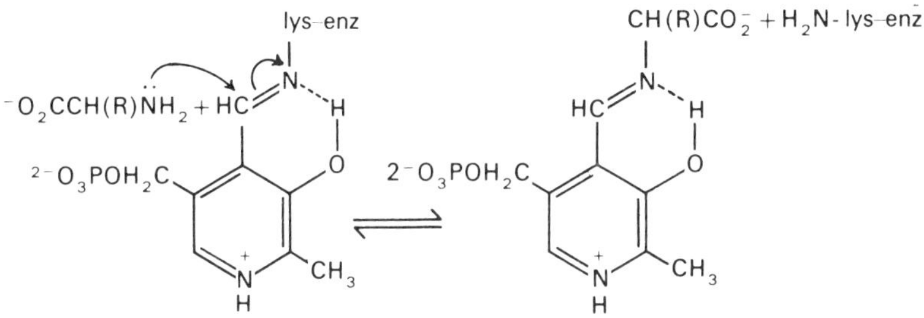

(9.56)

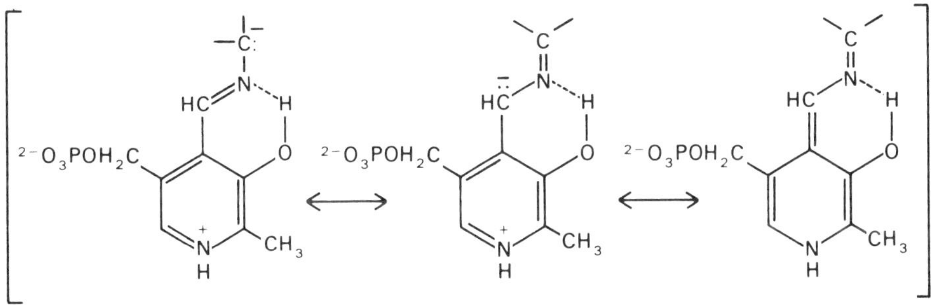

(9.57)

Clearly, prior orientation of the bond which is cleaved so that it is perpendicular to the plane containing the imino group and the pyridinium ring will facilitate bond cleavage by: (i) removing the necessity for rotation of groups to occur in the approach to the transition complex, (ii) by providing the geometry which will allow maximum overlap of the sp^3 orbital which, formally, describes the carbanionic electron pair, with the conjugated π oribtals of the imino group and the pyridinium ring. Notice that the carbanion contains two centres of high charge density, one at C-2 of the substrate residue and one at the aldehyde carbon of the PLP residue. The carbanion can therefore be re-protonated at either of these two sites and this, together with the fact that its structure is such as to allow charge to flow into, or out of the amino-acid side chain allows convincing mechanisms to be written for all types of reaction catalysed. These mechanisms are most logically classified in terms of the bond which is initially broken in the amino-acid residue and are discussed in this context below.

(i) Cleavage of the C-2–H bond

(a) Transamination. In transamination reactions (9.58) the carbanion which results from cleavage of the C-2–H bond is re-protonated by the apo-enzyme at the aldehyde carbon of the cofactor residue. The new Schiff base, which is a tautomer of that formed originally between substrate and co-enzyme is then

hydrolysed under the catalytic influence of the apo-enzyme. The products of hydrolysis are a 2-oxo-acid with the same carbon skeleton as the substrate amino-acid and enzyme bound **pyridoxamine phosphate** (PMP). After the oxo-acid has diffused away from the PMP–enzyme the first half reaction (9.59) of transamination is complete.

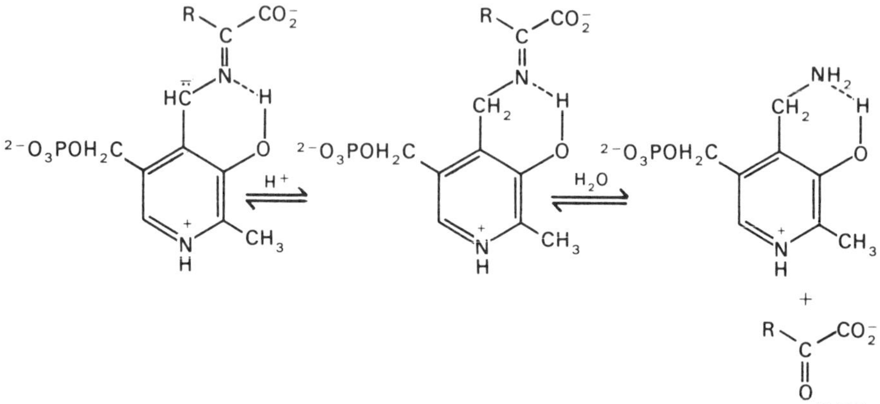

(9.58)

PLP–enz + amino-acid¹ ⇋ PMP–enz + oxo-acid¹ **(9.59)**

PMP–enz + oxo-acid² ⇋ PLP–enz + amino-acid² **(9.60)**

Sum: amino-acid¹ + oxo-acid² ⇌ oxo-acid¹ + amino-acid²

The overall process is then completed by occurrence of the second half reaction (9.60) which is, essentially, the reverse of the first. The PMP–enzyme binds a second 2-oxo-acid, the Schiff base which is formed is de-protonated at the 'aldehyde' carbon, and then re-protonated at C-2 of the amino acid residue. Finally a 2-amino-acid with the same carbon skeleton as the oxo-acid bound at the start of this second half reaction is liberated by transimination. PMP enzymes have been isolated as products of the first half reaction of a number of transaminases and the mechanism formulated above is commensurate with the 'ping-pong bi-bi' kinetics which have been observed for many transaminases. This type of kinetics (9.61), requires two forms of enzyme and the first product must be formed and released before the second substrate can bind.

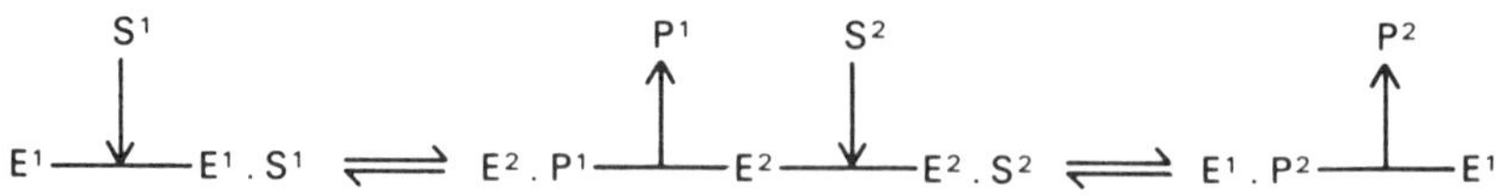

(9.61)

(*b*) *Racemisation.* Amino-acid racemases de-protonate C-2 of the amino-acid-PLP Schiff base and then re-protonate it non-stereospecifically. It is not known how these enzymes achieve non-stereospecific re-protonation.

(*c*) *Elimination from, and replacement in, amino-acid side chains. αβ-Elimination and β-replacement.* Amino acids which contain good leaving groups such as RCO_2–, –OPO_3^{2-}, RS– or indolyl or groups such as HO– which may be converted to a good leaving group attached to C-3 can undergo elimination which may, or may not, be followed by Michael type addition of a nucleophile across the (activated) double bond which is produced. Scheme (9.62) indicates that the negative charge in the Schiff base carbanion can act as an internal nucleophile towards C-3 of the amino-acid residue. The leaving group is expelled and a new Schiff base, corresponding to that which would result from PLP and a 2,3-unsaturated amino-acid is formed. This Schiff base can undergo one of two types of subsequent reaction, depending on the specificity of the apo-enzyme. It can undergo transimination, immediately, to produce the unsaturated amino-acid and resting PLP–enzyme. In fact, the unsaturated amino-acid does not appear as such because it immediately tautomerises and the imino tautomer is hydrolysed, rapidly and spontaneously, to yield an oxo-acid and ammonium ions. Alternatively the apo-enzyme might catalyse Michael type addition of a nucleophile across the newly formed double bond. Note that the imino group conjugated to the pyridinium ring activates the double bond towards nucleophilic attack in exactly the same way as do carbonyl groups (Section 4.2). Re-protonation at C-2 of the amino-acid residue followed by transimination then produces an amino-acid which is a β-substitution product of the substrate.

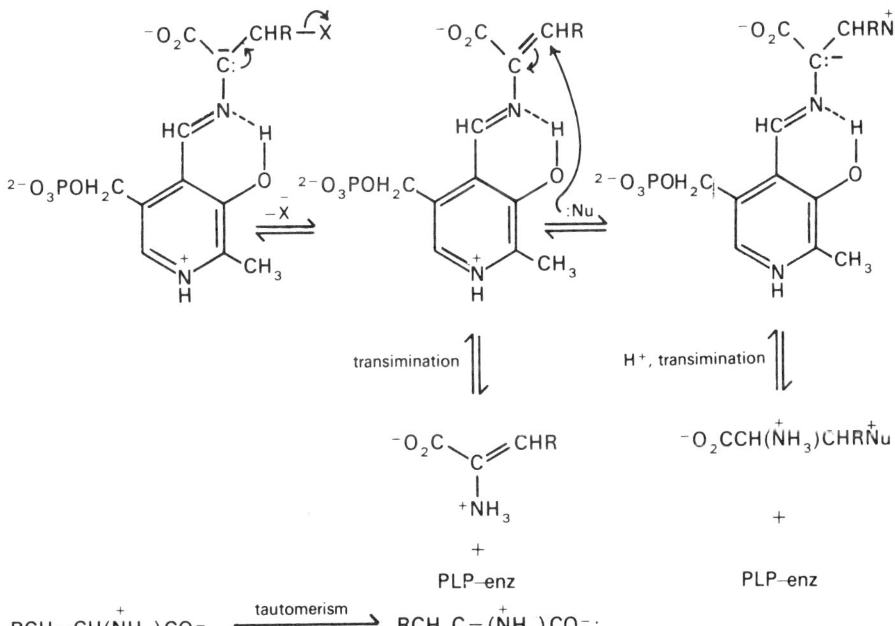

$$RCH{=}CH(\overset{+}{N}H_3)CO_2^- \xrightleftharpoons{\text{tautomerism}} RCH_2C{=}(\overset{+}{N}H_2)CO_2^-;$$

$$RCH_2C{=}(\overset{+}{N}H_2)CO_2^- + H_2O \rightleftharpoons RCH_2COCO_2^- + NH_4^+$$

$X = RCO_2^-, -OPO_3^{2-}, RS-, HO-$ etc. **(9.62)**

$\beta\gamma$-Elimination and γ-replacement These two types of reaction, which are comparatively rare, illustrate yet another aspect of the versatility of pyridoxal phosphate as a co-enzyme. As indicated in (9.63) the structure of the substrate–PLP Schiff base is such that not only can it delocalise negative charge resulting from abstraction of a proton from C-2 but it can also de-localise

negative charge resulting from abstraction of a proton from C-3 of the amino-acid residue.

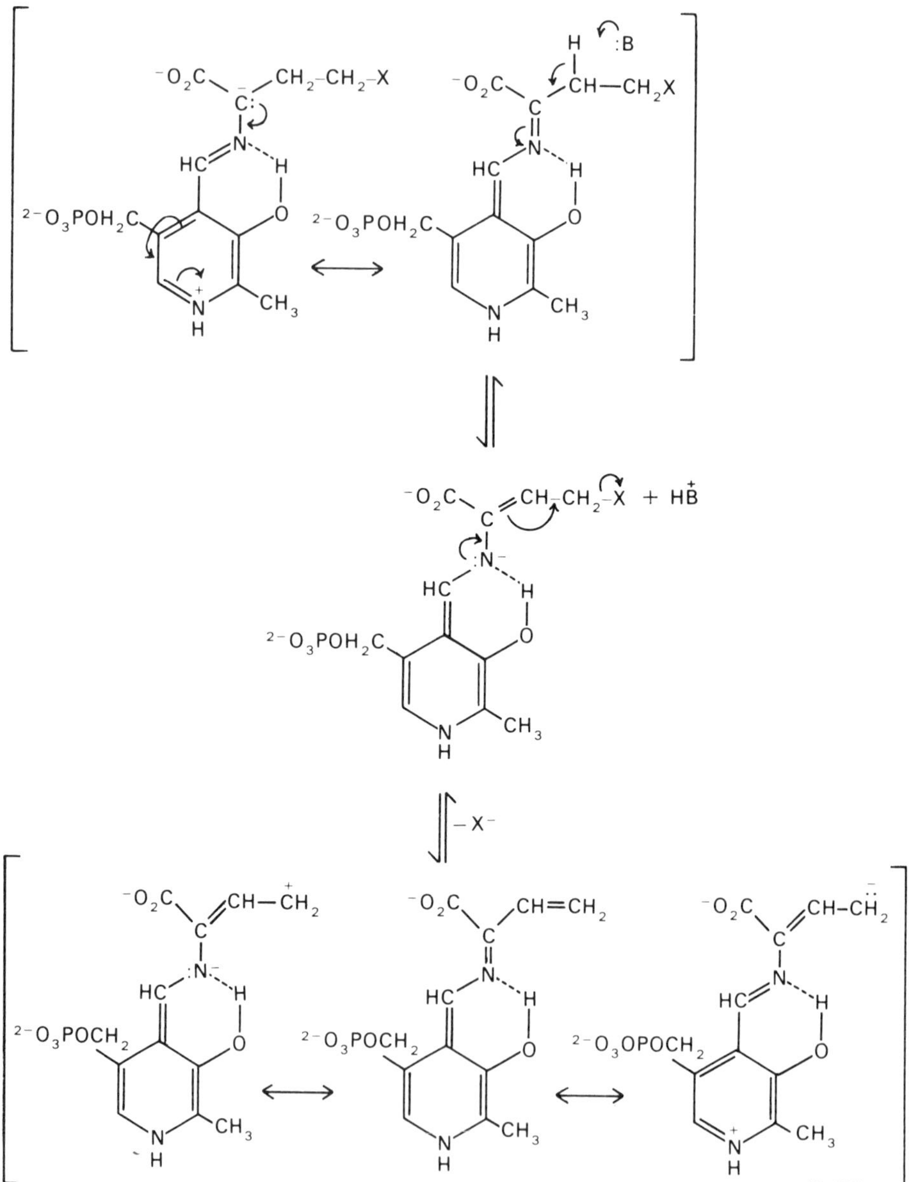

(9.63)

The negative charge delocalised over the di-anion then displaces the C-4-substituent of the amino-acid residue. The description of the elimination product as a resonance hybrid indicates that charge within it is mobile and that it could have three fates:

(i) The apo-enzyme reprotonates the Schiff base at C-4 of the amino-acid residue. Transimination then allows 2-aminobut-2-enoic acid to diffuse away from the PLP–enzyme. The unsaturated amino-acid immediately tautomerises and the tautomer is hydrolysed spontaneously to yield 2-oxobutanoate and ammonium ions (9.64).

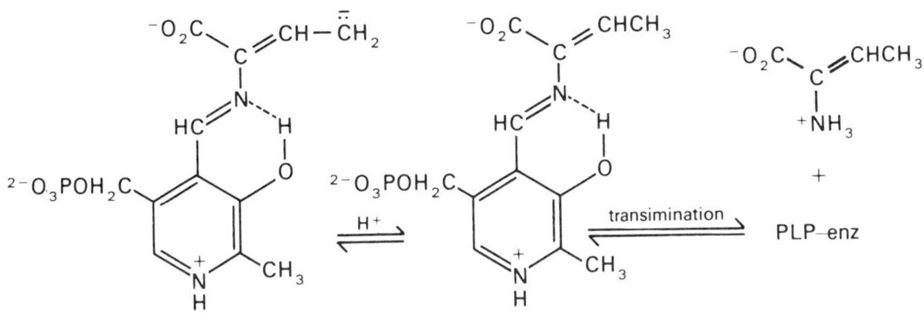

$$CH_3CH{=}C(\overset{+}{N}H_3)CO_2^- \underset{}{\overset{\text{tautomerism}}{\rightleftharpoons}} CH_3CH_2C({=}\overset{+}{N}H_2)CO_2^-;$$

$$CH_3CH_2C({=}\overset{+}{N}H_2)CO_2^- + H_2O \rightleftharpoons CH_3CH_2COCO_2^- + NH_4^+ \qquad \textbf{(9.64)}$$

(ii) Alternatively the apo-enzyme, after re-protonating the Schiff base carbanion at C-4 of the amino-acid residue may catalyse Michael type addition of a nucleophile over the 2,3 double bond. Transimination then releases a product which corresponds to replacement of the leaving group at C-4 by hydrogen coupled to replacement of hydrogen at C-3 by a nucleophilic group. (9.65).

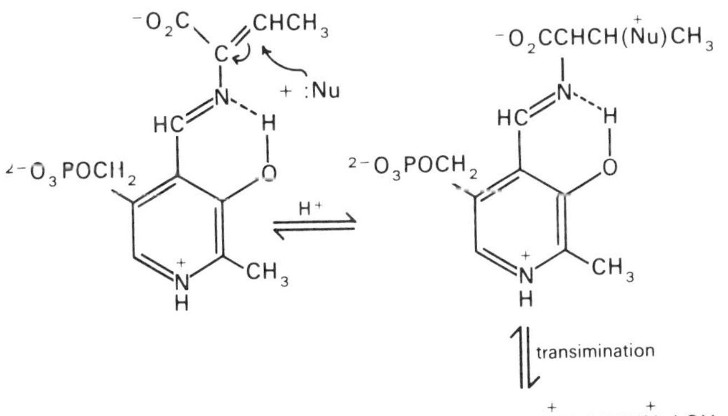

$$^-O_2CCH(\overset{+}{N}H_3)CH(\overset{+}{N}u)CH_3 + \text{PLP–enz} \qquad \textbf{(9.65)}$$

A specific example of this rare type of reaction is given by the biosynthesis of threonine in **Neorospora** (9.66).

$$^{2-}O_3POCH_2CH_2CH(\overset{+}{N}H_3)CO_2^- + H_2O \rightleftharpoons HPO_4^{2-} + CH_3CH(OH)CH(\overset{+}{N}H_3)CO_2^- \qquad \textbf{(9.66)}$$

O-phosphorylhomoserine

(iii) An appropriate apo-enzyme can catalyse Michael type addition of a nucleophile at C-4 followed by re-protonation at both C-2 and C-3 of the amino-acid residue. Transimination then liberates the product of γ-replacement (9.67).

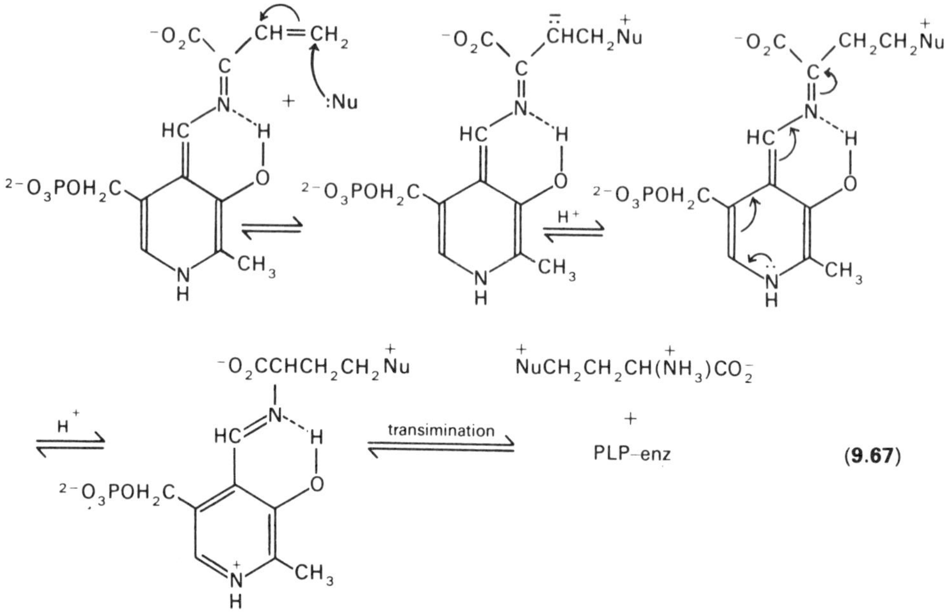

(9.67)

(*ii*) *Cleavage of the C-2–CO_2^- bond. De-carboxylation.* Enzymes which catalyse the de-carboxylation of 2-amino-acids induce cleavage of the C-2–CO_2^- bond in the substrate-PLP Schiff base (9.68). The carbanion is then protonated at C-2 of the amino-acid residue; subsequent transimination releases the amine derived by de-carboxylation of the substrate.

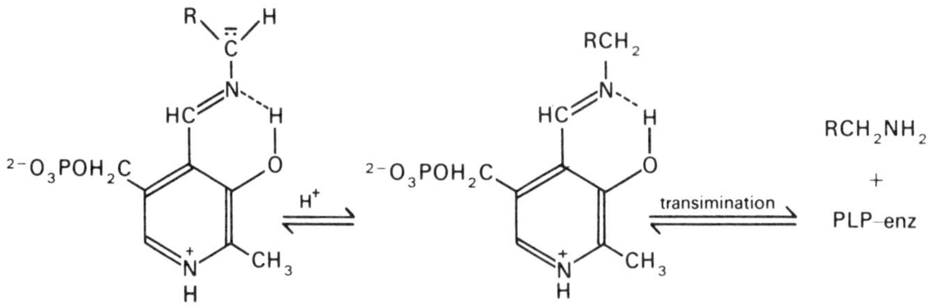

(9.68)

(*iii*) *Cleavage of the C-2–C-3 bond.* The reaction catalysed by serine trans-hydroxymethlase is probably the most thoroughly investigated reaction which involves cleavage of an amino-acid side chain mediated by a PLP dependant enzyme. The evidence which is available suggests that the reaction is essentially a reverse Aldol reaction (9.69). Formaldehyde does not, however, appear as such. The reaction only occurs at a significant rate in the presence of tetrahydrofolate (THF) which appears to act as a trap for the aldehyde (9.70) which, therefore, appears as 5,10-methylenetetrahydrofolate.

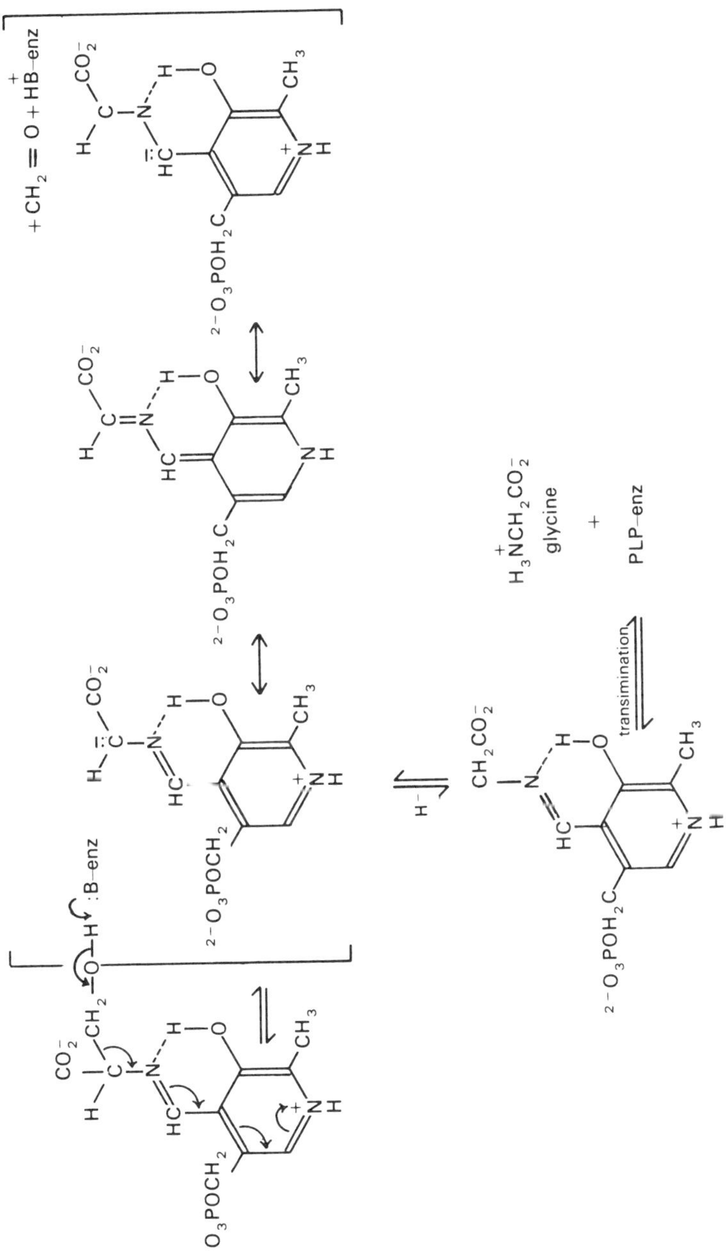

(9.69)

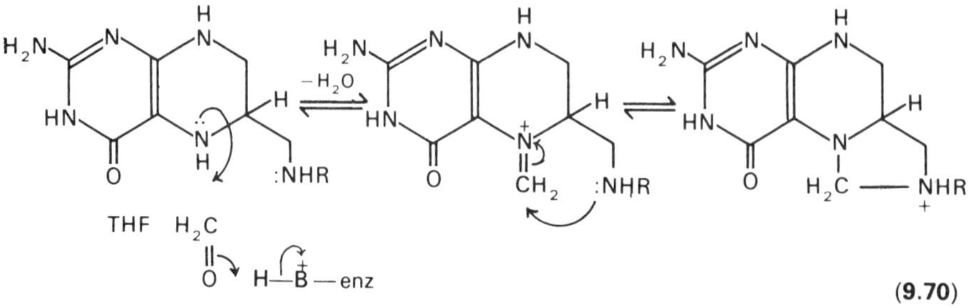

(9.70)

Thiamine pyrophosphate (TPP)

Thiamine pyrophosphate (9.71) is the co-enzyme which is concerned with both the oxidative and non-oxidative enzyme catalysed decarboxylation of 2-oxo-acids and with the formation of acyloins.

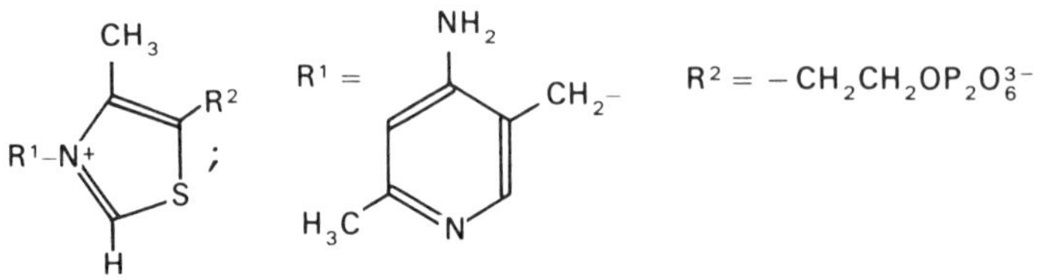

(9.71)

As noted in Section 6.2, decarboxylation of a carboxylic acid is facilitated if the side chain contains a structural feature which can de-localise negative charge in the carbanionic intermediate which is produced. The carbonyl group of 3-oxo-acids (*β*-keto-acids) is such a structural feature and it is found that this type of acid can easily be de-carboxylated in the laboratory. It therefore appears to be significant that the enzymes which catalyse the de-carboxylation of 3-oxo-acids do not require a specific co-enzyme. They either increase the efficiency of charge de-localisation in the anionic intermediate by formation of a Schiff base involving

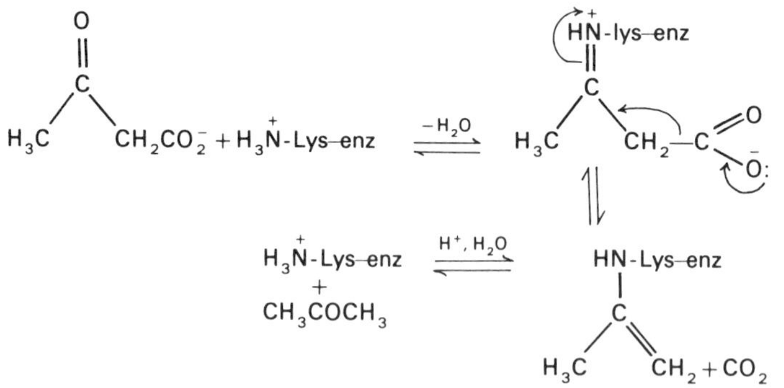

(9.72)

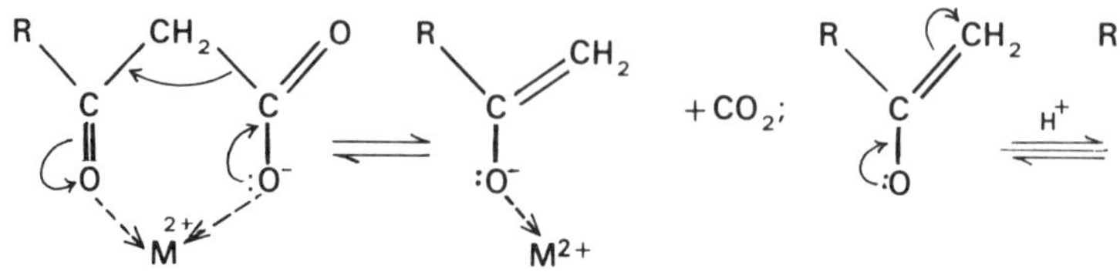

(9.73)

the 6-amino group of a lysine residue (9.72), or provide electrophilic catalysis using an enzyme bound metal ion (9.73).

The situation with 2-oxo-acids is different. The carbonyl groups in these acids is located in a position where it cannot de-localise developing negative charge and the enzymes which catalyse the de-carboxylation of this type of acid utilise a specific co-enzyme, thiamine pyrophosphate (TPP). The co-enzyme is bound tightly, but not co-valently, to its apo-enzymes and, as will be shown, its role is to delocalise charge produced during de-carboxylation.

Non-oxidative decarboxylation

The de-carboxylation of pyruvate, catalysed by yeast pyruvate decarboxylase (9.74), has been thoroughly investigated.

$$CH_3COCO_2^- + H^+ \underset{\text{decarboxylase, TPP}}{\overset{\text{pyruvate}}{\rightleftharpoons}} CH_3CHO + CO_2 \qquad \textbf{(9.74)}$$

The mechanism of the reaction appears to be as shown in (9.75). The C-2 hydrogen of TPP is demonstrably acidic (charge in the anion is delocalised to the positively charged nitrogen atom) and the co-enzyme appears to be present in the active-centre of its apo-enzymes as the C-2 anion. This anion acts as a nucleophile towards the 2-oxo-acid giving rise to an adduct which, unlike the keto-acid itself, can undergo facile de-carboxylation because the charge in the anionic intermediate (the anion of 2-hydroxyethyl thiamine pyrophosphate in this case) can be stabilised by charge delocalisation. Protonation of the anion, followed by regeneration of the resting holo-enzyme with concomitant production of acetaldehyde completes the reaction. It is clear that TPP functions by providing the substrate with a temporary substituent, at C-2, into which charge can flow during de-carboxylation.

Acyloin condensations.

The fate of the anion formed by decarboxylation of a 2-oxo-acid-TPP adduct is not always as described above. A number of enzymes catalyse reactions in which the anions which result from decarboxylation of the substrate act as nucleophiles towards an electrophilic centre, usually a carbonyl group or, in oxidative de-carboxylation, a disulphide group. For example in bacteria acetolactate, a precursor of valine, is produced from pyruvate as shown in (9.76); the reaction is catalysed by acetolactate synthetase.

A variant of this type of reaction is that catalysed by transketolase during operation of the pentose phosphate pathway (9.77).

It is known that the enzyme catalyses the transfer of a ketol or acyloin group from D-ribulose-5-phosphate to D-ribose-5-phosphate. The probable mechanism of the transfer is shown in (9.78). An anion is produced from a D-ribulose-5-phosphate-TPP adduct by removal of glyceraldehyde-3-phosphate. D-ribose-5-phosphate is then bound at the active site of the enzyme and the steps shown are reversed producing sedoheptulose-5-phosphate and resting enzyme.

It should now be clear that cyanide ions (Chapter 6) and the TPP C-2 anion mediate acyloin condensations in a similar manner. Both anions are sufficiently nucleophilic to be able to attack a carbonyl group, both can labilise a C–H or C–C bond at a carbonyl group by virtue of their ability to de-localise charge and both are good leaving groups.

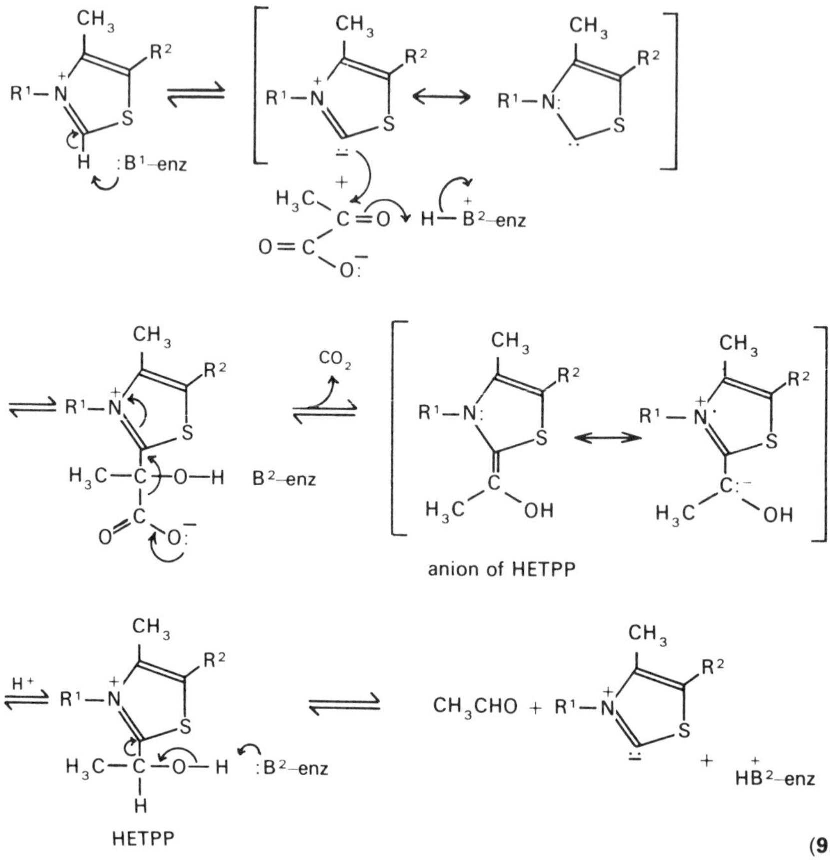

(**9.75**)

(HETPP = 2-hydroxyethylthiamine pyrophosphate)

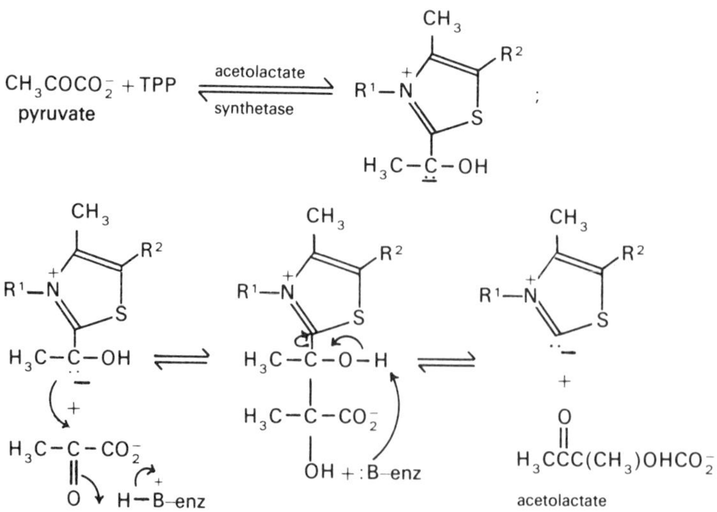

(**9.76**)

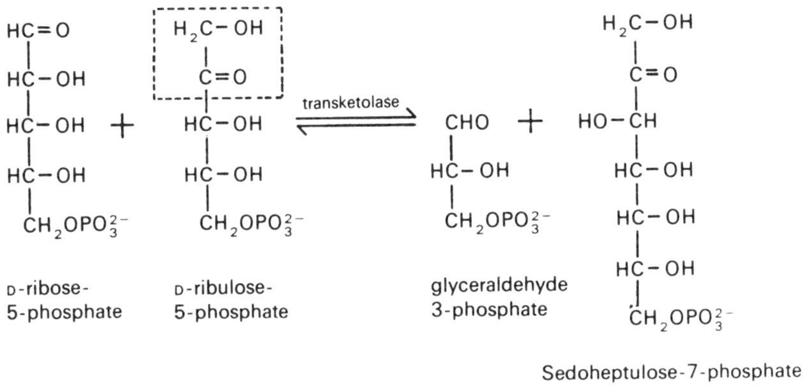

(9.77)

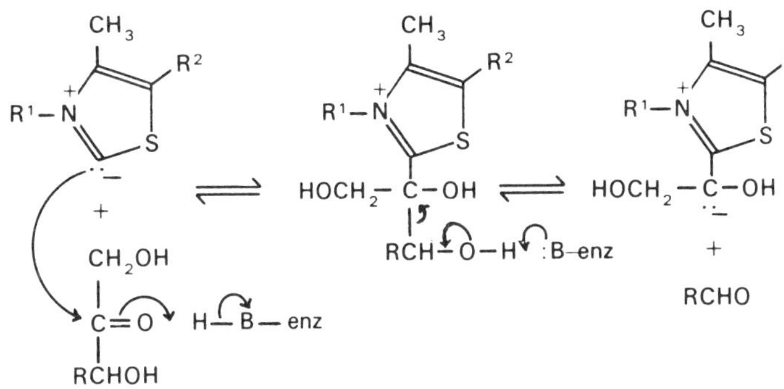

R = $^{2-}O_3POCH_2CH(OH)-$ **(9.78)**

Oxidative decarboxylations.

Both pyruvate and α ketoglutarate can undergo oxidative de-carboxylation in mitochondria (9.79 and 9.80); the reactions are important components of oxidative metabolism in aerobic cells. The two reactions proceed by the same basic pathway and are catalysed by analogous enzyme complexes. Both of the latter contain three component enzymes – a specific, TPP dependant α-keto-acid dehydrogenase, a specific dihydrolipoamide transacylase and a dihydrolipoamide dehydrogenase. The somewhat illogically named dehydrogenases catalyse the production of a carbanion from the α-keto-acid TPP adduct as described previously for yeast pyruvate decarboxylase (9.75 and 9.81).

$$\underset{}{CH_3COCO_2^-} + \overset{+}{NAD} + \underset{\text{co-enzyme A}}{CoASH} \xrightleftharpoons[\text{dehydrogenase complex}]{\text{pyruvate}} \underset{\text{acetyl-co-enzyme A}}{CH_3COSCoA} + CO_2 + NADH \qquad \textbf{(9.79)}$$

$$^-O_2CCH_2CH_2COCO_2^- + \overset{+}{NAD} + CoASH \xrightleftharpoons[\text{dehydrogenase complex}]{\alpha\text{-ketoglutarate}} \underset{\text{succinylco-enzyme A}}{^-O_2CCH_2CH_2COSCoA} + CO_2 + NADH \qquad \textbf{(9.80)}$$

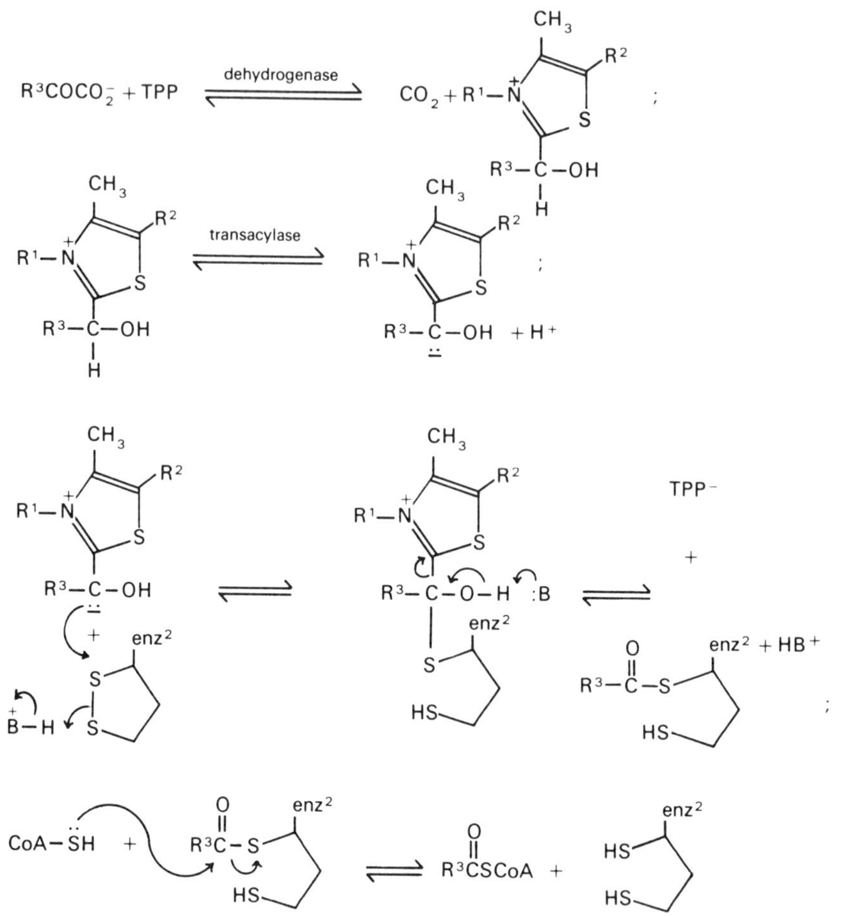

(enz^2 = transacylase). **(9.81)**

In contrast to the reactions involving TPP described previously the anion is protonated and the hydroxyalkylthiamine pyrophosphate which results is used as substrate for the second component of the enzyme complex, the dihydrolipoamide transacylase. These enzymes contain lipoic acid (9.82; oxidised form shown) bound covalently by an amide linkage involving the 6-amino group of a lysine residue. Like all disulphides lipoic acid readily undergoes 2 electron reduction to the dithiol form and this is a key element in the operation of the enzyme complexes. The transacylases labilise the C-2-hydrogen of their substrates and the carbanion attacks the C-8-sulphur atom of oxidised lipoic acid.

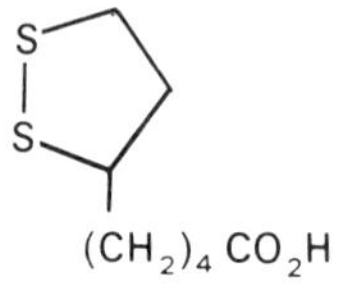

(9.82)

Expulsion of the TPP anion from the adduct produces an *S*-acylholo-enzyme and completes 2 electron oxidation of the de-carboxylated 2-keto-acid. Finally the transacylases catalyse transfer of the enzyme bound acyl group to co-enzyme A (a transesterification, Section 6.2). The third component of the enzyme complexes, the dihydrolipoamide dehydrogenase, is a flavo-enzyme which allows 'turn-over' to occur by regenerating the oxidised lipoyltransacylase as shown in (9.83), cf. (9.44).

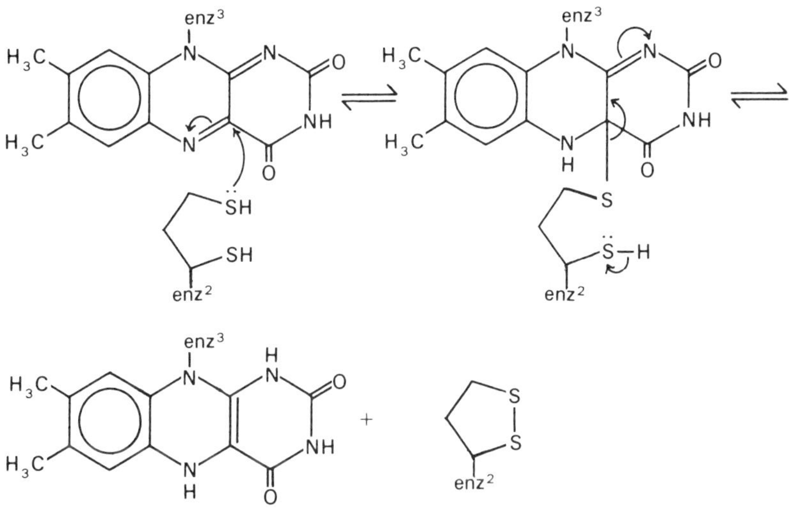

$$enz^3\text{-}FADH_2 + \overset{+}{NAD} \rightleftharpoons enz^3\text{-}FAD + NADH + H^+ \qquad \textbf{(9.83)}$$

enz^2 = transacylase, enz^3 = dihydrolipoamide dehydrogenase.

Biotin

Although a number of enzymes which mediate carboxylation reactions in cells do so using acid-base catalysis a small but well defined group of other enzymes require the presence of specific co-factor, biotin, together with a co-substrate, ATP, in order to catalyse carboxylation. Two examples of reactions catalysed by biotin dependant enzymes are described by (9.84) and (9.85).

$$ATP^{4-} + HCO_3^- + \underset{\text{pyruvate}}{CH_3COCO_2^-} \underset{\text{carboxylase}}{\overset{\text{pyruvate}}{\rightleftharpoons}} \underset{\text{oxaloacetate}}{^-O_2CCH_2COCO_2^-} + ADP^{3-} + HPO_4^{2-} + H^+ \qquad \textbf{(9.84)}$$

$$ATP^{4-} + HCO_3^- + \underset{\text{acetyl co-enzyme A}}{CH_3COSCoA} \underset{\text{carboxylase}}{\overset{\text{acetylcoenzyme A}}{\rightleftharpoons}} \underset{\text{malonyl co-enzyme A}}{^-O_2CCH_2COSCoA} + ADP^{3-} + HPO_4^{2-} + H^+ \qquad \textbf{(9.85)}$$

The structure of enzyme bound biotin is shown in (9.86). It consists of an imidazolone ring fused to a tetrahydrothiophene ring, the latter carrying a valeroyl side chain by which the co-enzyme is attached to specific apo-enzymes via an amide bond involving a lysine residue. It is now accepted that the role of the co-substrate, ATP, is to activate bicarbonate by formation of a mixed anhydride which then carboxylates biotin in the holo-enzyme. A carboxyl group is then transferred from the carboxybiotinylenzyme to an incipient nucleophile (9.87).

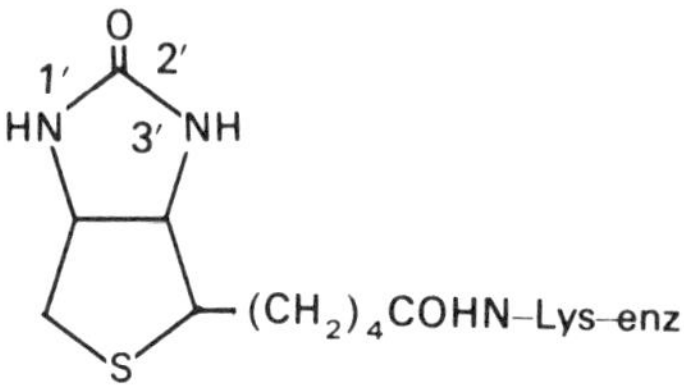

(9.86)

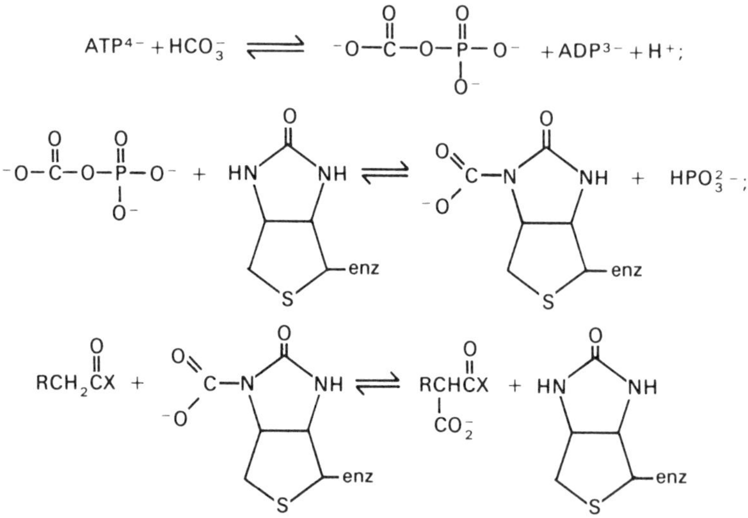

(9.87)

The evidence which is available suggests that biotin is carboxylated at N-1^1. An N-1^1 carboxylated enzyme has been isolated from the reaction catalysed by **Achromobacter** β-methylglutarylCoA carboxylase and also from the reaction catalysed by **E. Coli** acetylCoA carboxylase; the former carboxylated enzyme has been shown to be chemically competent to carboxylate β-methylglutarylCoA. However, it has been pointed out that amide nitrogens are only very weakly nucleophilic and unlikely to be capable of displacing phosphate from carbonyl-

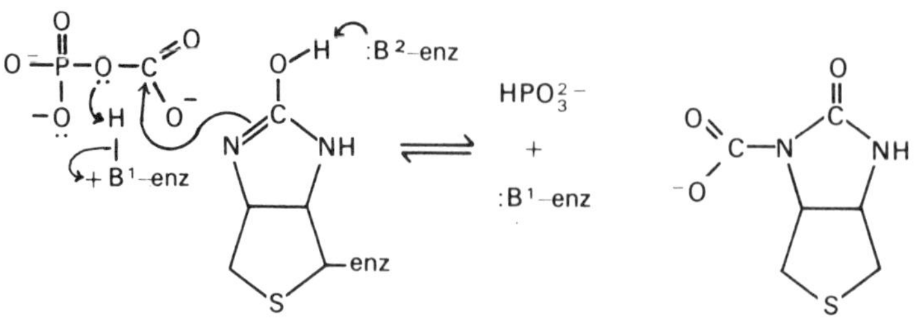

(9.88)

phosphate. The apo-enzyme might, however, promote enolisation of the amide function and the enolic isourea (a better nucleophile than the amide) might be responsible for displacing phosphate from carbonylphosphate in a base catalysed process (9.88).

Acid catalysed re-formation of the enolic isourea when the carboxyl group is transferred to an enolic anion would also effectively make the biotinylenzyme a good leaving group. Unfortunately, it seems that there are no simple *in vitro* reactions from which analogies can be drawn in order to explain the mode of action of biotin.

9.4. Tetrahydrofolate derivative and *S*-adenosyl methionine

Most enzymes which catalyse methylation, hydroxymethylation or formylation reactions *in vivo* utilise either a derivative of tetrahydrofolic acid (THF, 9.89) of *S*-adenosylmethionine (SAM, 9.90) as donor of the one carbon fragment.

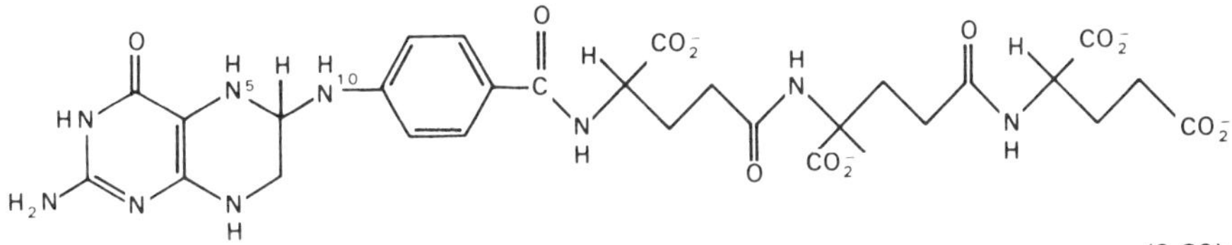

(9.89)

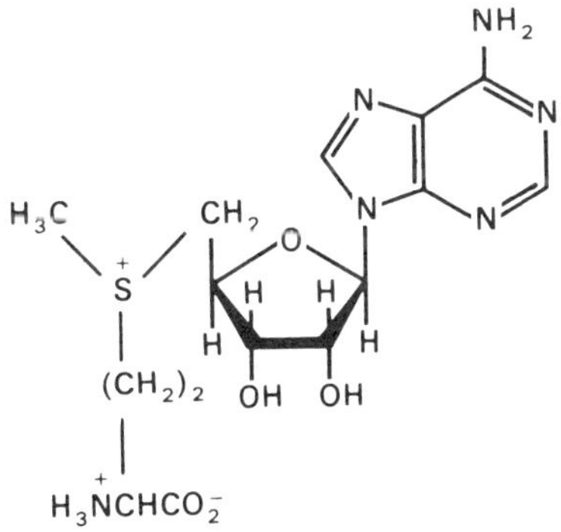

(9.90)

Derivatives of THF

Biochemical reactions involving derivatives of THF tend to be complex and, as in the case of the flavins and biotin, no obvious analogies in the form of simple *in vitro* reactions are available with which to explain the mode of action of the co-enzyme. THF derivatives are involved with the transfer of one carbon fragments at the oxidation levels of methanol, formaldehyde and formic acid. The interconversion of the three THF derivatives involved is shown in (9.91) and representative one carbon transfers at the oxidation levels noted are illustrated by, respectively, reactions (9.92), (9.93) and (9.94).

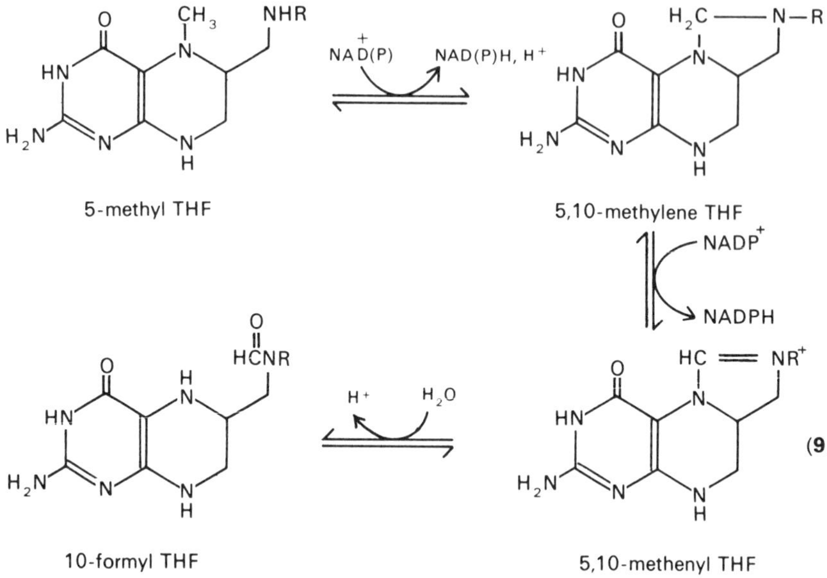

(**9.91**)

$$\text{5-methylTHF} + HSCH_2CH_2CH(\overset{+}{N}H_3)CO_2^- \xrightleftharpoons[\text{MTHF-HCMT}]{\text{methyl-cobalamine, SAM}} CH_3SCH_2CH_2CH(\overset{+}{N}H_3)CO_2^- + \text{THF}$$

homocysteine — methionine

MTHF-HCMT = liver N^5-methyltetrahydrofolate-homocysteine methyltransferase (**9.92**)

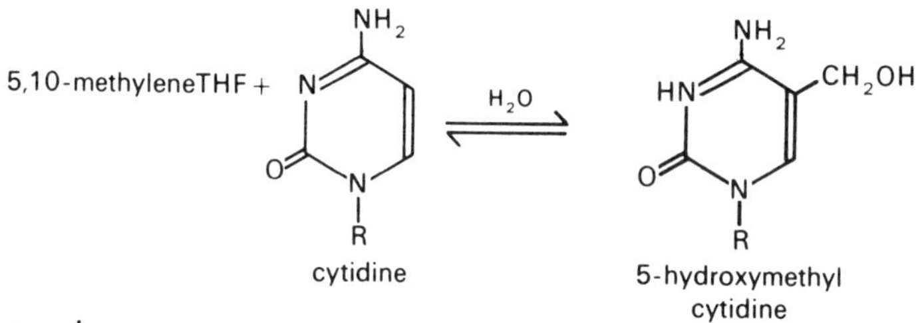

+ THF (**9.93**)

R = 1′-D-ribosyl

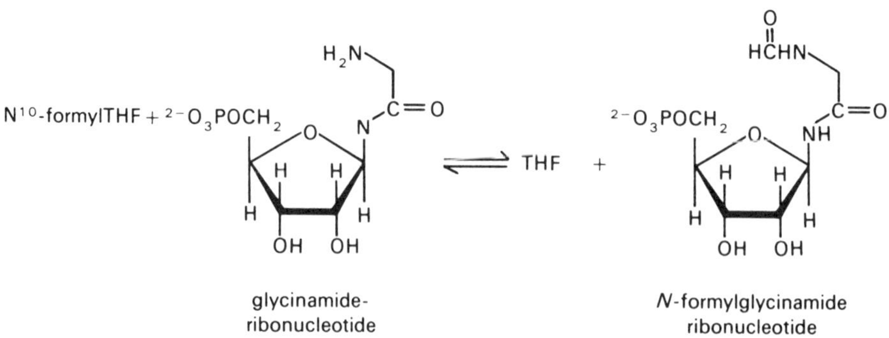

(**9.94**)

Reaction (9.92) is particularly complex. The enzyme catalysing it in liver requires the presence of both methylcobalamine and *S*-adenosylmethionine, whose precise functions are unknown. Furthermore it is not obvious how the methyl group in 5-methyl THF is activated towards transfer. Formally it is present as part of a secondary amino group. As noted in Section 6.1 amino groups are not good leaving groups, although it is conceivable that the enzyme might quaternize the nitrogen atom and thus convert it to a better leaving group. Similarly the mechanism of (9.93) is obscure; formally the reaction may proceed as shown by (9.95).

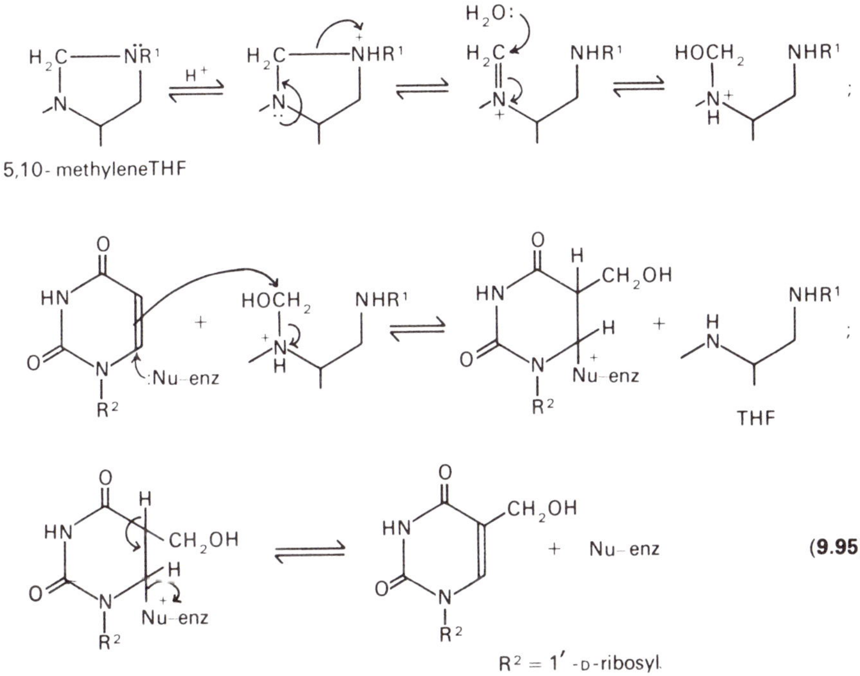

(9.95)

Reaction (9.94) and others similar to it are formally transamidations; they may proceed by a pathway which is similar to that given for other acylations (6.84).

S-Adenosylmethionine

Although 5-methyltetrahydrofolate is the co-enzyme involved in the formation of methionine by methylation of homocysteine (9.92) the majority of enzymes which catalyse methylations require *S*-adenosylmethionine (9.90) as co-enzyme. *S*-Adenosylmethionine is formed in a rather unusual reaction in which the nucleophilic sulphur of methionine attacks C-5[1] of ATP displacing tripolyphosphate (which actually appears as phosphate and pyrophosphate, (9.96).

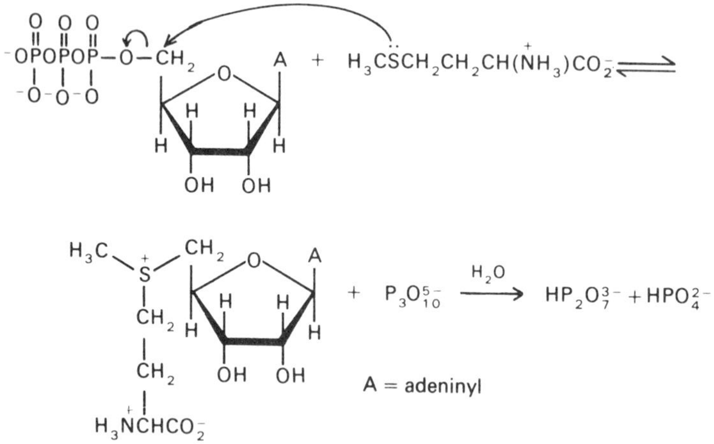

(9.96)

The co-enzyme is a (tertiary) sulphonium salt. As noted in Section 5.1 compounds of this type are efficient alkylating agents and SAM is no exception in this matter. It is usually, but not invariably, attacked by nucleophiles at the carbon atom of the methyl group; the thio-ether, *S*-adenosylhomocysteine (a good leaving group) is displaced in what appear to be typical S_N2 reactions. Several examples of this type of methylation have already been given in Section 5.1 (5.90, 5.91, 5.93) making further elaboration here unnecessary.

9.5. Co-enzyme A

Co-enzyme A is the co-enzyme that is concerned with the reactions of acyl groups, particularly acetyl groups, in living cells. Although its total structure is complex (9.97) the functional part of the molecule is the thiol group and for this reason the co-enzyme is usually represented as CoASH. Acyl groups are 'carried' by the co-enzyme attached to its thiol group. Thus, the reactions of acyl derivatives of co-enzyme A are those of thiol esters. Two types of reaction are of particular importance, namely those in which the co-enzyme acts as an acylating agent and those in which removal of a proton from C-2 of the thiol ester generates an enolate anion. The explanation of why thiol esters are effective acylating agents and effective sources of enolate anions was given in Section 6.1 as were examples of the two types of reaction involved (6.105, 6.130, 6.131, 6.132) and no further information will be presented here.

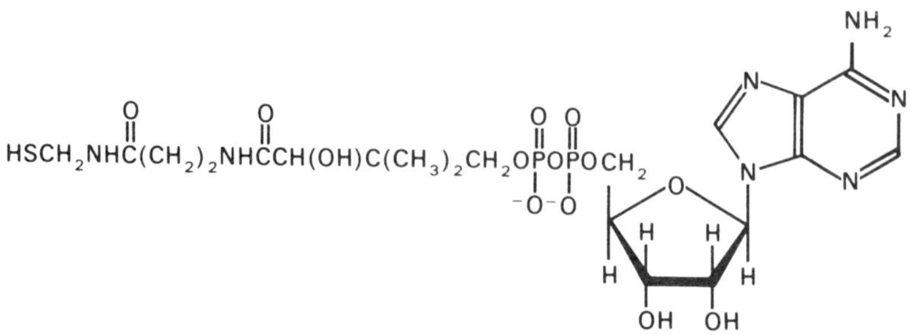

(9.97)

Further reading

Further information relating to the topics dealt with in this book is given in the following publications:—

General Chemistry Chemistry, Matter and The Universe, R. E. Dickerson and I. Geis, Benjamin/Cummins, New York, 1976. (Also contains useful summaries of structural and metabolic biochemistry).

Chemical Bonding Elementary Molecular Bonding Theory. P. G. Perkins, Methuen, London, 1969.

Principles of Chemistry, S. Dunston, Van Nostrand; London, 1968.

Physical Chemistry A Biologist's Physical Chemistry, J. Gareth Morris, Edward Arnold, London, 1968.

Organic Chemistry Organic Chemistry, Fourth Edition, R. T. Morrison and R. N. Boyd. Allyn & Bacon, Boston, 1983.

Organic Chemistry, Third Edition, J. B. Hendrickson, D. J. Cram and G. S. Hammond. McGraw-Hill, New York, 1970.

Organic Reaction Mechanisms A guidebook to Mechanism in Organic Chemistry, Fifth Edition. P. Sykes, Longman, London and New York, 1981.

Stereochemistry Organic Stereochemistry, H. Kagan, Edward Arnold, London, 1979.

Organic Stereochemistry, G. Hallas, McGraw-Hill, London, 1967.

Biochemistry Biochemistry, Second Edition. A. L. Lehninger, Worth; New York, 1975.

Biochemistry, D. E. Metzler, Academic Press, London, 1977.

Enzymes and their co-factors Enzyme Structure and Mechanism. A. Fersht, W. H. Freeman, Reading and San Francisco, 1977.

Enzymatic Reaction Mechanisms. C. Walsh, W. H. Freeman & Co., San Francisco, 1979.

Bio-organic Mechanisms, Vols I and II, T. C. Bruice and S. J. Benkovic; W. A. Benjamin, New York and Amsterdam, 1966.

Physical Methods Organic Spectroscopy, Second Edition, S. F. Dyke, A. J. Floyd, M. Sainsbury and R. S. Theobold. Longmans, London and New York, 1978.

A Biologist's Guide to Principles and Techniques of Practical Biochemistry. Second Edition, Ed. B. C. Williams and K. Wilson, Edward Arnold, London, 1981.

Index